AF594994

Advanced Powder Metallurgy Technologies

Advanced Powder Metallurgy Technologies

Special Issue Editor

Pavel Novák

MDPI • Basel • Beijing • Wuhan • Barcelona • Belgrade • Manchester • Tokyo • Cluj • Tianjin

Special Issue Editor
Pavel Novák
University of Chemistry and Technology
Czech Republic

Editorial Office
MDPI
St. Alban-Anlage 66
4052 Basel, Switzerland

This is a reprint of articles from the Special Issue published online in the open access journal *Materials* (ISSN 1996-1944) (available at: https://www.mdpi.com/journal/materials/special_issues/Powder_Metallurgy_Tech).

For citation purposes, cite each article independently as indicated on the article page online and as indicated below:

LastName, A.A.; LastName, B.B.; LastName, C.C. Article Title. *Journal Name* **Year**, *Article Number*, Page Range.

ISBN 978-3-03936-523-4 (Hbk)
ISBN 978-3-03936-524-1 (PDF)

Cover image courtesy of Pavel Novak.

Contents

About the Special Issue Editor

Pavel Novák, Ph.D., is working as an associate professor with the Department of Corrosion Engineering of University of Chemical Technology, Prague. Fields of scientific interests include powder metallurgy, intermetallics, tool materials, surface treatment of metals, and metallography. He serves as a member of editorial boards of Materials and Manufacturing Technology. P. Novak is the author and co-author of more than 140 publications on WoS and three accepted patent applications. The most important research results include the optimization of powder metallurgy technology for the synthesis of intermetallics and the development of novel high-temperature and wear-resistant intermetallic-based materials.

 *materials*

Editorial

Advanced Powder Metallurgy Technologies

Pavel Novák

University of Chemistry and Technology, Prague, Department of Metals and Corrosion Engineering, Technická 5, 166 28 Prague 6, Czech Republic; panovak@vscht.cz

Received: 31 March 2020; Accepted: 3 April 2020; Published: 8 April 2020

Abstract: Powder metallurgy is a group of advanced processes for the synthesis, processing, and shaping of various kinds of materials. Initially inspired by ceramics processing, the methodology comprising of the production of a powder and its transformation to a compact solid product has attracted great attention since the end of World War II. At present, there are many technologies for powder production (e.g., gas atomization of the melt, chemical reduction, milling, and mechanical alloying) and its consolidation (e.g., pressing and sintering, hot isostatic pressing, and spark plasma sintering). The most promising ones can achieve an ultra-fine or nano-grained structure of the powder, and preserve it during consolidation. Among these methods, mechanical alloying and spark plasma sintering play a key role. This Special Issue gives special focus to the advancement of mechanical alloying, spark plasma sintering and self-propagating high-temperature synthesis methods, as well as to the role of these processes in the development of new materials.

Keywords: powder metallurgy; mechanical alloying; spark plasma sintering; self-propagating high-temperature synthesis

---

Powder metallurgy is a relatively old discipline, as some trials date back to at least 3000 B.C. The technological importance of this branch starts at the beginning of the 20th century. To date, it has undergone significant development through the hot isostatic pressing process, which allows the achievement of an isotropic structure, metal injection molding and other methods for net shape production to the methods used to obtain nanostructured materials. Among these methods, mechanical alloying for powder production, and spark plasma sintering for its consolidation, probably play the most important role. These methods are applied for the manufacture of a Ni-Co-Cr-Fe-Ti high entropy alloy by Moravcik et al. [1]. The results of this paper show that Ti increases the strength of the alloys by solid solution strengthening, reaching an ultimate tensile strength of approx. 1600 MPa, together with the ductility of 9 %, even though there are oxide inclusions in the material. The presented results could lead to future, novel inclusion-tolerant materials. Three papers in the Special Issue are focused on the processing of Fe-Al-Si-based materials by the same set of methods [2–4]. The Fe-Al-Si alloys exhibit very good oxidation resistance against high-temperature oxidation. As well as the binary Fe-Al alloys and some other aluminide systems, these materials show the anomaly of not only the yield strength, but also of the ultimate tensile strength. This means that these mechanical properties increase with the temperature in some temperature intervals (for these alloys at 400–500 °C). This increase in strength is also coupled with considerable ductility, even though these alloys are brittle at room temperature. Together with oxidation resistance and thermal stability, this implies that these alloys could probably be applied, for example, as exhaust valves of internal combustion engines [2]. The properties of this alloy could be further improved by the addition of a combination of molybdenum with nickel or titanium. This leads to very good tribological properties, leading to a significantly lower wear rate than in the case of high-performance tool steels [3]. Cech et al. investigated how the feedstock powder composition influences the structure and properties of Fe-Al-Si alloys [4]. The authors compare the use of elemental powders (Fe,Al,Si) with pre-alloyed ones (Fe-Al, Fe-Si and Al-Si alloys). There are no significant differences in the phase composition and microstructure of the products, dependending on the feedstock powder composition. However, the use of softer elemental powders shortens the mechanical alloying process. The pre-alloyed Fe-Al powder improved the fracture toughness of the alloy. Ti-Al and Ti-Al-Si alloys are, as well as Fe-Al based alloys, known as high-temperature materials, which are problematically produced by common metallurgy methods. In this Special Issue, the possibility of the synthesis of Ti_3Al intermetallic phase by self-propagating high-temperature synthesis from elemental titanium and aluminium powder is described [5]. It was found that the reactions between titanium and aluminium powder start with the formation of a metastable Ti_2Al_5 phase. This phase then reacts with titanium, leading to the desired Ti_3Al phase. The series of Ti-Al-Si alloys with varying ratios between titanium, aluminium and silicon was prepared by the combination of mechanical alloying and spark plasma sintering [6]. As a reference, melting metallurgy was applied to alloys of the same chemical composition. The powder metallurgy could obtain a significantly finer microstructure and reach a correspondingly higher ultimate tensile strength. On the other hand, the fracture toughness of the melting-metallurgy-prepared samples was higher.

A significant part of the Special Issue is devoted to the functional materials, which are used for their shape memory properties, magnetic behaviour, optical properties, self-healing beahaviour or biodegradability in the organism. Salvetr et al. studied the synthesis of a NiTi shape memory alloy by self-propagating high-temperature synthesis, milling and spark plasma sintering [7]. The alloy reached almost the theoretical density, a very high ultimate compressive strength of more than 2000 MPa, but it lost the transformation behaviour. Further heat treatment at the temperature of 600–700 °C with slow cooling was needed to recover the transformation behaviour, which is crucial for the shape memory effect [7]. The magnetic materials were investigated by Skotnicova et al. [8]. This paper focused on the coercivity enhancement of Nd-Fe-B magnets by optimizing the microstructure by the addition of a $Dy_3Co_{0.6}Cu_{0.4}H_x$ alloy during mechanical activation. The whole-powder metallurgy process included strip casting, hydrogen decrepitation, milling (mechanical activation) and sintering. The results show the possibility of using a unified initial alloy for the manufacture of magnets with tailored magnetic characteristics.

Powder metallurgy was initially inspired by ceramics processing, while the spark plasma sintering was originally developed for the processing of metals. Nowadays, this method allows for strong development in the processing of special ceramic materials. In this Special Issue, magnesium aluminate spinels for optical applications, with the addition of LiF, MnF_2, and CoF_2 were consolidated this method [9]. Returning back to the world of metals, a composite material composed of aluminium alloy and nickel particles, prepared by a combination of gas atomization and spark plasma sintering, is presented [10]. This material reveals the functionality of self-healing, which is relatively new and unusual for metals. Kubasek et al. [11] presents a novel biodegradable composite material composed of zinc and magnesium, which can possibly be applied in future surgery for the temporary fixing implants. The problematics of composite materials are also solved from the viewpoint of the advancement of the mixing process for ternary powder mixtures [12].

Materials for cutting tools usually contain elements such as cobalt and tungsten, which are listed by the European Commission as critical raw materials (CRM). The review [13] shows the ways that these elements can be substituted in the manufacturing sector and recycled to minimize their economic impact, and how to improve the manufacturing process itself. One of the materials which are applicable as tool materials is maraging steel. This substitute for conventional materials is very popular because it can be processed by 3D-printing methods. The selective laser melting process using the atomized powder of the maraging steel was described, as well as the subsequent post-processing heat treatment [14].

As seen above, the Special Issue *Advanced Powder Metallurgy Technologies* provides insight into the recent trends in powder metallurgy from the viewpoints of both the advancement of the technology and the development of new materials using these methods. It is great to see that the modern powder metallurgy can move technology, economics, and also healthcare, forward.

Acknowledgments: I would like to thank all the authors for their hard work on the papers for this Special Issue, for bringing this fascinating science field forward and also for summarizing the state-of-the art in a nice review paper. The exposure and promotion of our work in this special issue has been facilitated by the editorial and production office of MPDI.

Conflicts of Interest: The author declares no conflict of interest.

References

1. Moravcik, I.; Gamanov, S.; Moravcikova-Gouvea, L.; Kovacova, Z.; Kitzmantel, M.; Neubauer, E.; Dlouhy, I. Influence of Ti on the Tensile Properties of the High-Strength Powder Metallurgy High Entropy Alloys. *Materials* **2020**, *13*, 578. [CrossRef] [PubMed]
2. Novak, P.; Vanka, T.; Nova, K.; Stoulil, J.; Prusa, F.; Kopecek, J.; Hausild, P.; Laufek, F. Structure and Properties of Fe-Al-Si Alloy Prepared by Mechanical Alloying. *Materials* **2019**, *12*, 2463. [CrossRef] [PubMed]
3. Průša, F.; Proshchenko, O.; Školáková, A.; Kučera, V.; Laufek, F. Properties of FeAlSi-X-Y Alloys (X,Y=Ni, Mo) Prepared by Mechanical Alloying and Spark Plasma Sintering. *Materials* **2020**, *13*, 292. [CrossRef]
4. Cech, J.; Hausild, P.; Karlik, M.; Boucek, V.; Nova, K.; Prusa, F.; Novak, P.; Kopecek, J. Effect of Initial Powders on Properties of FeAlSi Intermetallics. *Materials* **2019**, *12*, 2846. [CrossRef] [PubMed]
5. Skolakova, A.; Salvetr, P.; Novak, P.; Leitner, J.; Deduytsche, D. Mechanism of the Intermediary Phase Formation in Ti-20 wt.% Al Mixture during Pressureless Reactive Sintering. *Materials* **2019**, *12*, 2171. [CrossRef] [PubMed]
6. Knaislova, A.; Novak, P.; Kopecek, J.; Prusa, F. Properties Comparison of Ti-Al-Si Alloys Produced by Various Metallurgy Methods. *Materials* **2019**, *12*, 3084. [CrossRef] [PubMed]
7. Salvetr, P.; Dlouhy, J.; Skolakova, A.; Prusa, F.; Novak, P.; Karlik, M.; Hausild, P. Influence of Heat Treatment on Microstructure and Properties of NiTi46 Alloy Consolidated by Spark Plasma Sintering. *Materials* **2019**, *12*, 4075. [CrossRef] [PubMed]
8. Skotnicova, K.; Prokofev, P.A.; Kolchugina, N.B.; Burkhanov, G.S.; Lukin, A.A.; Koshkid'ko, Y.S.; Cegan, T.; Drulis, H.; Romanova, T.; Dormidontov, N.A. Application of a Dy3Co0.6Cu0.4Hx Addition for Controlling the Microstructure and Magnetic Properties of Sintered Nd-Fe-B Magnets. *Materials* **2019**, *12*, 4235. [CrossRef] [PubMed]
9. Talimian, A.; Pouchly, V.; Maca, K.; Galusek, D. Densification of Magnesium Aluminate Spinel Using Manganese and Cobalt Fluoride as Sintering Aids. *Materials* **2019**, *13*, 102. [CrossRef] [PubMed]
10. Michalcova, A.; Knaislova, A.; Kubasek, J.; Kacenka, Z.; Novak, P. Rapidly Solidified Aluminium Alloy Composite with Nickel Prepared by Powder Metallurgy: Microstructure and Self-Healing Behaviour. *Materials* **2019**, *12*, 4193. [CrossRef] [PubMed]
11. Kubasek, J.; Dvorsky, D.; Capek, J.; Pinc, J.; Vojtech, D. Zn-Mg Biodegradable Composite: Novel Material with Tailored Mechanical and Corrosion Properties. *Materials* **2019**, *12*, 3930. [CrossRef] [PubMed]
12. Olszowka-Myalska, A.; Wrzesniowski, P.; Myalska, H.; Godzierz, M.; Kuc, D. Impact of the Morphology of Micro- and Nanosized Powder Mixtures on the Microstructure of Mg-Mg2Si-CNT Composite Sinters. *Materials* **2019**, *12*, 3242. [CrossRef] [PubMed]
13. Rizzo, A.; Goel, S.; Grilli, M.L.; Iglesias, R.; Jaworska, L.; Lapkovskis, V.; Novak, P.; Postolnyi, B.O.; Valerini, D. The Critical Raw Materials in Cutting Tools for Machining Applications: A Review. *Materials* **2020**, *13*, 1377. [CrossRef] [PubMed]
14. Strakosova, A.; Kubasek, J.; Michalcova, A.; Prusa, F.; Vojtech, D.; Dvorsky, D. High Strength X3NiCoMoTi 18-9-5 Maraging Steel Prepared by Selective Laser Melting from Atomized Powder. *Materials* **2019**, *12*, 4174. [CrossRef] [PubMed]

Article

Influence of Ti on the Tensile Properties of the High-Strength Powder Metallurgy High Entropy Alloys

Igor Moravcik [1,*], Stepan Gamanov [1,2], Larissa Moravcikova-Gouvea [1], Zuzana Kovacova [3], Michael Kitzmantel [3], Erich Neubauer [3] and Ivo Dlouhy [1,2]

1. Institute of Materials Science and Engineering, Brno University of Technology, Technicka 2896/2, 61669 Brno, Czech Republic; gamanov@ipm.cz (S.G.); gouvea@fme.vutbr.cz (L.M.-G.); dlouhy@fme.vutbr.cz (I.D.)
2. Institute of Physics of Materials of the Czech Academy of Sciences, Žižkova 513/22, 61600 Brno, Czech Republic
3. RHP-Technology GmbH, Forschungs- und Technologiezentrum, 2444 Seibersdorf, Austria; zuzana.kovacova@rhp-technology.com (Z.K.); michaelkitzmantel@rhp-technology.com (M.K.); erichneubauer@rhp-technology.com (E.N.)

* Correspondence: igor.moravcik@vutbr.cz; Tel.: +421-911-566-030

Received: 17 December 2019; Accepted: 20 January 2020; Published: 26 January 2020

Abstract: The focus of this study is the evaluation of the influence of Ti concentration on the tensile properties of powder metallurgy high entropy alloys. Three $Ni_{1.5}Co_{1.5}CrFeTi_X$ alloys with X = 0.3; 0.5 and 0.7 were produced by mechanical alloying and spark plasma sintering. Additional annealing heat treatment at 1100 °C was utilized to obtain homogenous single-phase face centered cubic (FCC) microstructures, with minor oxide inclusions. The results show that Ti increases the strength of the alloys by increasing the average atomic size misfit i.e., solid solution strengthening. An excellent combination of mechanical properties can be obtained by the proposed method. For instance, annealed $Ni_{1,5}Co_{1,5}CrFeTi_{0.7}$ alloy possessed the ultimate tensile strength as high as ~1600 MPa at a tensile ductility of ~9%, despite the oxide contamination. The presented results may serve as a guideline for future alloy design of novel, inclusion-tolerant materials for sustainable metallurgy.

Keywords: multi principal element alloy; tensile strength; fracture; ductility; powder

---

1. Introduction

The equiatomic high entropy alloys (HEA) designed by Yeh and Cantor [1,2] are a new class of metallic materials, composed of at least five elements and with a common feature being a lack of single major element. Despite some ongoing arguments regarding the origins (as well as the existence) of their special properties [3], the mentioned core idea spawned a new direction in the development of metallic, as well as non-metallic materials [4,5]. Even though considerable resources has been dedicated to the study of a vast range of HEA properties, phase compositions as well as guiding principles for further alloy design [6–9], the properties of most HEAs are still not competitive in comparison to more classic materials [3,10]. One of the biggest issues with HEAs is the vast compositional space resulting from inherent chemical complexity, in which it is extremely hard to choose the compositions [11]. On top of this, the competitiveness of any selected composition is a matter of a combination of several properties, not strictly constrained to combination of strength and ductility [12,13], as it may seem from current studies. One of the new HEA alloy systems which possess a very interesting combination of mechanical properties, oxidation and corrosion resistance combined with exceptional wear resistance [14–18], is $Ni_{1.5}Co_{1.5}CrFeTi_X$. It exhibits a very high mechanical strength and ductility in a solutionized (single-phase FCC) state [17], while it can be heat treated to contain intermetallic strengthening phases.

Its strength in single-phase state is derived from relatively high atomic size misfit δ (5.04 % for X = 0.5) of its elements [15], which is a measure of the extent of solid solution strengthening [19] i.e., the magnitude of critical resolved shear stress required for the dislocation movement. The atomic size difference δ based on the Hume—Rothery rules is defined as [20]:

$$\delta = 100\sqrt{\sum_{i=1}^{n} c_i\left(1-\frac{r_i}{\bar{r}}\right)^2}. \quad (1)$$

In this equation, c_i is the atomic concentration (molar fraction), r_i presents the atomic radius of i-th element and $\bar{r}$ is the average atomic radius $\bar{r} = \sum_{i=1}^{n} c_i r_i$ of a given system.

The difference between the atomic sizes of Ni, Co, Cr and Fe elements present in $Ni_{1.5}Co_{1.5}CrFeTi_X$ HEA are relatively small [21,22]. The only exception is Ti, which possess much larger atoms, with respect to the latter. It is therefore intuitive that Ti concentration will largely dictate the extent of solid solution strength, and it would be advisable to keep it as high as possible. On the other hand, the binary mixing enthalpy ΔH_{AB} of Ti-Ni pair (−35 kJ/mol) and Ti-Co pair (−35 kJ/mol) is much more negative than for the other possible atomic pairs (minimum −7 kJ/mol for Ni-Cr) [21,23]. This means that increasing the Ti concentration will inevitably result in the thermodynamic drive towards the formation of intermetallic phases—decreasing the solid solution strengthening effect. To probe the effect of Ti concentration change on the mechanical properties as well as phase composition and Ti partitioning effects, the $Ni_{1,5}Co_{1,5}CrFeTi_X$ alloys have been tested. Altogether, three different HEAs with Ti atomic ratios; x = 0.3, 0.5 and 0.7 were prepared. To prevent the segregation effects which may occur during the casting of such complex systems [24] and formation of coarse-grained microstructures, an alternative powder metallurgical manufacturing route of mechanical alloying (MA) and spark plasma sintering (SPS) was used. The latter process enables us to produce fine-grained materials and to study the properties without the need for time- and energy-consuming additional hot and cold-working processes, which are sometimes required for cast materials [15,25].

2. Materials and Methods

Three alloys with chemical composition of $Ni_{1.5}Co_{1.5}Cr_1Fe_1Ti_X$ (where x = 0.3; 0.5; 0.7) were prepared by a powder metallurgy route and will be further referred to as Ti0.3, Ti0.5 and Ti0.7, respectively. The target composition in atomic % are $Ni_{28}Co_{28}Cr_{19}Fe_{19}Ti_6$, $Ni_{27}Co_{27}Cr_{18}Fe_{18}Ti_9$ and $Ni_{26}Co_{26}Cr_{17}Fe_{17}Ti_{12}$ for Ti0.3, Ti0.5 and Ti0.7 alloys, respectively (also presented latter is Section 3.3), whereas, in weight %, the compositions are $Ni_{30}Co_{29}Cr_{19}Fe_{19}Ti_5$, $Ni_{29}Co_{29}Cr_{17}Fe_{18}Ti_8$ and $Ni_{28}Co_{28}Cr_{16}Fe_{17}Ti_{11}$ in the same order.

It should be noted that the conditions used for mechanical alloying and spark plasma sintering are similar to our previous publication [15] on similar HEA. Alloys were prepared from commercial-grade purity, whereas the information on the powders given in Table 1. Powders were milled with 10:1 ball to powder weight ratio using balls with 15 mm and 20 mm diameter (1:1 ratio). Each starting powder mixture weighted 100 g. A ball and powder mixture was introduced into a steel milling container together. A milling container was then sealed and flushed with N_2 gas to limit oxidation during the milling process. Dry milling was performed in Pulverisette 6 ball mill (Fritsch GmbH, Idar-Oberstein, Germany) with milling speed of 250 rounds per minute for 35 h of milling (30 min pauses after each hour of milling were used to prevent overheating). In the end of the dry-milling process, powders were further wet-milled for 1 h with 100 mL of ethanol to remove powders stuck to milling balls. Extracted powders were then filtered and dried. Metallographic specimens were prepared from powders and bulk materials were prepared by standard metallographic procedures. Energy-dispersive X-ray spectroscopy (EDS) analyses were performed on every powder to ensure its chemical homogeneity before the SPS process.

Table 1. Manufacturers and product identification of used powders.

Element	Manufacturer	Product Identification
Co	Alfa Aesar	Particle size < 44 μm, purity 99.5%, LOT: W08B011
Cr	Alfa Aesar	Particle size < 44 μm, purity 99%, LOT: S18A034
Fe	Aldrich	Particle size 5–9 μm, purity ≥99.5% Lot # MKBS9265V
Ni	Aldrich	Particle size < 50 μm, purity 99.7%, Lot # MKBR6365V
Ti	Alfa Aesar	Particle size < 44 μm, purity 99.5%, LOT: W17A045

MA powders were subsequently sintered by the SPS method at RHP-Technology GmbH (city, country) into the form of 5 mm thick cylinders with 36 mm diameter using graphite dies. To prevent powder contamination by carbon, graphite dies were coated prior to sintering with chemically inert boron nitride. Sintering temperature of 1150 °C measured by thermocouple inside the die, with 30 MPa of pressure and 10 min. dwell time at sintering temperature was used. The heating rate of 100 °C/min. was used up to a temperature of 1000 °C. The heating rate was then reduced to 50 °C/min. up to 1150 °C/min. The whole process was performed in a vacuum atmosphere. At the end of sintering, the setup was left to naturally cool down, until it was opened at ~200 °C. Sintered samples were then annealed at 1100 °C for 24 h and water cooled, to obtain microstructural relaxation, homogenization and pore closure (in case of Ti0.3 alloy). These materials will be henceforth referred to as Ti0.3A, Ti0.5A and Ti0.7A, respectively. The porosity, as well as particle size measurement, was performed with image analysis in ImageJ software on images taken with optical microscope (Olympus GX51, Tokyo, Japan). Average grain size measurement was performed with a linear intercept method on FCC grains and oxide particle size was measured with special module embedded in ImageJ software. ULTRA PLUS SEM (Carl Zeiss AG, Oberkochen, Germany) and Lyra XMA FEG/SEM (TESCAN, Brno, Czech Republic) were used for SEM and energy-dispersive X-ray spectroscopy (EDS) analysis of bulk microstructures and fracture surfaces. A Smartlab instrument (Rigaku, Tokyo, Japan) with Co source was utilized for X-ray diffraction (XRD) analysis of phase composition and lattice parameters. A microhardness test with a Microharness tester LM247AT (Leco, St. Joseph, MI, USA) was carried out with 100 g load (HV0.1). Tensile tests were carried out on cylindrical tensile specimens with gauge length of 12.5 mm and diameter of 3.5mm were cut and machined from bulk materials. An Instron 8801 machine (Instron, Norwood, MA, USA) was used for tensile tests performed at room temperature and cross-head speed of 0.25 mm/min. Two tensile samples were tested for each of annealed TixA materials. Calculation of property diagrams (CALPHAD) were performed using Thermo-Calc software version 2019a (TCHEA3 database version 3.1, (Thermo-Calc, Solna, Sweden)). The thermodynamic modeling was performed for estimation of correct temperatures for annealing, as well as to get a better understanding of possible phase composition of the manufactured materials.

3. Results

3.1. Prepared Powder before SPS

Average particle size d of powders is given in Table 2. It is worth noting that each prepared powder had a large number of relatively small particles (above 75% of particles with d below 10 μm for all powders) and a low number of relatively large particles. This is especially evident in powder Ti0.3 with particles diameter up to 250 μm as can be seen in Figure 1. Therefore, Ti0.3 presents the coarsest powder among all of them, while Ti0.5 and Ti0.7 are relatively identical. The EDS analysis revealed homogenous chemical composition on powder particles; more precise analysis was later performed on sintered and annealed bulk samples.

Table 2. The average sizes of the Ni1.5Co1.5CrFeTix powder particles after mechanical alloying.

Powder	Average Size d [μm]	Deviation [μm]	d_{max} [μm]
Ti0.3	10.68	20.96	249.80
Ti0.5	9.08	15.95	154.36
Ti0.7	8.27	12.70	141.19

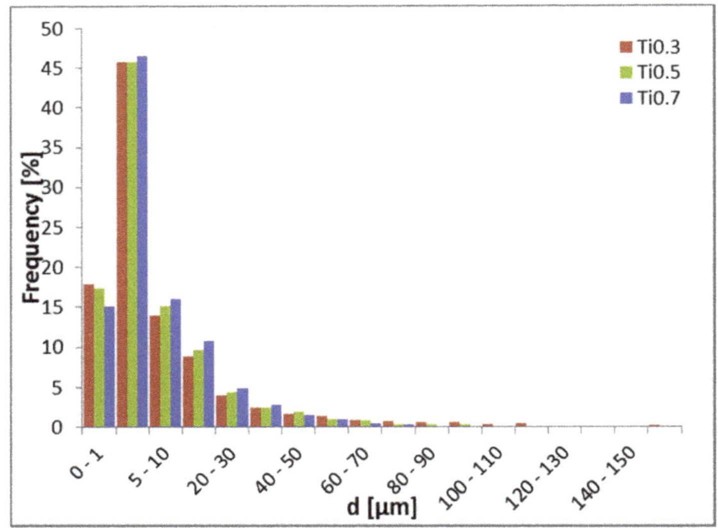

Figure 1. Histogram of powder particles distribution of Tix alloys. The largest fraction of particles have a size between 1 and 5 μm. Particles with size larger than 100 μm are significantly frequent only in the case of Ti0.3 powder.

3.2. Phase Diagram

The results of the CALPHAD predictions of $Ni_{1.5}Co_{1.5}CrFeTi_x$ alloys showing equilibrium phases are shown in Figure 2. It should be noted, in the start, that the accuracy of the CALPHAD predictions for such complex systems may not be perfect [26], and the real microstructures and phases fraction may show certain discrepancies. Nevertheless, it serves as useful general guidance for further phase characterization. The Ti0.3 material should be composed from a mixture the FCC solid solution phase and Ni_3Ti (HCP DO_{24}) ordered η phase, found commonly in Ni-base superalloys. The Ti0.5 alloy should contain an additional ordered ($L1_2$) FCC phase, precipitating at temperatures below ~900 °C. According to the prediction, this ordered FCC phase should have a $(Ni,Co)_3Ti$ chemistry. It has been shown before that the Ni_3Ti ordered η phase and ordered FCC $L1_2$ phases show similar morphologies and they can coexist in the microstructures of Ni-base alloys [27], due to their close chemistry and crystallography. With increasing Ti concentration, the melting temperature is slightly decreasing (Figure 2c) in Ti0.7 alloy, while the phase composition should be identical to that of Ti0.5 alloy, but with increasing volume fraction of η and ordered FCC $L1_2$ phase. The ordered FCC $L1_2$#2 phase (marked by red rectangle in Figure 2c) has probably not formed in the structure and should be identical to FCC $L1_2$. It probably appeared due to an uncertainty in the calculation, which has been observed before [14,28]. At the processing temperatures to which the materials were subjected to (1150 or 1100 °C), the microstructures of the alloys should contain only single FCC phase. This single-phase microstructure should be mostly retained due to relatively fast cooling after SPS [14,15]. Despite this, some extent of second phase precipitation may occur, especially in Ti0.7 alloy. To make the microstructures more uniform, annealing treatment was used. After annealing at 1100 °C and water quenching, all materials should show single FCC phase, ensuing the highest extent of solid solution strengthening.

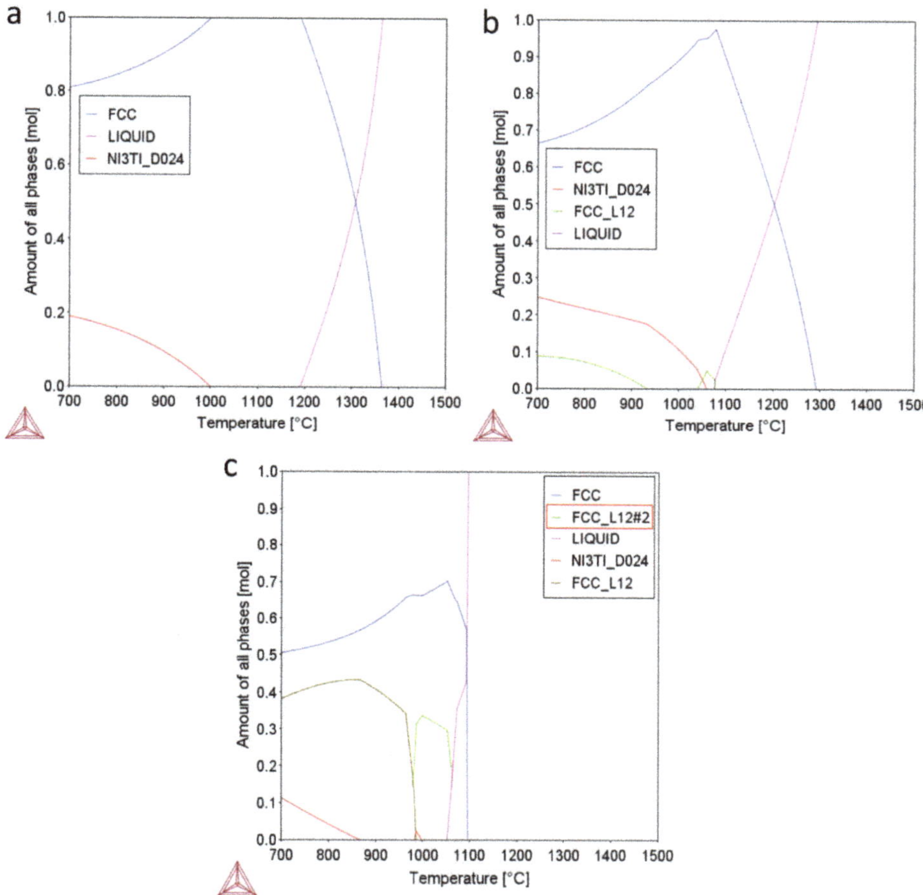

Figure 2. Results of the CALPHAD predictions of Ni1.5Co1.5CrFeTix alloys showing equilibrium phases and their respective fraction at different temperatures; (**a**) Ti0.3; (**b**) Ti0.5; (**c**) Ti0.7.

3.3. Microstructure and Phase Analysis after Sintering and Annealing

Table 3 shows data acquired via image and XRD analysis of sintered and annealed alloys. Significant porosity was measured only for Ti0.3 alloy, which had the coarsest powder, while the rest of the alloys obtained full-density already after SPS. The lower porosity (at same sintering temperature) of Ti0.5 and Ti0.7 materials compared to Ti0.3 can be also caused by lowering of melting temperature by Ti, as calculated before. In regards to porosity, additional annealing was successful since it reduced porosity of Ti0.3 from 3.77% to 0.43%. The representative microstructures of all materials are presented in Figure 3. The grain size of all alloys in sintered state was very fine i.e., below 1.7 μm, whereas it grew after annealing up to 2.13 μm for the Ti0.7 alloy. Differences in shades of gray between different FCC grains in the BSE images of Figure 3 are caused by differences in crystallographic orientation, as shown in Appendix A Figure A1, due to special setup used for the imaging of grains [29] without the need for electrochemical etching. A large fraction of oxides (visible as black dots) was present in all alloys even though the powders were prepared in an inert atmosphere. More detailed EDS analysis on the representative black particles present in all microstructures is given in the Appendix A in Figure A2 and Table A1, proving that they pertain to oxides and not to porosity. It should be noted that the small size of the oxides combined with the resolution and insensitivity of EDS to lighter elements did not enable us to measure the character of oxides precisely. Even the largest of the oxides are still too small for EDS and the signal coming out of oxides is contaminated by the signal from the matrix below the particles (Table A1). However, it is clear that the oxides are mostly enriched in Ti elements. The high concentration of light Ti and O (compared to other present elements) causes them to appear dark compared to the matrix in back-scattered electron (BSE) images. This is different to the pores which also appear black, but in the secondary electron micrographs [30,31]. The powders of commercial-purity already possessed an oxide layer on their surfaces prior to mechanical alloying, which got dispersed into the powder particles during the latter process. In addition, the contamination of the powders was enhanced due to an additional milling step in ethanol. These oxides get trapped inside of the specimen during the SPS-densification process. The average size of the oxides was below 100 nm for all sintered alloys and between 150 and 175 nm for annealed alloys. It should be noted that the small size of the oxides combined with the resolution and insensitivity of EDS to lighter elements did not enable us to measure oxides character precisely. XRD analysis of the produced alloys shown in Figure 4 revealed that all SPS-ed alloys are dominantly composed of FCC solid solution phase (in agreement with the CALPHAD predictions) with a minor fraction of oxide phase. As mentioned before, the oxide phase is a result of powder surface contamination. However, due to a small fraction of this oxide phase(s) (~3.5%), it would be imprecise to conclude its true nature only from XRD. It corresponds to a Ti oxide, which can also contain small concentrations of other elements. The fraction of the secondary phase did not decrease after annealing at 1100 °C, which is caused by high thermal stability of the oxide. The only exception to this is the Ti0.7 alloy, which showed a decrease of the secondary phase after annealing. Therefore, it can be concluded that Ti0.7 alloy prior to annealing contained an additional fraction of intermetallic phase as predicted by the CALPHAD with the same peaks as the oxide phase, while a decrease in their intensity was observed after annealing. This tertiary intermetallic phase was not observed by SEM before due to its low fraction.

Table 3. Various measured characteristics of Tix alloys in sintered and annealed state measured by Image and XRD analysis.

Measured Characteristics.	Sintered Alloys			Annealed Alloys		
Image Analysis	**Ti0.3**	**Ti0.5**	**Ti0.7**	**Ti0.3A**	**Ti0.5A**	**Ti0.7A**
Porosity [%]	3.77	0.03	0.02	0.43	0.06	0.01
Average grain size [μm]	0.44	1.69	1.14	1.30	1.74	2.13
Average oxide size [nm]	61.80	94.41	87.40	155.54	171.13	171.13
Area of oxides [%]	5.01	5.09	5.03	7.39	8.32	8.29
XRD Analysis	**Ti0.3**	**Ti0.5**	**Ti0.7**	**Ti0.3A**	**Ti0.5A**	**Ti0.7A**
Lattice parameter [Å]	3.58	3.58	3.59	3.58	3.59	3.60
FCC phase [%]	95.40	96.40	92.90	96.60	96.50	96.00
Oxides + intermetallics [%]	4.60	3.60	7.10	3.40	3.50	4.00

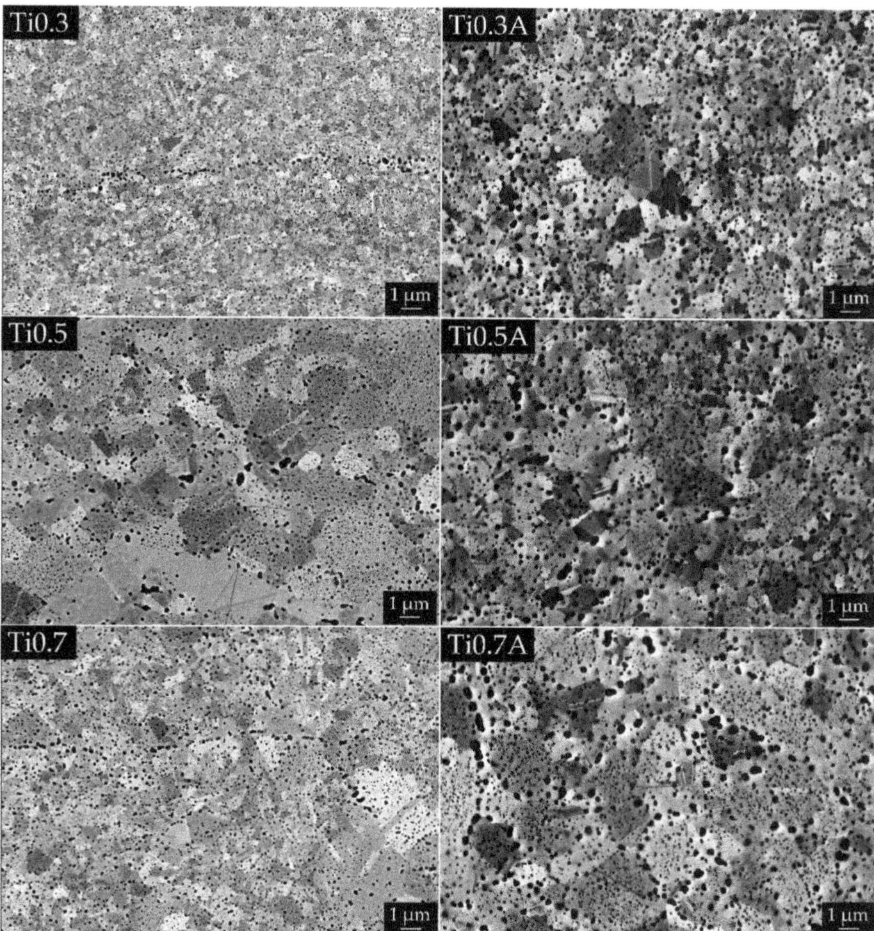

Figure 3. Structure of Tix alloys in sintered and annealed state, backscattered electron (BSE) SEM micrographs. It can be seen that FCC grains in all materials coarsened as well as the oxides (black dots). There are visible lines of coarsest oxides in the case of (**Ti0.3**) and (**Ti0.7**); these oxides probably copy the shape of original powder particles. There are also very fine particles inside the grains (best seen in case of (**Ti0.7A**); these particles are suspected to not be the oxides but intermetallic particles. Please note that the light grey–dark grey contrast between FCC grains is caused by different crystallographic orientation.

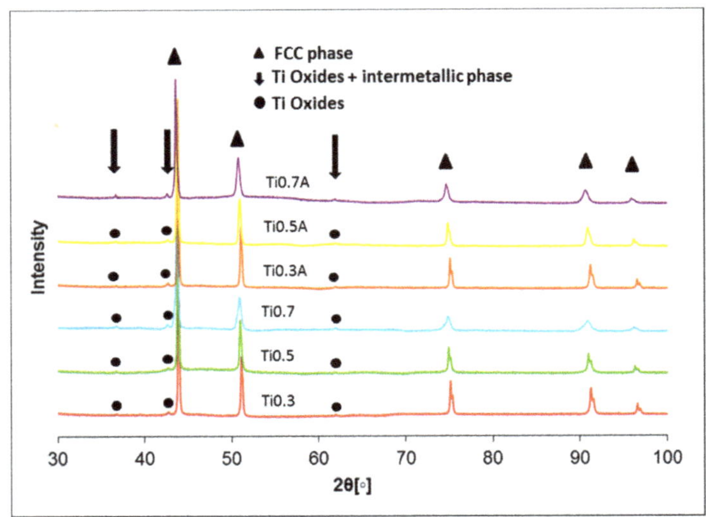

Figure 4. XRD spectrums of sintered an annealed Tix alloys.

EDS analysis of matrix of the Tix alloys are given in Table 4 with a representative EDS map in Figure 5 (including the oxygen map).

Table 4. Chemical compositions of the respective HEA FCC matrix obtained by EDS analysis of materials in as-sintered and annealed conditions.

	Sintered Alloys		Annealed Alloys	
Ti0.3	**Measured Composition [at.%]**	**Target Composition [at.%]**	**Measured Composition [at.%]**	**Target Composition [at.%]**
Co	27.97 ± 0.53	28.3	27.63 ± 0.16	28.3
Cr	17.28 ± 0.44	18.87	19.33 ± 0.12	18.87
Fe	20.11 ± 0.51	18.87	21.09 ± 0.16	18.87
Ni	30.05 ± 0.69	28.3	27.17 ± 0.18	28.3
Ti	4.59 ± 0.83	5.66	4.79 ± 0.61	5.66
Ti0.5	**Measured Composition [at.%]**	**Target Composition [at.%]**	**Measured Composition [at.%]**	**Target Composition [at.%]**
Co	27.36 ± 0.70	27.27	26.60 ± 0.22	27.27
Cr	16.82 ± 0.31	18.18	18.46 ± 0.10	18.18
Fe	20.60 ± 1.03	18.18	21.14 ± 0.19	18.18
Ni	28.56 ± 0.34	27.27	26.23 ± 0.26	27.27
Ti	6.66 ± 1.81	9.09	7.58 ± 0.76	9.09
Ti0.7	**Measured Composition [at.%]**	**Target Composition [at.%]**	**Measured Composition [at.%]**	**Target Composition [at.%]**
Co	25.90 ± 0.54	26.32	25.57 ±0.47	26.32
Cr	16.95 ± 0.94	17.54	17.90 ± 0.19	17.54
Fe	19.45 ± 1.04	17.54	19.76 ± 0.26	17.54
Ni	27.82 ± 1.78	26.32	25.17 ± 0.77	26.32
Ti	9.88 ± 1.42	12.28	11.60 ± 1.50	12.28

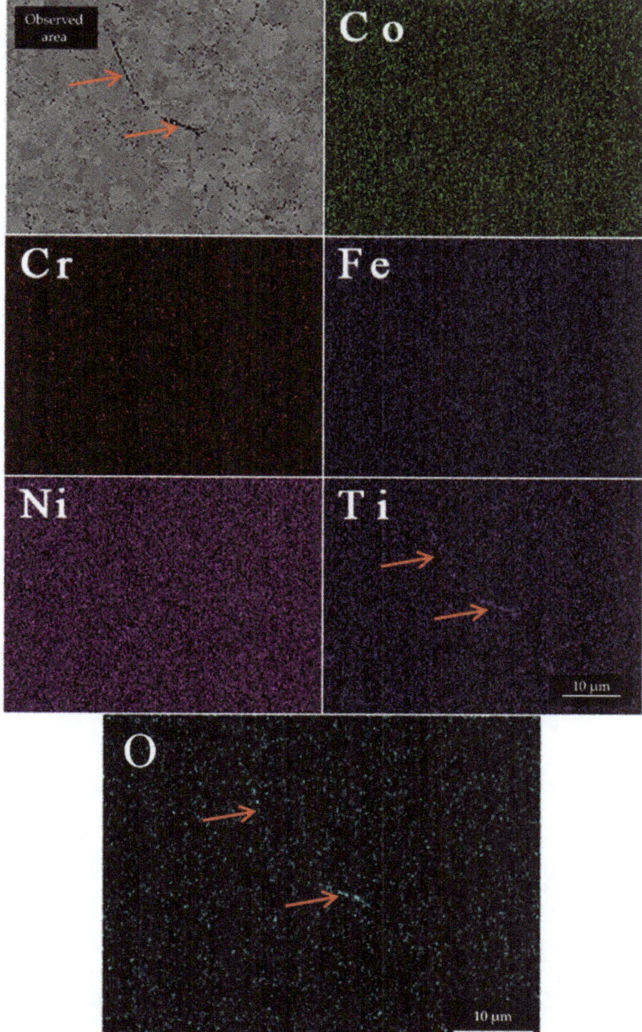

Figure 5. Representative EDS elemental maps of Ti0.5 alloy. The elements are evenly distributed in the alloy. The only exception is Ti, which concentrates in the oxide particles. The Ti- enriched oxide strings highlighted by red arrows denominate powder particle boundaries prior to SPS densification, but they are also present inside of former particles. The signal of O is strongest in the same regions as Ti, most noticeably in the region highlighted by red arrows of oxide strings.

Statistical data on overall chemical compositions of Table 4 are composed of point EDS spectrums (points were always placed farther away from the oxides). The alloy compositions present a good fit to the target chemical compositions. The content of Ti in matrix of Ti0.7 alloy increased after annealing. This indirectly suggests us that some intermetallic compounds were present before annealing, as evidenced before by XRD. As these phases dissolved during annealing, the content of Ti in matrix increased. The elements in FCC phase of materials show even distribution (Figure 5), while only the Ti is localized also in the oxide particles (denoted by red arrow).

3.4. Mechanical Testing

The results of microhardness test are presented in Table 5 The $Ni_{1.5}Co_{1.5}CrFeTi_x$ materials exhibit relatively high hardness values, due to very fine grain size and the presentation of hard particles. It should be noted that the influence of porosity on the hardness of the Ti0.3 materials was negligible, since the size of the micro-indents was significantly smaller than the size of the pores. Therefore, it was possible to measure hardness on areas with full-density at a considerable distance from the pores. As expected, the average hardness of the materials is increasing with increasing the Ti concentration in both conditions (SPS-ed and annealed). The annealing treatment decreased the average hardness and led to higher hardness uniformity (smaller values scatter). Compared to Ti0.3 and Ti0.5, the annealing of Ti0.7 sample led to much lower hardness decrease.

Table 5. Results of microhardness testing on the sintered and annealed bulk materials.

Material	Ti0.3	Ti0.5	Ti0.7	Ti0.3A	Ti0.5A	Ti0.7A
Microhardness HV 0.1	448 ± 20.1	524 ± 25.6	556 ± 22.0	355 ± 4.5	379 ± 3.9	500 ± 7.4

The results of tensile tests are displayed in Figure 6 and Table 6 The worst values of tensile yield strength ($R_{p0.2}$) of 781 MPa, ultimate tensile strength (R_m) of 845 MPa and only ~2% of ductility (A_t) were observed for the Ti0.3A samples due to premature fracture caused by porosity. On the other hand, full density samples Ti0.5A showed very good combination of $Rp_{0.2} \cong 930$ MPa, $R_m \cong 1200$ MPa and $A_t \cong 14\%$. Slight necking was also observed (plastic deformation after plastic instability threshold R_m). The highest strength properties of $Rp_{0.2} \cong 1220$ MPa, $R_m \cong 1680$ MPa were measured for Ti0.7A material, with slightly lower ductility of $A_t \cong 9\%$. It is interesting to note that the highest strain hardening (R_m–$R_{p0.2}$) was observed in the strongest material Ti0.7A, which is contradictory to usual observations [32].

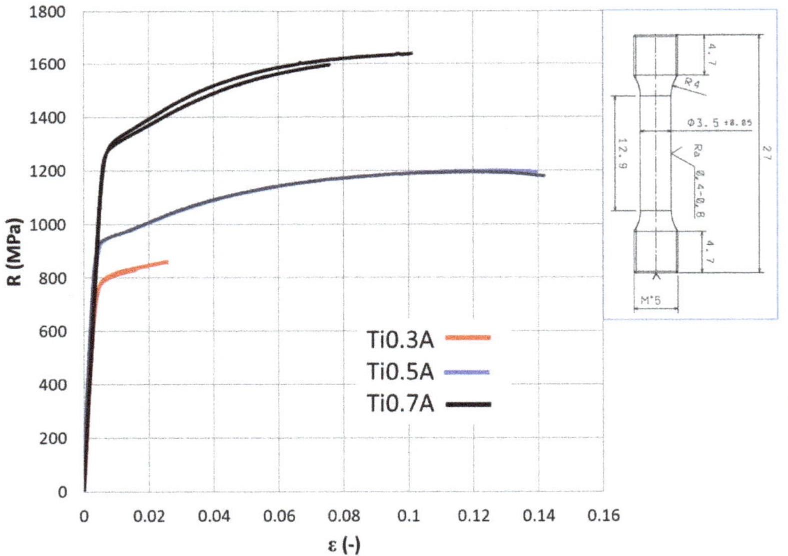

Figure 6. The tensile curves obtained from the $Ni_{1.5}Co_{1.5}CrFeTi_x$ alloy specimens after annealing. The sample geometry is presented on the right-hand side, dimensions in mm.

Table 6. The average results of the tensile tests carried out on $Ni_{1,5}Co_{1,5}CrFeTi_x$ alloys.

Alloy	E (GPa)	$R_{p0.2}$ (MPa)	R_m (MPa)	A_t (%)	Z (%)	Strain Hardening R_m–$R_{p0.2}$ (MPa)
Ti0.3A	229.8	781.5	845	2.1	2.8	63.5
Ti0.5A	265.6	930.5	1199	14.1	24.8	268.5
Ti0.7A	228.7	1281.5	1618.5	8.8	12.1	337

Figure 7 shows representative fracture morphologies of broken tensile specimens. All of the samples exhibited ductile fracture dimples with the size between of 200–800 nm. On fracture surface of Ti0.3A material, surfaces of pores present in the fracture surface of the samples are exposed, denoted by a red arrow with original pore surface (matrix-pore interphase) marked by red spline. Despite the low measured porosity in Ti0.3A alloy after annealing (<0.5%), the pores appeared on fracture surfaces of Ti0.3A in much larger contents. This happens because the forming crack tip is following areas with the largest porosity levels, as they locally act as stress concentrators. The fine oxide particles are found at the ductile dimple centers, which suggests their role as a nucleation size for dimple formation. In comparison to Ti0.3A, no porosity was observed on the fracture surface of Ti0.5A, which agrees well with its largest ductility. The dimple size of the Ti0.5A is slightly smaller with lower number density of oxides in their centers. The pore-like formations denoted by the green arrow (Figure 7c) correspond to areas with larger oxides that were pulled from the matrix during fracture process. Larger internally cracked oxide particles can also be observed on the fracture surfaces (Figure 7d). The fracture surface of the Ti0.7A material is much flatter with lower surface roughness compared to Ti05A material, corresponding to its lower ductility. This type of fracture corresponds to low-energy ductile tearing observed in severely strengthened ductile metals [33]. Ductile dimples of Ti0.7 material are largely free of oxides and very shallow, which suggest that dimple nucleation started inside the FCC matrix due to its higher strength and consequential lower plasticity, compared to Ti0.3A and Ti0.5A.

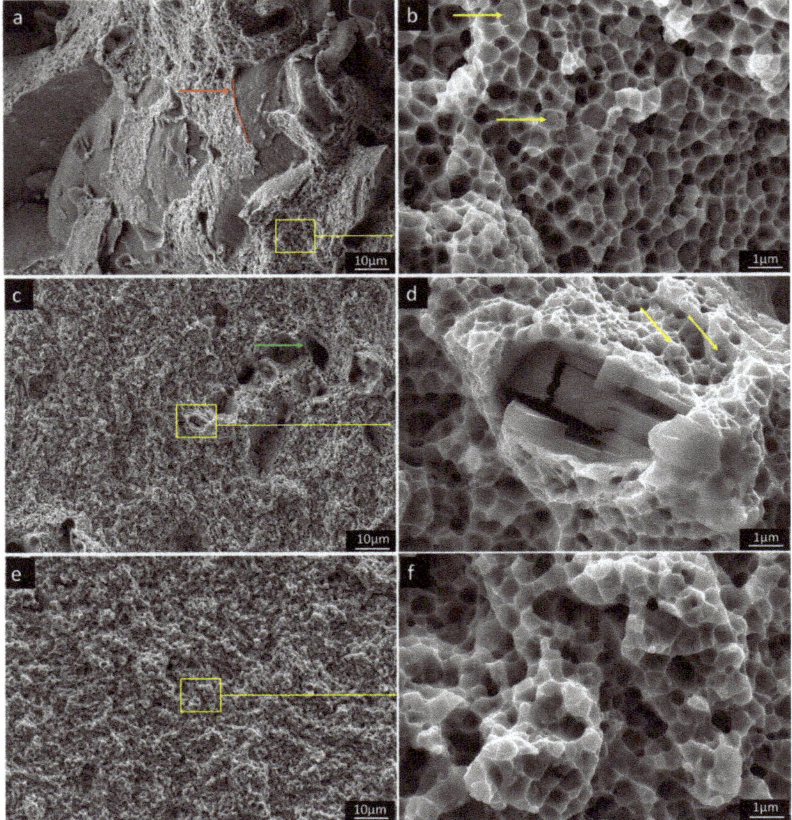

Figure 7. Fracture surfaces of $Ni_{1.5}Co_{1.5}CrFeTi_XA$ material; oxide particles inside the ductile dimples are marked by yellow arrows. (**a**,**b**) Ti0.3A, showing ductile dimples and residual porosity—the edge of the former pore is denoted by a red arrow and spline; (**c**,**d**) Ti0.5A showing ductile dimples and holes are denoted by green arrow with larger fractured oxide particles; (**e**,**f**) Ti0.7A exhibiting very flat ductile fracture surface and small dimples lacking oxide particles. Oxide particles inside the ductile dimples are marked by yellow arrows.

4. Discussion

Considering the influence of the Ti concentration on the properties of the milled powders, with its increasing concentration, the powder particle size is decreasing. This can be associated with increasing the intrinsic strength of the alloy by Ti. The extent of solid solution strengthening given approximately by atomic size difference in Table 7 is increasing with increasing Ti. This causes strengthening of the FCC matrix but decreases in its plasticity. Consequently, the powders have a lower tendency to agglomerate, resulting in smaller particle size.

Table 7. The calculated atomic size difference of produced $Ni_{1.5}Co_{1.5}CrFeTi_X$ alloys in a single-phase state.

Alloy	Ti0.3	Ti0.5	Ti0.7
Atomic size difference δ (%)	4.08	5.03	5.71

The increase in the Ti concentration has a positive influence on the densification rate of $Ni_{1.5}Co_{1.5}CrFeTi_X$ alloy (at the same sintering temperatures) since an alloy with the least Ti (Ti0.3) alloy showed the largest porosity after SPS. This can be caused by a decreasing of melting temperature by Ti (Figure 2), which promotes faster densification.

Considering only the comparison of TiXA alloys between each other, the increase in strength associated with grain boundary strengthening can be neglected due to a relatively narrow range of obtained grain sizes (from 1.3 μm to 2.13 μm in annealed state). In the same way, Orowan strengthening by the oxide inclusions can also be omitted since the volume fraction of oxides is similar (~3.5%) in all compositions. Consequently, the aforementioned increase of strength of the solid solution by Ti is directly observed from the increase of lattice parameter [34] from 3.58 Å in Ti0.3A material, to 3.6Å in Ti0.7A caused by Ti. This is in good agreement with the increasing hardness and tensile strength by Ti, especially in annealed conditions. On the other hand, the ductility of the materials seems to be decreasing with increasing Ti concentration. It should be noted that low ductility of the Ti0.3A material was caused by the porosity, rather than by the inherent properties of the FCC solid solution.

All the feedstock powders were probably contaminated on their surfaces by oxygen prior to MA. After the MA and SPS densification, even though it was problematic to measure their precise chemical composition, we can safely say that the oxides contained mostly Ti, due to its highest affinity to oxygen with respect to other present elements. This suggests that all the oxides of different elements recombined during MA (in situ) or subsequent SPS sintering into Ti-rich oxides. This also exhorts the possibility of modification of these oxides by different elements with even higher affinity to oxygen than Ti (for example Y, Ce) as shown before in [35], in order to form more favorable and finer oxide dispersion. Compared to the previous study [15], the oxide contamination was in the present case significantly larger due to use of ethanol wet-milling after dry milling.

It is interesting to point out that, despite the significant contamination by oxides, which is to some extent inherent to the utilized processing route, the mechanical properties were not significantly deteriorated.

With increase in the Ti concentrations, a significant increase in propensity to form ordered intermetallic phases ($L1_2$ or DO_{24} type) in the FCC matrix phase. At the same time, Ti decreases the melting temperature of the $Ni_{1,5}Co_{1,5}CrFeTi_X$ alloys, due to a eutectic-type of transformation. However, the intermetallic phases predicted by the CALPHAD calculations were not formed even in the SPS-ed materials, or only in the minor fractions (Ti0.7). This occurred due to their slow formation, in combination with lower processing temperature of the PM process and relatively fast cooling after SPS [14].

Table 8 presents the comparison of our tensile results with other materials produced by the same method. Note that there is only a very limited amount of studies showing tensile testing on PM high entropy materials, due to certain problems with preparation of sufficient material volume. Table 8 shows that our results are equivalent, or even superior to other powder metallurgy (PM) HEAs in terms of tensile properties. The full-density Ti0.5A and Ti0.7A alloys even exhibit comparable properties to commercial FCC steels with similar strength levels.

The presented results show that HEAs such as $Ni_{1,5}Co_{1,5}CrFeTi_X$ show a very good tolerance for inclusions and changes of chemical compositions. With increasing use of scrap metal and energy saving technologies aiming at the reduction of the consumption of primary raw materials, such HEAs may play a role in future sustainable metallurgy production [36].

Table 8. The comparison of mechanical properties of produced $Ni_{1,5}Co_{1,5}CrFeTi_X$ alloys with other similar materials. The values for different PM materials are taken from Refs. [9,37]. The values of wrought steels are taken from https://www.materials.sandvik/cz.

Material	$R_{p0.2}$ (MPa)	R_m (MPa)	A (%)
G-Ti0.3	781.5	845	4.4
G-Ti0.5	930.5	1199	11.9
G-Ti0.7	1281.5	1618.5	8.2
Fe30Ni30Co29Mn5.5Cu5.5 + TiC	495	710	11
Fe30Ni30Co29Mn5.5Cu5.5	1125	1276	10
CoCrNi + Boride	1425	1432	1.86
X 7 CrNiAl 17-7—precipitation strengthened	1150	1300	12
SANDVIK 316LVM cold rolled	800	1100	12
X 10 CrNi 18-8 cold rolled	1150	1300	15

5. Conclusions

- The increase in the Ti concentration decreases the average particle size of the powder during mechanical alloying of $Ni_{1.5}Co_{1.5}CrFeTi_X$.
- The thermodynamic calculation with the ThermoCalc TCHEA3 database showed satisfactory prediction accuracy in relation to the obtained experimental data.
- The Ti increases the strength of the $Ni_{1.5}Co_{1.5}CrFeTi_X$ alloys by increasing average atomic size misfit i.e., solid solution strengthening.
- An excellent combination of high strength and ductility can be obtained in $Ni_{1.5}Co_{1.5}CrFeTi_x$ alloys by combining mechanical alloying, SPS and solution annealing.
- The mechanical alloying in ethanol resulted in the formation of oxide dispersion in the SPS-ed bulks.
- Despite the oxide formation, the mechanical properties were not significantly impeded.

Author Contributions: Conceptualization, I.M. and S.G.; methodology, I.M. and L.M.-G.; software, L.M.-G.; investigation, I.M. and S.G.; material preparation, Z.K.; E.N. and M.K.; writing—original draft preparation, I.M. and S.G.; writing—review and editing, L.M.-G. and I.D.; visualization, I.M.; supervision, I.D.; funding acquisition, I.D., E.N. and M.K. All authors have read and agreed to the published version of the manuscript.

Funding: This research was funded by Czech Science Foundation Project No. 19-22016Sand RHP-Technology GmbH.

Acknowledgments: The authors would like to acknowledge the kind support of Jozef Zapletal and Zdenek Spotz for their contribution to the analysis. This work was financially supported by Czech Science Foundation Project No. 19-22016S.

Conflicts of Interest: The authors declare no conflict of interest.

Appendix A

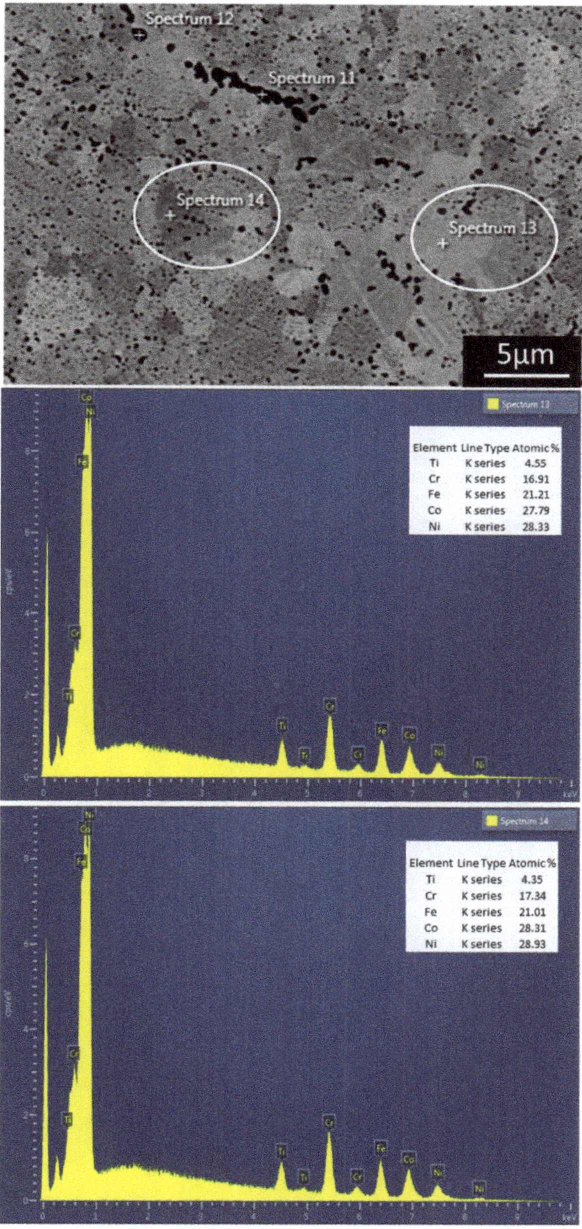

Figure A1. Representative EDS points in FCC matrix of Ti0.5 alloys in the SPS-ed state marked on the BSE micrograph. The presented points possess very close chemical composition despite the different shades of grey in the respective grains that spectras were taken from. Therefore, the contrast between different FCC grains in BSE mode is caused by differences in crystallographic orientation and not in chemistry.

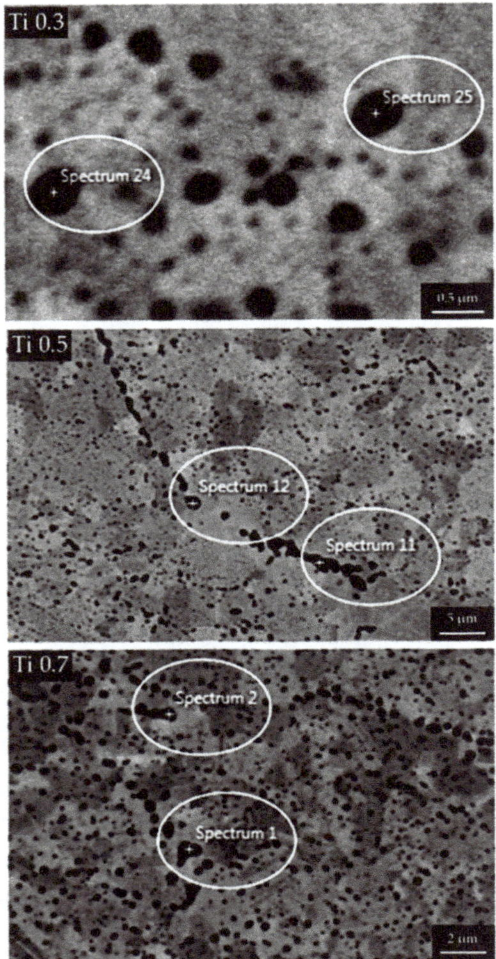

Figure A2. Representative EDS analysis of oxides on Tix alloys in SPS-ed state. The largest of the black dots were chose for point spectrum analysis to ensure that the majority of detected signal is coming from the particle and not from matrix below. The exact chemical compositions measured are presented in the Table A1 below. Micrograph of (**Ti0.5**) corresponds to an area also shown in Figure 5 in the main text.

Table A1. Chemical compositions of the oxides from Figure A1. The black dots represent oxide particles enriched mostly in Ti. The point EDS analysis cannot be considered as perfectly accurate measurement of chemical composition because even the largest of the oxides are still very small and the EDS signal coming out of them is contaminated by the signal from the Ni, Co, Cr and Fe rich FCC matrix below the particles. Therefore, residual traces of Ni, Co, Cr and Fe are probably a background noise.

Element	Ti0.3 Spectrum 24 (at.%)	Ti0.3 Spectrum 25 (at.%)	Ti0.5 Spectrum 11 (at.%)	Ti0.5 Spectrum 12 (at.%)	Ti0.7 Spectrum 1 (at.%)	Ti0.7 Spectrum 2 (at.%)
O	**30.21**	**41.46**	**56.89**	**57.20**	**37.6**	**26.51**
Ti	**50.50**	**30.95**	**40.55**	**40.11**	**44.9**	**38.98**
Cr	11.83	8.41	0.96	1.01	5.83	8.26
Fe	2.25	5.83	0.71	0.75	4.44	6.95
Co	2.67	6.87	0.33	0.56	4.63	10.01
Ni	2.41	5.65	0.56	0.37	2.61	9.29

References

1. Yeh, J.W.; Chen, S.K.; Lin, S.J.; Gan, J.Y.; Chin, T.S.; Shun, T.T.; Tsau, C.H.; Chang, S.Y. Nanostructured High-Entropy Alloys with Multiple Principal Elements: Novel Alloy Design Concepts and Outcomes. *Adv. Eng. Mater.* **2004**, *6*, 299–303. [CrossRef]
2. Cantor, B.; Chang, I.T.H.; Knight, P.; Vincent, A.J.B. Microstructural development in equiatomic multicomponent alloys. *Mater. Sci. Eng. A* **2004**, *375*, 213–218. [CrossRef]
3. Miracle, D.B.; Senkov, O.N. A critical review of high entropy alloys and related concepts. *Acta Mater.* **2017**, *122*, 448–511. [CrossRef]
4. George, E.P.; Raabe, D.; Ritchie, R.O. High-entropy alloys. *Nat. Rev. Mater.* **2019**, *4*, 515–534. [CrossRef]
5. Wei, X.F.; Liu, J.X.; Li, F.; Qin, Y.; Liang, Y.C.; Zhang, G.J. High entropy carbide ceramics from different starting materials. *J. Eur. Ceram. Soc.* **2019**, *39*, 2989–2994. [CrossRef]
6. Ding, J.; Yu, Q.; Asta, M.; Ritchie, R.O. Tunable stacking fault energies by tailoring local chemical order in CrCoNi medium-entropy alloys. *Proc. Natl. Acad. Sci. USA* **2018**, *115*, 8919–8924. [CrossRef]
7. Luo, H.; Li, Z.; Lu, W.; Ponge, D.; Raabe, D. Hydrogen embrittlement of an interstitial equimolar high-entropy alloy. *Corros. Sci.* **2018**, *136*, 403–408. [CrossRef]
8. Guo, S.; Liu, C.T. Phase stability in high entropy alloys: Formation of solid-solution phase or amorphous phase. *Prog. Nat. Sci. Mater. Int.* **2011**, *21*, 433–446. [CrossRef]
9. Moravcik, I.; Gouvea, L.; Cupera, J.; Dlouhy, I. Preparation and properties of medium entropy CoCrNi/boride metal matrix composite. *J. Alloys Compd.* **2018**, *748*, 979–988. [CrossRef]
10. Wei, S.; Kim, J.; Tasan, C.C. Boundary micro-cracking in metastable Fe45Mn35Co10Cr10 high-entropy alloys. *Acta Mater.* **2019**, *168*, 76–86. [CrossRef]
11. Gorsse, S.; Miracle, D.B.; Senkov, O.N. Mapping the world of complex concentrated alloys. *Acta Mater.* **2017**, *135*, 177–187. [CrossRef]
12. Luo, H.; Li, Z.; Mingers, A.M.; Raabe, D. Corrosion behavior of an equiatomic CoCrFeMnNi high-entropy alloy compared with 304 stainless steel in sulfuric acid solution. *Corros. Sci.* **2018**, *134*, 131–139. [CrossRef]
13. Moravcik, I.; Hadraba, H.; Li, L.; Dlouhy, I.; Raabe, D.; Li, Z. Yield strength increase of a CoCrNi medium entropy alloy by interstitial nitrogen doping at maintained ductility. *Scr. Mater.* **2020**, *178*, 391–397. [CrossRef]
14. Moravcikova-Gouvea, L.; Moravcik, I.; Omasta, M.; Vesely, J.; Cizek, J.; Minarik, P.; Cupera, J.; Zadera, A.; Jan, V.; Dlouhy, I. High-strength $Al_{0.2}$ $Co_{1.5}CrFeNi_{1.5}Ti$ high-entropy alloy produced by powder metallurgy and casting: A comparison of microstructures, mechanical and tribological properties. *Mater. Charact.* **2019**, *159*, 110046. [CrossRef]
15. Moravcik, I.; Cizek, J.; Zapletal, J.; Kovacova, Z.; Vesely, J.; Minarik, P.; Kitzmantel, M.; Neubauer, E.; Dlouhy, I. Microstructure and mechanical properties of $Ni_{1.5}Co_{1.5}CrFeTi_{0.5}$ high entropy alloy fabricated by mechanical alloying and spark plasma sintering. *Mater. Des.* **2017**, *119*, 141–150. [CrossRef]
16. Chuang, M.H.; Tsai, M.H.; Wang, W.R.; Lin, S.J.; Yeh, J.W. Microstructure and wear behavior of $AlxCo_{1.5}CrFeNi_{1.5}Tiy$ high-entropy alloys. *Acta Mater.* **2011**, *59*, 6308–6317. [CrossRef]

17. Moravcik, I.; Gouvea, L.; Hornik, V.; Kovacova, Z.; Kitzmantel, M.; Neubauer, E.; Dlouhy, I. Synergic strengthening by oxide and coherent precipitate dispersions in high-entropy alloy prepared by powder metallurgy. *Scr. Mater.* **2018**, *157*, 24–29. [CrossRef]
18. Chang, Y.J.; Yeh, A.C. The evolution of microstructures and high temperature properties of $AlxCo_{1.5}CrFeNi_{1.5}Tiy$ high entropy alloys. *J. Alloys Compd.* **2015**, *653*, 379–385. [CrossRef]
19. Toda-Caraballo, I.; Rivera-Díaz-del-Castillo, P.E.J. Modelling solid solution hardening in high entropy alloys. *Acta Mater.* **2015**, *85*, 14–23. [CrossRef]
20. Hume-Rothery, W.; Coles, B.R. The transition metals and their alloys. *Adv. Phys.* **1954**, *3*, 149–242. [CrossRef]
21. Yang, X.; Zhang, Y. Prediction of high-entropy stabilized solid-solution in multi-component alloys. *Mater. Chem. Phys.* **2012**, *132*, 233–238. [CrossRef]
22. Chen, H.; Kauffmann, A.; Laube, S.; Choi, I.C.; Schwaiger, R.; Huang, Y.; Lichtenberg, K.; Müller, F.; Gorr, B.; Christ, H.J.; et al. Contribution of Lattice Distortion to Solid Solution Strengthening in a Series of Refractory High Entropy Alloys. *Metall. Mater. Trans. A* **2017**, *49*, 772–781. [CrossRef]
23. Takeuchi, A.; Inoue, A. Classification of Bulk Metallic Glasses by Atomic Size Difference, Heat of Mixing and Period of Constituent Elements and Its Application to Characterization of the Main Alloying Element. *Mater. Trans.* **2005**, *46*, 2817–2829. [CrossRef]
24. Varalakshmi, S.; Kamaraj, M.; Murty, B.S. Processing and properties of nanocrystalline CuNiCoZnAlTi high entropy alloys by mechanical alloying. *Mater. Sci. Eng. A* **2010**, *527*, 1027–1030. [CrossRef]
25. Dieter, G.E.; Bacon, D. *Mechanical Metallurgy*, 3rd ed.; Mc Graw-Hill Book Co.: New York, NY, USA, 1986.
26. Christofidou, K.A.; McAuliffe, T.P.; Mignanelli, P.M.; Stone, H.J.; Jones, N.G. On the prediction and the formation of the sigma phase in CrMnCoFeNix high entropy alloys. *J. Alloys Compd.* **2019**, *770*, 285–293. [CrossRef]
27. Heo, Y.U.; Takeguchi, M.; Furuya, K.; Lee, H.C. Transformation of DO24 η-Ni3Ti phase to face-centered cubic austenite during isothermal aging of an Fe–Ni–Ti alloy. *Acta Mater.* **2009**, *57*, 1176–1187. [CrossRef]
28. Pickering, E.J.; Muñoz-Moreno, R.; Stone, H.J.; Jones, N.G. Precipitation in the equiatomic high-entropy alloy CrMnFeCoNi. *Scr. Mater.* **2016**, *113*, 106–109. [CrossRef]
29. Zaefferer, S.; Elhami, N.N. Theory and application of electron channelling contrast imaging under controlled diffraction conditions. *Acta Mater.* **2014**, *75*, 20–50. [CrossRef]
30. Shishkin, A.; Drozdova, M.; Kozlov, V.; Hussainova, I.; Lehmhus, D. Vibration-Assisted Sputter Coating of Cenospheres: A New Approach for Realizing Cu-Based Metal Matrix Syntactic Foams. *Metals* **2017**, *7*, 16. [CrossRef]
31. Shishkin, A.; Hussainova, I.; Kozlov, V.; Lisnanskis, M.; Leroy, P.; Lehmhus, D. Metal-Coated Cenospheres Obtained via Magnetron Sputter Coating: A New Precursor for Syntactic Foams. *JOM* **2018**, *70*, 1319–1325. [CrossRef]
32. Mecking, H.; Kocks, U.F. Kinetics of flow and strain-hardening. *Acta Metall.* **1981**, *29*, 1865–1875. [CrossRef]
33. Noell, P.J.; Carroll, J.D.; Boyce, B.L. The Mechanisms of Ductile Rupture. *Acta Mater.* **2018**, *161*, 83–98. [CrossRef]
34. Labusch, R. A Statistical Theory of Solid Solution Hardening. *Phys. Status Solidi* **1970**, *41*, 659–669. [CrossRef]
35. Hadraba, H.; Chlup, Z.; Dlouhy, A.; Dobes, F.; Roupcova, P.; Vilemova, M.; Matejicek, J. Oxide dispersion strengthened CoCrFeNiMn high-entropy alloy. *Mater. Sci. Eng. A* **2017**, *689*, 252–256. [CrossRef]
36. Rombach, G. Raw material supply by aluminium recycling—Efficiency evaluation and long-term availability. *Acta Mater.* **2013**, *61*, 1012–1020. [CrossRef]
37. Fu, Z.; MacDonald, B.E.; Dupuy, A.D.; Wang, X.; Monson, T.C.; Delaney, R.E.; Pearce, C.J.; Hu, K.; Jiang, Z.; Zhou, Y.; et al. Exceptional combination of soft magnetic and mechanical properties in a heterostructured high-entropy composite. *Appl. Mater. Today* **2019**, *15*, 590–598. [CrossRef]

Article

Structure and Properties of Fe–Al–Si Alloy Prepared by Mechanical Alloying

Pavel Novák [1,*], Tomáš Vanka [1], Kateřina Nová [1], Jan Stoulil [1], Filip Průša [1], Jaromír Kopeček [2], Petr Haušild [3] and František Laufek [2,4]

1. Department of Metals and Corrosion Engineering, University of Chemistry and Technology, Prague, Technická 5, Prague 6, 166 28 Prague, Czech Republic
2. Institute of Physics of the ASCR, v. v. i., Na Slovance 2, Prague 8, 182 21 Prague, Czech Republic
3. Department of Materials, Faculty of Nuclear Sciences and Physical Engineering, Czech Technical University in Prague, Trojanova 13, Prague 2, 120 00 Prague, Czech Republic
4. Czech Geological Survey, Geologická 6, Prague 5, 152 00 Prague, Czech Republic

* Correspondence: panovak@vscht.cz

Received: 3 July 2019; Accepted: 31 July 2019; Published: 2 August 2019

Abstract: Fe–Al–Si alloys have been previously reported as an interesting alternative to common high-temperature materials. This work aimed to improve the properties of FeAl20Si20 alloy (in wt.%) by the application of powder metallurgy process consisting of ultrahigh-energy mechanical alloying and spark plasma sintering. The material consisted of Fe_3Si, FeSi, and $Fe_3Al_2Si_3$ phases. It was found that the alloy exhibits an anomalous behaviour of yield strength and ultimate compressive strength around 500 °C, reaching approximately 1100 and 1500 MPa, respectively. The results also demonstrated exceptional wear resistance, oxidation resistance, and corrosion resistance in water-based electrolytes. The tested manufacturing process enabled the fracture toughness to be increased ca. 10 times compared to the cast alloy of the same composition. Due to its unique properties, the material could be applicable in the automotive industry for the manufacture of exhaust valves, for wear parts, and probably as a material for selected aggressive chemical environments.

Keywords: iron silicide; Fe–Al–Si alloy; mechanical alloying; spark plasma sintering; characterization

1. Introduction

Materials based on Fe–Al ordered phases have been subjected to extensive research and development over the last decades, due to their low cost, low density, high specific strength, high creep resistance, as well as excellent high-temperature oxidation and sulphidation resistance [1]. The positive effect of aluminium on the heat-resistance of iron-based alloys was reported already in 1894 [2]. However, because of room temperature brittleness and manufacturing problems, these alloys were only sporadically applied for a long time. A significant advance in the production of these materials was achieved in former Czechoslovakia in the 1950s, where a Fe–Al–C alloy called Pyroferal was developed [3], being composed of FeAl phase and particles of Al_4C_3 carbide [3]. This material with excellent oxidation resistance and good high-temperature mechanical properties was designed to be produced by casting, which made it easily producible and cost-efficient [3]. However, several problems arose during application of this material. In acidic solutions or high-temperature environments with water vapour, aluminium carbide was hydrolysed to methane, which destroyed the material. Therefore, the extensive search for adequate carbon-free Fe–Al alloys began. Several alloy types containing Nb, Ta, Cr and other elements have been developed [4–10]. However, due to the economic situation of the EU, most of these elements are now listed as critical raw materials (CRMs) [11]. Hence, the development of these alloys is not preferred.

Recently, Fe–Al–Si alloys have been developed by our team as another carbon-free alternative to iron-based intermetallic alloys. Even though silicon is also listed as a CRM, the problem can be easily solved by the use of silicon recovered from recycled electronics, because a lower purity is required for these alloys than that used in electronics. These alloys are characterised by extremely good high-temperature oxidation resistance in air and hardness [12]. The oxidation resistance was found to be given by aluminium oxide layer on the surface and also additionally by the sub-layer enriched by iron silicides as a result of the consumption of aluminium upon formation of the oxide layer [12]. Further published papers also demonstrated the exceptional oxidation resistance of Fe–Si–Al ternary alloy in CO_2–H_2O gaseous atmospheres [13]. Since the processing of iron-based intermetallics by conventional metallurgical routes is relatively problematic [14–16], powder metallurgy using reactive sintering has previously been tested for the Fe–Al–Si alloys [17]. Even though a deep optimisation of the process was carried out [17], there was still a significant portion of residual porosity which affected the results of the high-temperature oxidation tests, as internal oxidation was observed in the case of some Fe–Al–Si alloys [12]. In addition, the microstructure was relatively coarse, and the alloys were determined to be very brittle at the room temperature, having a fracture toughness of ca. 0.35 MPa·$m^{1/2}$. Due to high residual porosity of the samples produced by reactive sintering, it was not possible to measure the intrinsic mechanical and corrosion properties (e.g., tensile/compressive strength, wear resistance, corrosion behaviour in electrolytes, etc.) and, therefore, they remain unknown for this alloy.

Since it has been reported for several alloy systems that structure refinement could lead to an increase in fracture toughness [18,19] and, also, in order to overcome the abovementioned manufacturing obstacles, a different powder metallurgical route was tested for the preparation of this alloy—a combination of ultrahigh-energy mechanical alloying and spark plasma sintering. Mechanical alloying, as one of the most efficient techniques applicable to achieve fine-grained structure [20,21], is in fact high-energy milling, usually using ball mills. In this process, the high kinetic energy of balls causes the following phenomena: crushing of particles leading to the reduction of the particle size; local mechanical joining and welding of particles by plastic deformation, friction forces and diffusion; structure refinement due to severe plastic deformation; and formation of solid solutions and chemical compounds (intermetallics) [22]. The process of mechanical alloying usually takes tens of hours [23–25], which is its main disadvantage. In our previous work [26] we developed the ultrahigh-energy mechanical alloying process, which enables intermetallics to be obtained from metallic powders already in the time between 120 and 240 min. Our process uses high ball-to-powder ratio (over 50:1) and high rotational velocity (at least 400 rpm). No lubrication medium is added in our technology in order to maximise the friction forces between the balls, powder, and wall of the milling vessel [26].

Spark plasma sintering (SPS) is a modern compaction method which uses uniaxial pressing accompanied by passage of the electric current through the sample. In the SPS process, rapid heating of the sample is caused by Joule heat, which accompanies the passage of high electric current and electric discharges (spark or plasma) between particles [27–29]. The SPS method is especially suitable for the consolidation of nanocrystalline materials and phases with lower thermal stability due to high sintering rate and a corresponding reduction of thermal exposure of the consolidated material [30].

In this work, the combination of unique ultrahigh-energy mechanical alloying and spark plasma sintering was tested in order to improve the mechanical properties of FeAl20Si20 alloy (in wt %). This alloy composition has previously been determined to be the most resistant against high-temperature oxidation among Fe–Al–Si alloys [12], but also highly brittle at room temperature when being produced by reactive sintering powder metallurgy or casting. Complex characterisation of the alloy prepared by the abovementioned process was carried out in order to find the future application range for this alloy.

2. Materials and Methods

FeAl20Si20 alloy (in wt %) was prepared by the combination of mechanical alloying (MA) and subsequent spark plasma sintering (SPS). Mechanical alloying was carried out in a planetary ball mill (PM 100 CM, Retsch, Haan, Germany)) under the following conditions, as optimised in our previous paper dealing with the synthesis of intermetallics [26]:

- milling duration: 240 min;
- change of rotation direction each 30 min;
- rotation speed: 400 rpm;
- atmosphere: argon;
- powder batch: 5 g;
- ball-to-powder weight ratio: 70:1.

For mechanical alloying, the following elemental powders were applied: iron (purity 99.9%, particle size <44 µm, supplied by Strem Chemicals, Newburyport, MA, USA), aluminium (purity 99.7%, particle size <44 µm, supplied by Strem Chemicals). and silicon (purity 99.5%, particle size <44 µm, supplied by Alfa Aesar, Haverhill, MA, USA).

The powder was consolidated by SPS method using a HP D10 device (FCT Systeme GmbH, Rauenstein, Germany). A pressure of 48 MPa was applied at 1000 °C for 10 min with a previous heating rate of 300 K/min and a cooling rate of 50 K/min. The weight of the batch for sintering was 5 g.

The microstructure of the alloys produced by a combination of ultrahigh-energy mechanical alloying and spark plasma sintering was studied using a VEGA 3 LMU scanning electron microscope (TESCAN, Brno, Czech Republic) after etching using modified Kroll's reagent (5 mL HNO_3, 10 mL HF, and 85 mL H_2O). Phase composition was identified by X-ray diffraction analysis (XRD) using a X'Pert Pro X-ray diffractometer (PANalytical, Almelo, The Netherlands). The amounts of phases, and their lattice parameters and crystallite sizes were calculated from the XRD data using Rietveld pattern refinement by the means of Topas 5 software.

The distribution of the phases in the sample was imaged using an electron backscatter diffraction (EBSD) analyser (manufactured by EDAX) on a FERA III scanning electron microscope (TESCAN, Brno, Czech Republic). Porosity was measured by image analysis (ImageJ 1.48) on the polished non-etched samples as the area fraction of pores.

Mechanical properties of the SPS-consolidated material were determined by the means of hardness measurement, determination of fracture toughness, and compression tests. Hardness was measured using the Vickers method with a load of 30 kg. Fracture toughness was determined by the indentation test (Vickers indenter FM-700 (Future-Tech, Kawasaki-City, Japan) with the load of 1 kg) and evaluated from the length of cracks by Palmqvist Equation (1) [31]:

$$K_C = 0.016 \times \left(\frac{E}{HV}\right)^{1/2} \times \left(\frac{F}{c^{3/2}}\right) \tag{1}$$

where E is the Young´s modulus (Pa), HV is the Vickers hardness, F is the applied load (N), and c is the half length of the crack after indentation (m).

Compression tests were carried out using LabTest 5.250SP1-VM universal loading machine (produced by LaborTech, Kateřinky, Czech Republic) at the following temperatures: room temperature, 400, 500, 600, and 700 °C with the initial deformation rate of 0.001 s^{-1}.

The wear resistance was measured using modified pin-on-disc method, where the "pin" was the tested sample and "disc" was a grinding paper P1200. The normal force used in the test was 5.8 N. The wear rate was calculated from the measured weight losses by the Equation (2) [32]:

$$w = \frac{\Delta m \times 1000}{\rho \times l} \tag{2}$$

where w, Δm, ρ, and l are wear rate ($mm^3 \cdot m^{-1} \cdot N^{-1}$), weight loss (g), density (5.91 $g \cdot cm^{-3}$), and sliding distance on the grinding paper (2500 m), respectively. The density of the material was determined by the Archimedes method.

Cyclic and isothermal oxidation tests were carried out at 800 and 1000 °C in air. Oxidation rate was determined from weight gains caused by the oxide formation on the surface of thermally exposed samples. In isothermal oxidation tests, the samples were heated continuously in alumina crucibles for 100, 200, 300, and 400 h, followed by air-cooling.

Cyclic oxidation tests were applied in order to reveal the susceptibility of the oxide layers to the spallation due to thermally induced stresses. The duration of one oxidation cycle was 48 h. After each cycle, samples were air-cooled, weighed, and heated again to the test temperature.

The microstructure of the oxide layers was documented using a VEGA 3 LMU scanning electron microscope (TESCAN, Brno, Czech Republic), and phase composition was determined by XRD. Glow discharge optical emission spectroscopy (GDOES, Horiba JobinYvon GD Profiler II) was applied for a depth profile chemical analysis of the oxide layers.

To evaluate the oxidation kinetics, the parabolic rate constant was calculated for all oxidation durations according to Equation (3) [33]:

$$k_p = \frac{\left(\frac{\Delta m}{A}\right)^2}{t} \tag{3}$$

where k_p, Δm, A, and t are parabolic rate constant ($g^2 \cdot m^{-4} \cdot s^{-1}$), weight gain (g), exposed area (m^2), and duration of oxidation (s), respectively.

The thermal stability of the alloy was tested by the evaluation of microstructure and measurement of Vickers hardness after annealing at 800 and 1000 °C. The hardness was measured in the core of the material on a crosscut.

In order to test the behaviour of the material in water-based electrolytes, the FeAl20Si20 alloy was characterised by means of electrochemical impedance spectroscopy (EIS) on FAS2 potentiostat (Gamry Instruments, Warminster, PA, USA). A disc sample 20 mm in diameter and 5 mm thick was polished by grinding paper P220 prior to each measurement. Samples were exposed in a pressure cell and the O-ring defined the tested area to 0.8 cm^2. Conditions were settled for 60 min in the testing solution and then the measurement was started. EIS spectra were measured within the range of frequencies from 10 kHz to 1 mHz for less aggressive media and from 10 kHz to 10 mHz in more aggressive media. Sampling was conducted using 5 points per decade and a testing amplitude 20 mV according to open circuit potential (EOC). Saturated silver-silver chloride electrode (ACLE) was used as reference and Pt wire as a counter electrode.

Study of the material corrosion behaviour was carried out in the series of sulphuric acid solutions with pH 0, 1, 2, 3, 4, and 5.6 (which is demineralised water with dissolved carbon dioxide), in a water-based solution of 0.5 $mol \cdot dm^{-3}$ NaCl and in a solution of 2.2 $g \cdot dm^{-3}$ NaF in demineralised water. X-ray photoelectron spectroscopy (XPS) analysis using ESCAProbeP device (Omicron Nanotechnology, Abingdon, UK) equipped with an Al Kα (λ = 1486.7 eV) X-ray source was used to determine the chemical composition of the passive layer after exposure in a water-based environment. The spectra were measured with an energy step of 0.05 eV and normalised to the binding energy of C1s peak (285.0 eV). Measured spectra were evaluated in CasaXPS 2.3.15 software (IMFP, RSf etc are part of the software library). The data for the chemical state evaluation were obtained from the NIST (Aithersburg, MD, USA) X-ray Photoelectron Spectroscopy Database [34].

3. Results

3.1. Microstructure and Phase Composition of FeAl20Si20 Alloy

Microstructure of the FeAl20Si20 alloy prepared by mechanical alloying and spark plasma sintering is shown in Figure 1a. The alloy exhibits very low porosity (0.1 vol %). The alloy is composed of two types of iron silicides (FeSi and Fe_3Si) and the $Fe_3Al_2Si_3$ ternary phase. The distribution of individual microstructure constituents and overall phase composition were characterised by EBSD (Figure 1b), XRD (Figure 2), and EDS (Table 1).

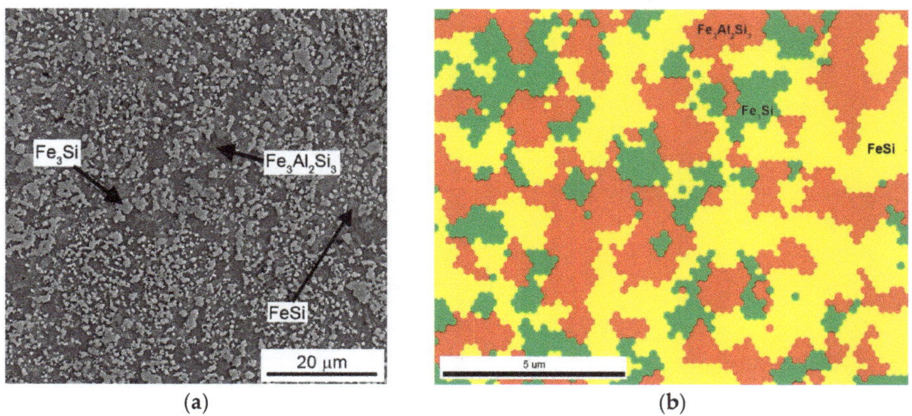

Figure 1. (**a**) Microstructure of FeAl20Si20 alloy produced by mechanical alloying and spark plasma sintering; (**b**) EBSD phase map of the alloy.

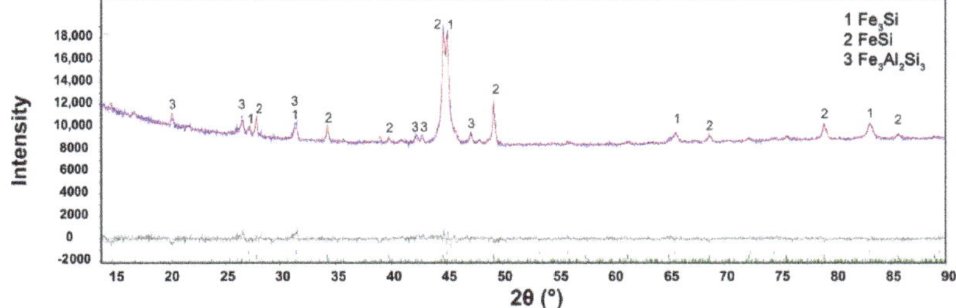

Figure 2. Rietveld plot of FeAl20Si20 alloy produced by mechanical alloying and spark plasma sintering.

The amount of Fe_3Si phase (Fm3m) was determined as 33.0 wt %. The lattice parameter was determined as 5.693×10^{-10} m, which is higher than the value presented in the PDF2 database (5.6700×10^{-10} m). The same behaviour was observed for FeSi phase ($P2_13$ structure), where the measured value of the lattice parameter reached 4.533×10^{-10} m, while the table value was 4.485×10^{-10} m. This indicates possible supersaturation of the phase by silicon or partial substitution or enrichment by aluminium. The presence of aluminium dissolved in silicides was proved by the EDS chemical microanalysis (Table 1). In the case of materials prepared by mechanical alloying, this kind of supersaturation is relatively common because the mutual solubility of elements is usually strongly increased by mechanical alloying. This is even observed in the examples, where mechanical alloying produced a solid solution of normally non-miscible elements, such as Mg and Fe [35].

Table 1. Average chemical composition of individual phases in FeAl20Si20 alloy (determined by EDS).

Phase	Chemical Composition (wt %)		
	Fe	Al	Si
Fe_3Si	83.9 ± 0.6	7.2 ± 0.6	8.9 ± 0.5
FeSi	65.5 ± 1.4	12.1 ± 3.2	22.4 ± 3.4
$Fe_3Al_2Si_3$	46.0 ± 4.3	27.6 ± 4.2	27.4 ± 4.3

The crystallite size of present phases is in the interval of 20–45 nm (Table 2), even though the size of individual particles of silicides reached approximately 0.2–5 μm (Figure 1a). This shows that the silicide particles observed by SEM are, in fact, polycrystals constituted of small nanosized grains. The crystallite size of $Fe_3Al_2Si_3$ ternary phase is ca. two times higher than this parameter of silicides. The probable reason is in the fact that the ternary phase was not present in the as-milled state [36] and thus formed during sintering, probably by the reaction of silicide phase and iron aluminide (FeAl with B2 structure), which disappeared during the sintering process. The determined phase composition corresponds well qualitatively with the Fe–Al–Si equilibrium phase diagram at 1000 °C [37]. However, a higher ratio between ternary phase and silicides could be expected, because the chemical composition of the alloy is very close to the ternary phase (Table 1).

Table 2. Weight fraction, lattice parameters, and crystallite size of individual phases in FeAl20Si20 alloy (determined by Rietveld refinement of XRD pattern).

Phase	Weight Fraction (wt %)	Space Group	Lattice Parameters ($\times 10^{-10}$ m)	Crystallite Size (nm)
Fe_3Si	33.0 ± 0.4	Fm3m	5.6934 ± 0.0004	23 ± 1
FeSi	37.5 ± 0.4	$P2_13$	4.5337 ± 0.0001	45 ± 1
			4.6512 ± 0.0001	
$Fe_3Al_2Si_3$	29.5 ± 0.4	P-1	6.3261 ± 0.0002	39 ± 1
			7.4990 ± 0.0002	

3.2. Mechanical and Tribological Properties of FeAl20Si20 Alloy

The hardness of the alloy reaches 811 ± 19 HV30 and the fracture toughness achieves the value of 3.50 ± 0.33 MPa $m^{1/2}$. The fracture toughness reaches rather low value, comparable mostly with ceramic materials (e.g., corundum-based ceramics) [38]. However, when the same alloy is produced by casting, it exhibits nearly ten times lower fracture toughness (0.35 ± 0.02 MPa $m^{1/2}$). This demonstrates that the structure refinement by mechanical alloying helps to reduce the room temperature brittleness of intermetallics.

The mechanical properties of the FeAl20Si20 alloy are summarised in Table 3. At room temperature, the alloy exhibits yield strength (YS) and ultimate compressive strength (UCS) of 1071 and 1085 MPa, respectively. The temperature dependence of yield strength and UCS shows anomalous behaviour. Both of these mechanical characteristics increase strongly at 500 °C and then rapidly decrease with temperature (Table 3). The abrasive wear rate of the FeAl20Si20 alloy was determined to be 4.5 ± 0.13 $mm^3\ m^{-1}$. For comparison, the AISI D2 cold work tool steel after the recommended regime of heat treatment achieves a value of approximate 15 $mm^3\ m^{-1}$ under the same test conditions. The results show that without the need for any heat treatment, the FeAl20Si20 alloy exhibits more than three times better wear resistance than properly heat-treated cold work tool steel with comparable compressive strength. Due to the absence of heat treatment, it can be expected that the wear resistance would not degrade strongly even when the temperature will increase, as is common during high-speed machining.

Table 3. Mechanical properties (yield strength and ultimate compressive strength) of FeAl20Si20 alloy in compression.

Temperature (°C)	Yield Strength (MPa)	UCS (MPa)
25	1071 ± 110	1085 ± 115
400	1101 ± 129	1140 ± 138
500	1163 ± 78	1508 ± 239
600	597 ± 103	1079 ± 236
700	348 ± 7	644 ± 1

3.3. High-Temperature Oxidation

Isothermal oxidation tests at 800 and 1000 °C revealed that a protective layer of aluminium oxide with minor admixtures of Fe_2O_3 and SiO_2 was formed (Figures 3 and 4). In addition, FeSi andFe_3Si phases were detected, which originate from the matrix below the oxide layer. The reason why the ternary Fe–Al–Si phase was not detected is discussed below. The main difference between the oxidation at 800 and 1000 °C is in the allotropic modification of aluminium oxide, which covers the surface of the material. At 1000 °C, α-Al_2O_3 (corundum, trigonal) is formed. On the other hand, γ-Al_2O_3 (cubic) is formed at 800 °C.

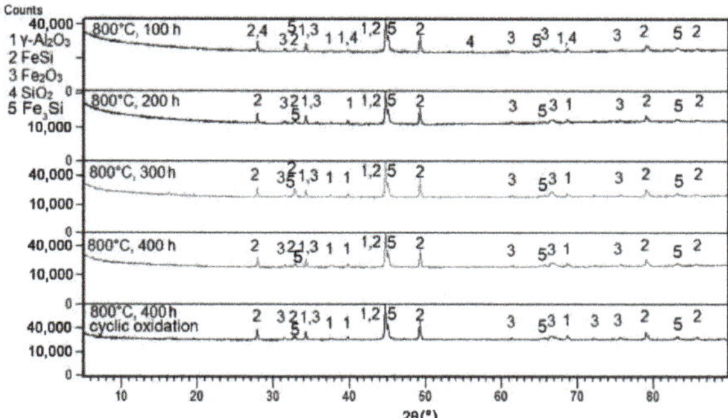

Figure 3. XRD patterns of FeAl20Si20 alloy after various durations of oxidation at 800 °C in air.

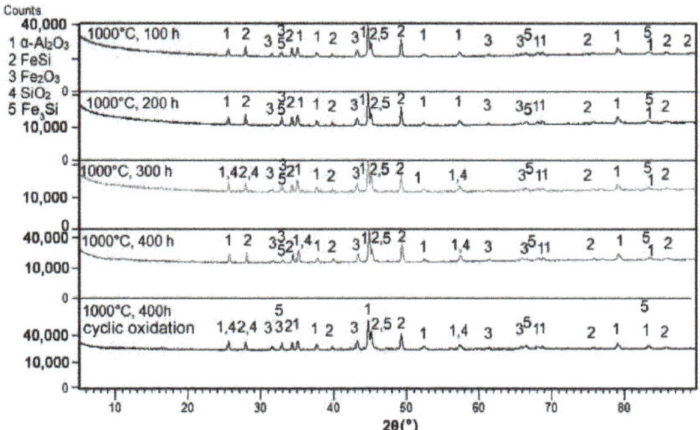

Figure 4. XRD patterns of FeAl20Si20 alloy after various durations of oxidation at 1000 °C in air.

At the first stage of oxidation, a mixture of Al_2O_3 and Fe_2O_3 is formed at 1000 °C. After 300 h, the formation of SiO_2 was detected (Figure 5). At a temperature of 800 °C, SiO_2 was detected by XRD after only 100 h of oxidation (Figure 4).

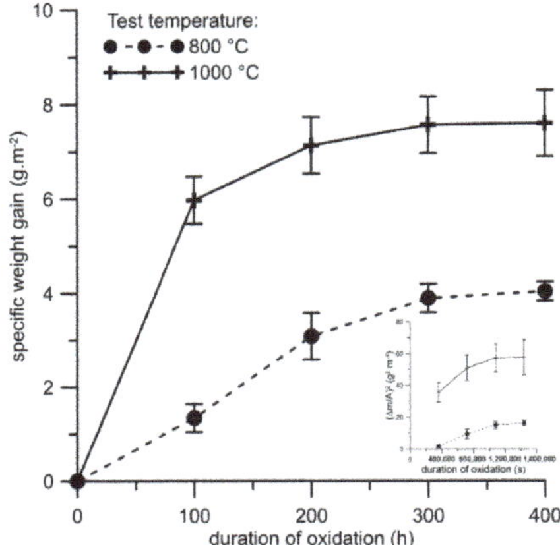

Figure 5. Dependence of specific weight gain (g m^{-2}) on the duration of isothermal oxidation at 800 and 1000 °C in air for FeAl20Si20 alloy, with the squared value of specific weight gain vs. time as insert.

The oxidation rate of the alloy, represented by weight gains caused by isothermal oxidation, is ca. two times lower at 800 °C than at 1000 °C (Figure 5). The growth of the oxide layer at 800 °C almost follows the parabolic law between 100 and 300 h, as seen on the insert in Figure 5, and it slows down after 300 h. This means that the oxidation is controlled by the diffusion of oxygen or aluminium through the oxide layer, and thus follows parabolic law. From the growth rate, it cannot be determined whether the layer grows inwards by oxygen diffusion or outwards by diffusion of aluminium to the surface because the diffusion rates of aluminium and oxygen in aluminium oxide are comparable [39]. The lowering of the oxidation rate at the end of the test is probably caused by the changes in chemical and phase composition below the oxide layer, as discussed below. At 1000 °C, the oxidation was much faster at the beginning, which is probably caused by reaction-controlled oxidation. After reaching a certain thickness of the oxide layer, the process changes to the diffusion-controlled mode. At the end of the oxidation test, the oxidation is slowed down due to the same effects as at 800 °C.

This shows that the oxide layer containing α-Al_2O_3 has superior protective effect, which was also confirmed by spallation of the oxide layer during the test. In the case of samples tested at 800 °C, the amount of delaminated oxides up to 0.4 g·m^{-2} was detected after 300–400 h of oxidation (Figure 6). On the contrary, no spallation was detected during isothermal oxidation at 1000 °C. The Pilling-Bedworth ratio [40] was calculated for this material and α-Al_2O_3 oxide layer as ca. 1.7, which indicates a layer with protective effect (a Pilling-Bedworth ratio below 1 indicates that no continuous oxide layer is formed on the surface, while a ratio above 2 implies layer spallation [40]).

Cyclic oxidation was also tested to prove the adherence of the oxide layer during cooling and heating up to the test temperature.

The weight gains due to cyclic oxidation (Figure 7) were slightly higher than during the isothermal test (Figure 5). At both temperatures, the oxidation follows parabolic law with just minor deviations, as can be seen in the insert in Figure 7. The calculated parabolic rate constants are slightly higher than during the isothermal oxidation (Table 4), probably due to defects in the oxide layers caused by the stresses induced during thermal cycling. The spallation of the oxide layer was observed at both test temperatures, but more significantly at 800 °C (Figure 8). However, the amounts of delaminated oxides were very low (less than 0.2 g m^{-2} at 1000 °C and 0.5 g·m^{-2} at 800 °C).

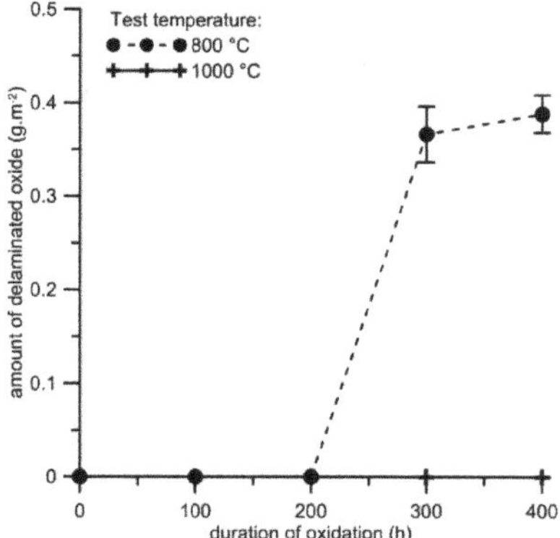

Figure 6. Dependence of the amount of delaminated oxide (g 006^{-2}) on the duration of isothermal oxidation at 800 and 1000 °C in air for FeAl20Si20 alloy.

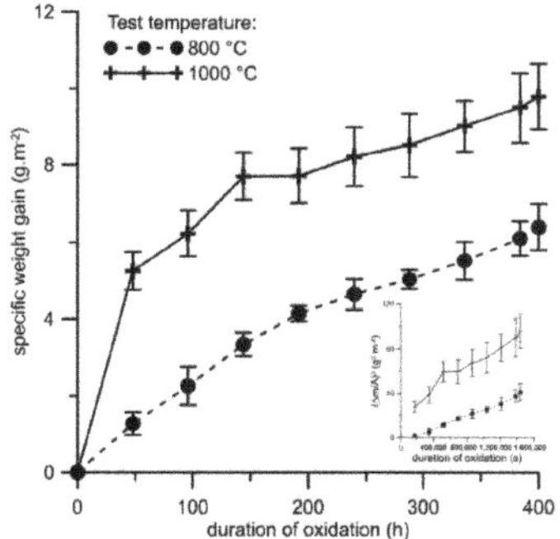

Figure 7. Dependence of specific weight gain (g m^2) on duration of cyclic oxidation at 800 and 1000 °C in air for FeAl20Si20 alloy, with the squared value of specific weight gain vs. time as insert.

Table 4. Parabolic rate constants (k_p) of isothermal and cyclic oxidation at 800 and 1000 °C.

Test Mode	Cyclic Oxidation Test		Isothermal Oxidation Test	
Temperature (°C)	800 °C	1000 °C	800 °C	1000 °C
k_p ($g^2 \cdot m^{-4} \cdot s^{-1}$)	2.58×10^{-5}	7.22×10^{-5}	1.09×10^{-5}	5.46×10^{-5}

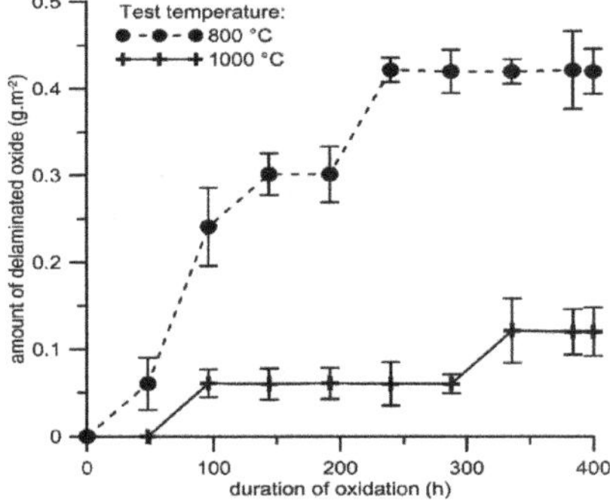

Figure 8. Dependence of the amount of delaminated oxide ($g \cdot m^{-2}$) on duration of cyclic oxidation at 800 and 1000 °C in air for FeAl20Si20 alloy.

To describe the chemical composition of the oxide layers, concentration-depth profiles of the present elements on the samples after cyclic oxidation were measured by GDOES (Figures 9 and 10). The high content of aluminium in the whole surface layer confirms that aluminium oxide is the main constituent of the oxide layer. The GDOES elemental profile also shows that the material is enriched by silicon and depleted by aluminium below the oxide layer. This implies that oxidation probably proceeds on the outer surface, controlled by the diffusion of aluminium through the oxide layer.

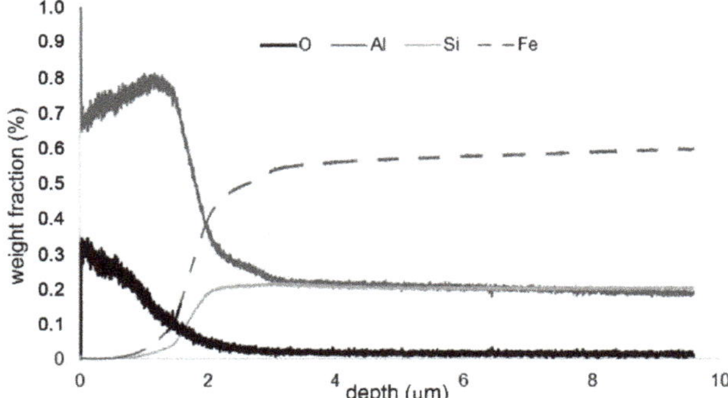

Figure 9. Concentration-depth profile of FeAl20Si20 alloy after cyclic oxidation at 800 °C for 400 h.

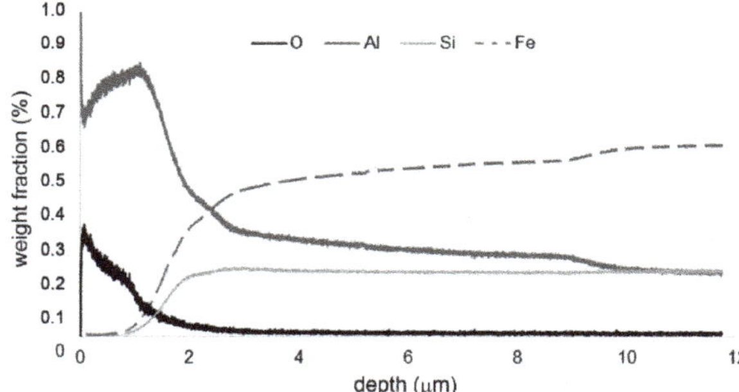

Figure 10. Concentration-depth profile of FeAl20Si20 alloy after cyclic oxidation at 1000 °C for 400 h.

Thermal stability was evaluated by the measurement of Vickers hardness (HV 30) after annealing at 800 and 1000 °C (Figure 11). The measurements were carried out in the core of the material on a crosscut. Only minor variations in the hardness can be seen after annealing at both temperatures, mostly lower than the standard deviation of the results. This indicates that the alloy is highly thermally stable, and that no significant changes in the structure occur during high-temperature exposure.

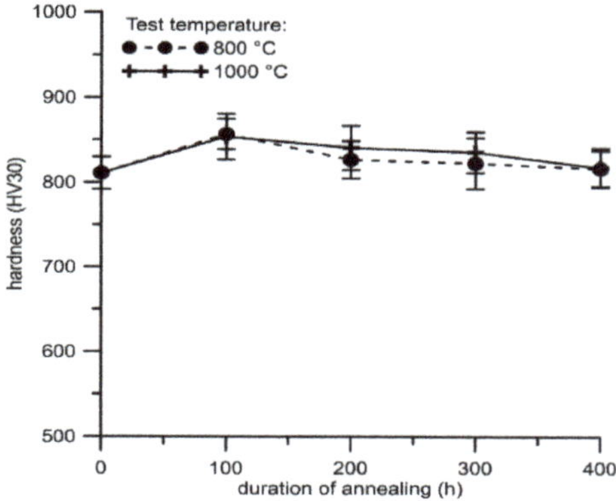

Figure 11. Hardness of FeAl20Si20 alloy after annealing at 800 and 1000 °C vs. duration of annealing.

3.4. Corrosion in Water-Based Electrolytes

The surface of samples shows two variations of behaviour in water-based environments. One is completely active, without passive layer, and is described by an equivalent circuit in Figure 12a. The second one is a surface with passive layer, as shown in Figure 12b.

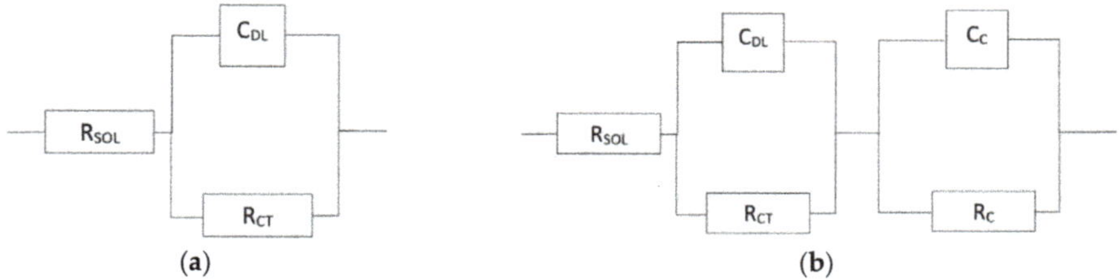

Figure 12. (**a**) Equivalent circuit with one RC couple (active surface), (**b**) Equivalent circuit with two RC couples (passive surface), (RSOL—resistance of the solution; RC—resistance of the passive layer; RCT—charge transfer resistance; CDL—capacitance of electrical double layer; CC—capacitance of passive layer).

The corrosion behaviour of the FeAl20Si20 alloy is presented in Figure 13. Phase shift (Figure 13b) shows two time constants in spectra of demi water, chloride solution, and sulphuric acid with pH 2. Both time constants in demineralised water and chloride solution are close to each other and with high values of impedance modulus (Figure 13a). This means that there is thick natural passive layer based on aluminium oxide, as shown by XPS (Table 5, Figure 14). Low pH dissolves the aluminium oxide passive layer and it is probably replaced by a very thin passive layer based on silicon oxide. The spectra recorded in low pH solutions still show two time constants, but significantly separated, and the impedance modulus is low. The fluoride solution does not allow for the formation of any passive layer, and the surface is completely active with only one time constant in the EIS spectra (Figure 13).

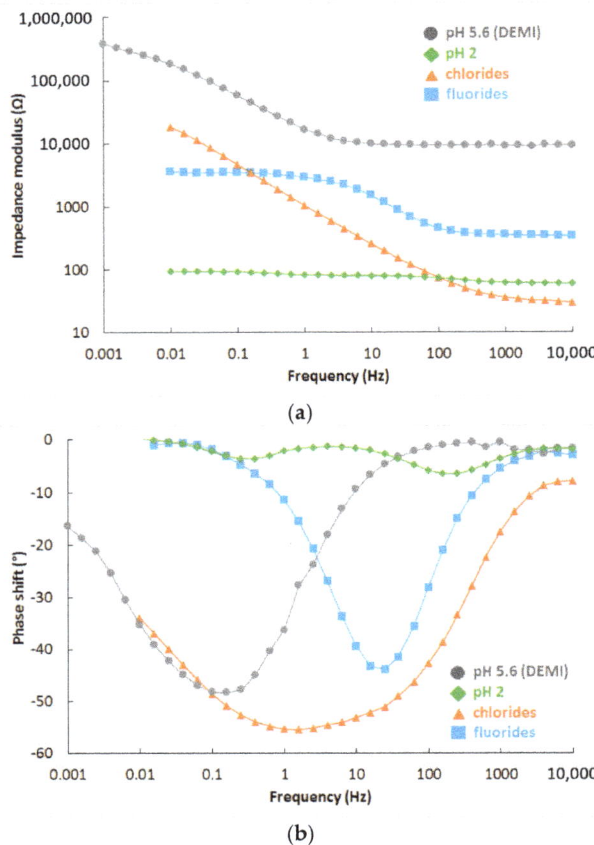

Figure 13. Bode plots of EIS spectra for different environments: (**a**) impedance modulus and (**b**) phase shift.

Table 5. Chemical composition of passivation layer on FeAl20Si20 formed in the electrolyte of pH = 5.6.

Element	Fe	Al-Metallic	Al-Oxide	Si	C	O
Concentration (atom %)	3.0	5.9	15.8	17.1	29.2	29.0

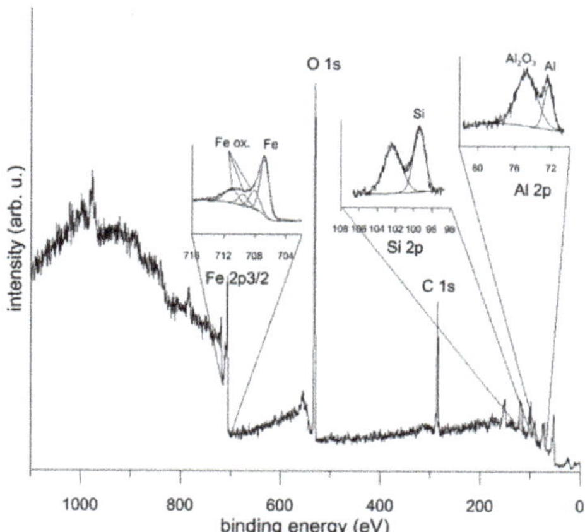

Figure 14. XPS spectrum of passivation layer on FeAl20Si20 formed in the electrolyte of pH = 5.6.

A summary of the results in sulphuric acid solution is given in Figure 15. The surface was covered by a thin layer of silicon oxide in the range of pH 0–2. There is a transition value of pH 3, where some aluminium oxide passive layer was present on the surface, but it has poorer quality when compared to spectrum at pH 4, which is the same as spectrum in non-aggressive demineralised water (pH 5.6).

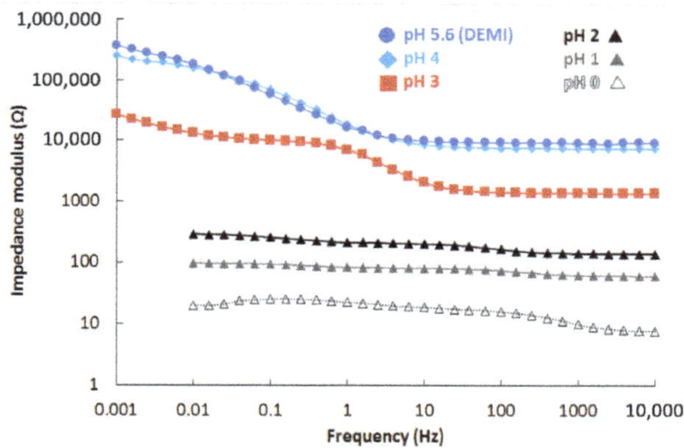

Figure 15. Bode plot (impedance modulus) in sulphuric acid solutions.

The complete results of spectra fitting are given in Table 6. The values of charge transfer resistance (RCT) are high for pH 3–5.6 and for chloride solution, when aluminium-based passive layer is present on the surface. Values corresponding to low pH are much lower since the silicon oxide-based passive layer has poor protective properties. The RCT is even lower than in fluorides when no passive layer is present on the surface, but material dissolution is driven by slow oxygen cathodic reduction unlike the fast hydrogen cathodic reaction in low pH environments. The passive layer resistance (RC) shows the conductivity of the layer, which is high for natural aluminium passive layer at pH 4 and 5.6 and lower for environments whereby the passive layer starts to be attacked (pH 3 and chloride solution), and it is the lowest for the silicon oxide-based passive layer.

Table 6. Results of EIS spectra fitting (RSOL—resistance of the solution; RC—resistance of the passive layer; RCT—charge transfer resistance; CDL—capacitance of electrical double layer; CC—capacitance of passive layer).

Parameter	pH 0	pH 1	pH 2	pH 3	pH 4	pH 5.6
R_{SOL} ($\Omega \cdot m^2$)	5.91×10^{-4}	4.88×10^{-3}	1.10×10^{-2}	1.11×10^{-1}	6.03×10^{-1}	7.36×10^{-1}
R_{CT} ($\Omega \cdot m^2$)	7.76×10^{-4}	1.19×10^{-3}	7.92×10^{-3}	6.28	9.54	23.54
R_C ($\Omega \cdot m^2$)	7.83×10^{-4}	1.57×10^{-3}	5.40×10^{-3}	5.88×10^{-1}	9.19	9.36
C_{DL} ($S \cdot s^{\alpha} \cdot m^{-2}$)	3.99×10^{2}	4.25×10^{2}	2.05×10^{2}	8.97	1.24	9.67×10^{-1}
α_{DL}	0.64	0.85	0.77	0.53	0.77	0.77
C_C ($S \cdot s^{\alpha} \cdot m^{-2}$)	2.28	1.21	1.22	2.42×10^{-1}	2.42×10^{-1}	4.17×10^{-1}
α_C	0.82	0.90	0.84	0.94	0.78	0.74

4. Discussion

The presented results for mechanical properties revealed the anomalous temperature dependence of both the yield strength and ultimate compressive strength. Among Fe–Al and Fe–Si phases, the anomalous behaviour of mechanical properties has been already reported for Fe_3Al, FeAl, and Fe_3Si phases [41–45]. However, only a weak effect has been described for Fe_3Si, corresponding to lower temperatures [45]. In addition, the silicides formed isolated particles in the investigated alloy. This implies that the matrix—$Fe_3Al_2Si_3$ ternary phase—probably also exhibits strong anomalous behaviour of the yield strength and ultimate tensile strength. This was not expected due to the crystal structure of this phase, which is triclinic (P-1). For definitive proof of the behaviour of this phase, samples containing pure $Fe_3Al_2Si_3$ phase would have to be prepared and tested.

The anomaly of YS and UCS at temperatures around 500 °C give rise to an interesting range of applications. These temperatures are common for exhaust valves of diesel internal combustion engines. During normal operation, the temperature reaches approximately 400 °C [45]. However, during the cleaning procedure of the filter of solid particles, the temperature increases to ca. 500 °C [46]. These conditions were demonstrated to be optimal for the use of this alloy. As the other results presented above show, the other characteristics (wear resistance, cyclic oxidation behaviour) also favour this material for this kind of application.

The oxidation tests at high temperatures were consistent with the oxidation mechanism previously published for this alloy prepared by reactive sintering [12]. During oxidation, a layer of aluminium oxide is formed. The reason for the predominance of aluminium oxide is its high thermodynamic stability, as compared with iron oxide and silicon oxide (Table 7) [47].

Table 7. Gibbs energy (ΔG_f) of formation of oxides at 800 °C (calculated on the basis of published data [47]).

Oxide Formula	ΔG_f (800 °C) ($kJ \cdot mol^{-1}$)
Al_2O_3	−1778
Fe_2O_3	−982
SiO_2	−983

At 800 °C, the oxidation product is γ-Al_2O_3 with cubic structure, which is then transformed to δ-Al_2O_3 during long-term exposure [48]. Such a transformation causes internal stresses which support the cracking and delamination of the oxide layer. On the other hand, the layer of α-Al_2O_3 formed at 1000 °C is stable and does not undergo any change during further exposure. Therefore, the spallation of the oxide layer is lower at 1000 °C than at 800 °C. Due to the formation of aluminium oxide, the zone below the oxide layer is depleted by aluminium. In this alloy, it leads to following transformation of the ternary phase to FeSi, which was detected below the surface (Figures 4 and 5):

$$Al_2Fe_3Si_3 + \frac{3}{2}O_2 \rightarrow Al_2O_3 + 3FeSi \quad (4)$$

Iron silicides are known to be highly oxidation-resistant [49]. Therefore, this silicon-enriched zone acts as a secondary protection when defects in the oxide layer occurs. This is probably also a reason why the oxidation slows down with oxidation duration (see the insert in Figure 5). Due to aluminium depletion, the source of aluminium decreases and is replaced by silicides.

Silicon-enriched zones also help to stabilise the mechanical properties in the near-surface area in high-silicon Fe–Al–Si alloys, as previously shown [12]. In binary Fe–Al alloys, the softer aluminium-deficient layer of Fe_3Al, or even Fe, can be expected. Therefore, other mechanical properties (wear resistance, fatigue life, creep limit) can also probably be negatively affected in the near-surface area in Fe–Al and Fe–Al–Si alloys with a lower amount of silicon.

A passive layer based on aluminium oxide is also responsible for exceptional corrosion resistance in electrolytes. This layer is stable at pH above 3. Below this value, an oxide film based on SiO_2 is probably formed with a lower protective effect. Therefore, this material should be used in environments with pH values above 3. The use of fluoride solutions is not recommended because the passive layer does not form in this kind of electrolyte due to formation of water-soluble $(AlF_6)^{3-}$ complexes [50].

Based on the factors discovered above, this alloy can be one of the solutions for a problem currently being addressed by the European Commission—substitution of critical raw materials [11]. In certain applications, it could possibly substitute chromium-containing alloys (corrosion-resistant and heat-resistant steels) or chromium-, molybdenum- and tungsten-alloyed tool steels.

5. Conclusions

In this work, the FeAl20Si20 alloy (in wt %) was prepared by the combination of unique ultrahigh-energy mechanical alloying and spark plasma sintering. The alloy exhibited very fine structure, composed of silicides (Fe_3Si and FeSi) and ternary phase ($Fe_3Al_2Si_3$). At room temperature, the ultimate compressive strength of the alloy was nearly 1100 MPa, while it increased to ca. 1500 MPa at 500 °C. The applied technology enabled achievement of a fracture toughness of 3.50 ± 0.33 MPa·m$^{1/2}$ at room temperature, which is almost ten times higher than the value determined for the material of the same composition prepared by casting. This value is almost comparable to ceramic materials. The abrasive wear rate of the alloy is a lower than in the case of the heat-treated AISI D2 cold work tool steel, which is considered as highly wear-resistant tool material. The FeAl20Si20 alloy exhibits excellent oxidation resistance at 800 and 1000 °C in air. From this viewpoint, the performance of the material at 1000 °C is better than at 800 °C, due to the formation of a highly protective α-Al_2O_3 layer. In water-based electrolytes, the material behaves passively at pH values above 3. The passivity is a result of the presence of an aluminium oxide layer on the surface. Under this pH value, the aluminium oxide layer dissolves and it is replaced by silicon oxide, which has a lower protective effect. Due to its superior chemical resistance, wear resistance, and mechanical behaviour, the material could be applicable in the automotive industry for the manufacture of exhaust valves, wear parts, or as a material for aggressive environments.

Author Contributions: Conceptualisation, P.N. and K.N.; Methodology, P.N. and J.K.; Formal Analysis, P.N. and F.P.; Investigation, T.V., K.N., J.S., J.K., and P.H.; Data Curation, P.N., K.N. and F.L.; Writing—Original Draft Preparation, P.N.; Writing—Review and Editing, K.N. and P.H.; Funding Acquisition, F.P.

Funding: This research was financially supported by Czech Science Foundation, project No. 17-07559S.

Conflicts of Interest: The authors declare no conflict of interest. The funders had no role in the design of the study; in the collection, analyses, or interpretation of data; in the writing of the manuscript, or in the decision to publish the results.

References

1. Zhu, X.; Yao, Z.; Gu, X.; Cong, W.; Zhang, P. Microstructure and corrosion resistance of Fe-Al intermetallic coating on 45 steel synthesized by double glow plasma surface alloying technology. *Trans. Nonferrous Met. Soc. China* **2009**, *19*, 143–148. [CrossRef]
2. Borsig, A. Zusatz von Aluminium zum Roheisen. *Stahl Eisen* **1894**, *14*, 6.

3. Kratochvíl, P. The history of the search and use of heat resistant Pyroferal alloys based on FeAl. *Intermetallics* **2008**, *16*, 587–591. [CrossRef]
4. Zamanzade, M.; Vehoff, H.; Barnoush, A. Effect of chromium on elastic and plastic deformation of Fe3Al intermetallics. *Intermetallics* **2013**, *41*, 28–34. [CrossRef]
5. Ferreira, P.I.; Couto, A.A.; de Paola, J.C.C. The effects of chromium addition and heat treatment on the microstructure and tensile properties of Fe-24Al (at.%). *Mater. Sci. Eng. A* **1995**, *192–193*, 165–169. [CrossRef]
6. Balasubramaniam, R. On the role of chromium in minimizing room temperature hydrogen embrittlement in iron aluminides. *Scr. Mater.* **1996**, *34*, 127–133. [CrossRef]
7. Zhang, Z.; Sun, Y.; Guo, J. Effect of niobium addition on the mechanical properties of Fe3Al-based alloys. *Scr. Metall. Mater.* **1995**, *33*, 2013–2017. [CrossRef]
8. Janda, D.; Fietzek, H.; Galetz, M.; Heilmaier, M. The effect of micro-alloying with Zr and Nb on the oxidation behavior of Fe3Al and FeAl alloys. *Intermetallics* **2013**, *41*, 51–57. [CrossRef]
9. Sundar, R.S.; Deevi, S.C. High-temperature strength and creep resistance of FeAl. *Mater. Sci. Eng. A* **2003**, *357*, 124–133. [CrossRef]
10. Klein, O.; Baker, I. Effect of chromium on the environmental sensitivity of FeAl at room temperature. *Scr. Metall. Mater.* **1992**, *27*, 1823–1828. [CrossRef]
11. Grilli, M.L.; Bellezze, T.; Gamsjager, E.; Rinaldi, A.; Novak, P.; Balos, S.; Piticescu, R.R.; Ruello, M.L. Solutions for Critical Raw Materials under Extreme Conditions: A Review. *Materials (Basel)* **2017**, *10*, 285. [CrossRef]
12. Novák, P.; Zelinková, M.; Šerák, J.; Michalcová, A.; Novák, M.; Vojtěch, D. Oxidation resistance of SHS Fe-Al-Si alloys at 800 °C in air. *Intermetallics* **2011**, *19*, 1306–1312. [CrossRef]
13. Li, H.; Zhang, J.; Young, D.J. Oxidation of Fe–Si, Fe–Al and Fe–Si–Al alloys in CO_2–H_2O gas at 800 °C. *Corros. Sci.* **2012**, *54*, 127–138. [CrossRef]
14. Kratochvíl, P.; Schindler, I. Hot rolling of iron aluminide Fe28.4Al4.1Cr0.02Ce (at%). *Intermetallics* **2007**, *15*, 436–438. [CrossRef]
15. Schindler, I.; Kratochvíl, P.; Prokopčáková, P.; Kozelský, P. Forming of cast Fe—45 at.% Al alloy with high content of carbon. *Intermetallics* **2010**, *18*, 745–747. [CrossRef]
16. Šíma, V.; Kratochvíl, P.; Kozelský, P.; Schindler, I.; Hána, P. FeAl-based intermetallics cast in an ultrsonic field. *Int. J. Mater. Res.* **2009**, *100*, 382–385. [CrossRef]
17. Novák, P.; Michalcová, A.; Voděrová, M.; Šíma, M.; Šerak, J.; Vojtěch, D.; Wienerová, K. Effect of reactive sintering conditions on microstructure of Fe–Al–Si alloys. *J. Alloy Compd.* **2010**, *493*, 81–86. [CrossRef]
18. Xia, Y.; Li, L.; Li, L. Effect of grain refinement on fracture toughness and fracture mechanism in AZ31 magnesium alloy. *Procedia Mater. Sci.* **2014**, *3*, 1780–1785. [CrossRef]
19. Schwarz, K.T.; Kormout, K.S.; Pippan, R.; Hohenwarter, A. Impact of severe plastic deformation on microstructure and fracture toughness evolution of a duplex-steel. *Mater. Sci. Eng. A* **2017**, *703*, 173–179. [CrossRef]
20. Vaidya, M.; Prasad, A.; Parakh, A.; Murty, B.S. Influence of sequence of elemental addition on phase evolution in nanocrystalline AlCoCrFeNi: Novel approach to alloy synthesis using mechanical alloying. *Mater. Des.* **2017**, *126*, 37–46. [CrossRef]
21. Fang, Q.; Kang, Z.; Gan, Y.; Long, Y. Microstructures and mechanical properties of spark plasma sintered Cu–Cr composites prepared by mechanical milling and alloying. *Mater. Des.* **2015**, *88*, 8–15. [CrossRef]
22. Froes, F.H.; Suryanarayan, C.; Russell, K.; Li, C.-G. Synthesis of Intermetallics by Mechanical Alloying. *Mater. Sci. Eng. A* **1995**, *192–193*, 612–623.
23. Al-Joubori, A.; Suryanarayana, C. Synthesis of metastable NiGe2 by mechanical alloying. *Mater. Des.* **2015**, *87*, 520–526. [CrossRef]
24. Naghiha, H.; Movahedi, B.; Asadabad, M.A.; Mournani, M.T. Amorphization and nanocrystalline Nb3Al intermetallic formation during mechanical alloying and subsequent annealing. *Adv. Powder Technol.* **2017**, *28*, 340–345. [CrossRef]
25. Molladavoudi, A.; Amirkhanlou, S.; Shamanian, M.; Ashrafizadeh, F. Synthesis and characterization of nanocrystalline CoTi intermetallic compound prepared by mechanical alloying. *Mater. Lett.* **2012**, *81*, 254–257. [CrossRef]
26. Novák, P.; Kubatík, T.; Vystrčil, J.; Hendrych, R.; Kříž, J.; Mlynár, J.; Vojtěch, D. Powder metallurgy preparation of Al-Cu-Fe quasicrystals using mechanical alloying and Spark Plasma Sintering. *Intermetallics* **2014**, *52*, 131–137. [CrossRef]

27. Chuvildeev, V.N.; Panov, D.V.; Boldin, M.S.; Nokhrin, A.V.; Blagoveshchensky, Y.V.; Sakharov, N.V.; Shotin, S.V.; Kotkov, D.N. Structure and properties of advanced materials obtained by Spark Plasma Sintering. *Acta Astronaut.* **2015**, *109*, 172–176. [CrossRef]
28. Zhang, Z.; Liu, Z.; Lu, J.; Shen, X.; Wang, F.; Wang, Y. The sintering mechanism in spark plasma sintering–Proof of the occurrence of spark discharge. *Scr. Mater.* **2014**, *81*, 56–59. [CrossRef]
29. Marder, R.; Estourne, C.; Chevallier, G.; Chaim, R. Plasma in spark plasma sintering of ceramic particle compacts. *Scr. Mater.* **2014**, *82*, 57–60. [CrossRef]
30. Liu, J.; Liang, C. Microstructure characterization and mechanical properties of bulk nanocrystalline aluminium prepared by SPS and followed by hightemperature extruded techniques. *Mater. Lett.* **2017**, *206*, 95–99. [CrossRef]
31. Peters, C.T. The relationship between Palmqvist indentation toughness and bulk fracture toughness for some WC-Co cemented carbides. *J. Mater. Sci.* **1979**, *14*, 1619–1623. [CrossRef]
32. Novák, P.; Vojtěch, D.; Šerák, J. Wear and corrosion resistance of a plasma-nitrided PM tool steel alloyed with niobium. *Surf. Coat. Technol.* **2006**, *200*, 5229–5236. [CrossRef]
33. Mrowec, S.; Stoklosa, A. Calculations of Parabolic Rate Constants for Metal Oxidation. *Oxid. Met.* **1974**, *8*, 379–391. [CrossRef]
34. *NIST X-ray Photoelectron Spectroscopy Database*; Version 4; National Institute of Standards and Technology: Gaithersburg, MD, USA, 2008. Available online: http//srdata.nist.gov/xps2 (accessed on 10 April 2019).
35. Hightower, A.; Fultz, B.; Bowman, R.C. Mechanical alloying of Fe and Mg. *J. Alloy Compd.* **1997**, *252*, 238–244. [CrossRef]
36. Novák, P.; Průša, F.; Nová, K.; Bernatiková, A.; Salvetr, P.; Kopeček, J.; Haušild, P. Application of mechanical alloying in synthesis of intermetallics. *Acta Phys. Pol. A* **2018**, *134*, 720–723. [CrossRef]
37. de Farias Azevedo, C.R.; Flower, H.M. Calculated ternary diagram of Ti–Al–Si system. *Mater. Sci. Technol.* **2000**, *16*, 372–381. [CrossRef]
38. Farzin-Nia, F.; Sterrett, T.; Sirney, R. Effect of machining on fracture toughness of corundum. *J. Mater. Sci.* **1990**, *25*, 2527–2531. [CrossRef]
39. Heuer, A.H. Oxygen and aluminum diffusion in α-Al2O3: How much do we really understand? *Eur. Ceram. Soc.* **2008**, *28*, 1495–1507. [CrossRef]
40. Proff, C.; Abolhassani, S.; Lemaignan, C. Oxidation behaviour of zirconium alloys and their precipitates—A mechanistic study. *J. Nucl. Mater.* **2013**, *432*, 222–238. [CrossRef]
41. Morris, D.G.; Munoz-Morris, M.A. A re-examination of the pinning mechanisms responsible for the stress anomaly in FeAl intermetallics. *Intermetallics* **2010**, *18*, 1279–1284. [CrossRef]
42. George, E.P.; Baker, I. Thermal vacancies and the yield anomaly of FeAl. *Intermetallics* **1998**, *6*, 759–763. [CrossRef]
43. Morris, D.G.; Munoz-Morris, M.A. The stress anomaly in FeAl–Fe3Al alloys. *Intermetallics* **2005**, *13*, 1269–1274. [CrossRef]
44. Brinck, A.; Neuhäuser, H. Yield stress and dislocation mechanisms in the D03 ordered intermetallic phase Fe3Al in the temperature range 240–500K. *Mater. Sci. Eng. A* **2004**, *387–389*, 969–972. [CrossRef]
45. Schaefer, H.-E.; Frenner, K.; Wurschum, R. High-temperature atomic defect properties and diffusion processes in intermetallic compounds. *Intermetallics* **1999**, *7*, 277–287. [CrossRef]
46. Castellano, J.; Chaudhari, A.; Bromham, J. Adaptive Temperature Control for Diesel Particulate Filter Regeneration. *SAE Tech. Pap.* **2013**, *1*, 517. [CrossRef]
47. Barin, I. *Thermochemical Data of Pure Substances*, 3rd ed.; VCH Verlagsgesellschaft mbH: Weinheim, Germany, 1995; ISBN 9783527287451.
48. Souza Santos, H.; Souza Santos, P. Pseudomorphic formation of aluminas from fibrillar pseudoboehmite. *Mater. Lett.* **1992**, *13*, 175–179. [CrossRef]

49. Roy, S.K.; Fasasi, A.; Pons, M.; Galerie, A.; Caillet, M. High-temperature oxidation behaviour of laser-surface alloyed iron-silicon coatings on iron. *J. Phys. IV* **1993**, *3*, 625–633. [CrossRef]
50. Corbillon, M.S.; Olazabal, M.A.; Madariaga, J.M. Potentiometric Study of Aluminium-Fluoride Complexation Equilibria and Definition of the Thermodynamic Model. *J. Solut. Chem.* **2008**, *37*, 567–579. [CrossRef]

Article

Properties of FeAlSi-X-Y Alloys (X,Y=Ni, Mo) Prepared by Mechanical Alloying and Spark Plasma Sintering

Filip Průša [1,*], Olga Proshchenko [1], Andrea Školáková [1], Vojtěch Kučera [1] and František Laufek [2,3]

1 Department of Metals and Corrosion Engineering, University of Chemistry and Technology, Technická 5, 166 28 Prague, Czech Republic; proscho@vscht.cz (O.P.); skolakoa@vscht.cz (A.Š.); kucerao@vscht.cz (V.K.)

2 Institute of Physics, The Czech Academy of Sciences, Na Slovance 1999/2, 182 00 Prague, Czech Republic; frantisek.laufek@geology.cz

3 Czech Geological Survey, Geologická 6, 152 00 Prague, Czech Republic

* Correspondence: prusaf@vscht.cz

Received: 25 November 2019; Accepted: 6 January 2020; Published: 8 January 2020

Abstract: Short-term mechanical alloying and compaction by spark plasma sintering was used for the production of FeAl20Si20Mo20-XNiX (X corresponds to 5–15 wt %) alloy, which showed an ultrafine-grained microstructure with dimensions of phases around 200 nm or smaller. It was found that the addition of Mo and Ni to the FeAl20Si20 alloy results in the formation of the AlMoSi phase compared to the three-phase FeAl20Si20 alloy, which initially contained FeSi, Fe_3Si, and $Fe_3Al_2Si_3$ phases. All the investigated alloys increased their hardness, reaching up to 1401 HV 1 for the FeAl20Si20Mo5Ni15 alloy, which contained in total 58.5% of the FeSi and $Fe_3Al_2Si_3$ phases. As a result, all the prepared alloys showed one order magnitude lower wear rates ranging from 3.14 to $5.97 \cdot 10^{-6}$ $mm^3 \cdot N^{-1} \cdot m^{-1}$ as well as significantly lower friction coefficients compared to two reference tool steels. The alloys achieved high compressive strengths (up to 2200 MPa); however, they also exhibited high brittleness even after long-term annealing, which reduced the strengths of all the alloys below approximately 1600 MPa. Furthermore, the alloys were showing ductile behavior when compressively tested at elevated temperature of 800 °C. The oxidation resistance of the alloys was superior due to the formation of a compact Al_2O_3 protective layer that did not delaminate.

Keywords: mechanical alloying; spark plasma sintering; hardness; compressive strength; oxidation resistance; wear

1. Introduction

The Fe–Al–Si-based alloys belong to a perspective group of materials that was developed by joining two binary alloy systems, namely Fe–Al and Fe–Si. These alloys might find their utilization as a much cheaper and lighter substitution of heat-resistant steels or even nickel superalloys [1,2]. The main advantages of these alloys are their excellent thermal stability and resistance against high-temperature reactions in oxygen or sulfate-bearing atmospheres [3–7]. Such behavior has been initially described for the Fe–Al system, which is capable of maintaining its mechanical properties up to 500 °C [8,9]. Simultaneously, the Fe–Al alloys create a compact and protective layer made of either of α-Al_2O_3 or γ-Al_2O_3, whose formation is temperature-dependent [2,10]. Comparing these two modifications, the latter mentioned provides significantly better protection, since it does not contain pores as the α-Al_2O_3 does [1].

The present oxidic layer acts as a shielding bipolar membrane that decreases the diffusion of metal atoms through the layer toward the environment while also blocking the gases' transport in the opposite direction. The α-Al_2O_3 modification provides lower protection since it contains many pores and is often prone to the formation of microcracks. Notably, the temperatures between 800 and 900 °C are known to be responsible for these kinds of defects. In this temperature interval, a metastable θ-Al_2O_3 is often formed and further transforms during cooling into a stable α-Al_2O_3 form, which is accompanied by a volume change [11]. As a result, tensile stresses are induced into the layer, allowing the formation of microcracks while simultaneously decreasing the adhesion of the layer to the alloy surface [12].

The negative effect of high porosity can be overcome by the addition of Si due to a formation of far more complex and dense compounds called spinels. Especially, the low addition of Si can effectively reduce the porosity of the protection layer. However, when some critical content of Si is exceeded, the porosity increases again, e.g., the Fe–Al–Si alloys containing over 30 wt % Si are exhibiting enormous porosity, making them unusable [10,13–17]. Simultaneously, the addition of Si toward the Fe–Al alloys suppresses the formation of aluminides in favor of silicides or alumino–silicides. Thus, e.g., the FeAl20Si20 (wt %) alloy is composed of FeSi, Fe_3Si and $Fe_3Al_2Si_3$ [15], or $FeAl_2Si$ phases [13,18]. The substitution of aluminides by silicides increases the hardness and strength of alloy as well as the wear resistance and thermal stability, although at the expense of toughness [9,10,13,15,16,19–22]. It has been found that the content of various forms of silicides determines the hardness of these alloys. The major contribution to the resulting hardness is mainly caused by the FeSi phase, which is known for its high hardness of 958 HV 1 [19]. Accordingly, to the work of Wu et al. [23], the FeSi phase exhibits the third-highest calculated hardness among the phases present in the Fe–Si binary system. According to this, the FeAlSi alloys were reported to exhibit the hardness of 730 HV 5 in case of FeAl10Si30 alloy [13–15]. On the other hand, the increase in Al content at the expense of Si results in a hardness decrease down to 440 HV 5 in the FeAl30Si10 alloy.

A further increase in the hardness and strength of the FeAlSi alloys may be achieved by the alloying of transition elements, especially Ni and Mo. Up to now, almost no reports have mentioned the influence of these elements on the microstructure and properties of quaternary and foremostly quinary alloys. Novák et al. [24] reported that the addition of Ni in the FeAl20Si20 (wt %) alloy prepared by reactive synthesis resulted in a decrease of porosity while increasing the hardness and wear resistance as well as thermal stability and oxidation resistance at 800 °C. The same alloy, although prepared by mechanical alloying (MA) and compacted via spark plasma sintering (SPS), showed quite similar results, increasing the hardness and compressive strength from 1049 HV 1 up to 1376 HV 1 and from 1085 MPa up to over 1800 MPa [25].

The Fe–Al–Si-based alloys are in general prepared by powder metallurgy, which overcomes the problems experienced during cast-metallurgy processes. Especially, the MA is capable of providing and retaining the beneficial microstructural refinement and homogeneity, which further improves the mechanical properties [26–30]. During the process, a highly localized cold welding allowing only a limited diffusion of elements; continuous fracturing and deformation strengthening is responsible for the ultrafine or even nanocrystalline microstructure of the alloys [27,31,32]. As a compaction method, various techniques including uniaxial pressing and isostatic pressing, both done at elevated temperatures, seem to be failing to deliver expected results. The main setback is a long duration of the processes allowing microstructural coarsening, which deteriorates the desirable properties gained by the MA [33]. Thus, the fast compaction via SPS is capable of providing almost full-density compacts and is especially of interest.

Thus, the present work describes the influence of Ni and Mo addition onto the complex properties of FeAl20Si20 alloy prepared by a combination of mechanical alloying and spark plasma sintering. The aim was to describe the effects of the different amounts of Ni and Mo as alloying elements on the microstructure, mechanical properties, and thermal stability, including oxidation resistance.

2. Materials and Methods

The FeAl20Si20Mo20-XNiX (X = 5–15 wt %) alloys were prepared from pure elements, which were mixed in appropriate amounts forming 20 g powder batches for mechanical alloying (MA). For this purpose, powders of Fe (purity of 99.9%, Strem Chemicals, Newburyport, US), Al (purity of 99.7%, Strem Chemicals, Newburyport, MA, US), Si (purity of 99.5%, Alfa Aesar, Lancashire, UK), Mo (purity of 99.5%, Alfa Aesar, Lancashire, UK), and Ni (purity of 99.5%, Merck, Darmstadt, United Kingdom) were used. The powders were placed into a milling jar together with milling balls, which were both made from AISI 420 stainless steel. Afterwards, the jar was sealed and flushed with Ar (purity of 99.996%) for 2 min with a constant flow of 2 l/min. Mechanical alloying was done in a milling device Retsch PM100 CM (Retsch, Haan, Germany) for 10.5 h while for each 30 min of the process, a short 10-min pause was maintained to suppress the excessive cold welding.

Then, prepared powders were compacted via spark plasma sintering (SPS, FCT Systeme, HP-D 10, Rauenstein, Germany) using a heating rate of 300 °C/min until reaching 900 °C, after which the heating rate was reduced to only 100 °C/min. The samples were compacted at a temperature of 1000 °C with a pressure of 48 MPa and remained at this temperature for 10 min. Afterwards, the samples were slowly cooled down to 300 °C with a speed of 50 °C/min to reduce the thermal stress–strains within the sample. Prepared samples were cut using a diamond blade cutting device (Leco Precision VC-50 Vari-Cut, St. Joseph, US) into samples which were either used for microstructural investigations of for mechanical testing.

Present phases were determined by powder X-ray diffraction (XRD, Bruker D8 Advance, Karlsruhe, Germany, Cu$K\alpha$ radiation and LynxEye-XE detector), while the actual chemical composition of the prepared samples was determined by X-ray fluorescence analysis (XRF, ARL 9400 XP, Thermo ARL, Switzerland). Semi-quantitative phase analysis, as well as the calculation of lattice and microstructural parameters were performed by the Rietveld method using the Topas 5 program (Bruker AXS, 2014). The microstructure of the prepared cross-sections was investigated using scanning electron microscopy (SEM, Tescan Lyra, Brno, Czech Republic) equipped with an energy-dispersive spectrometer (EDS, Oxford Instruments, 80 mm^2, High Wycombe, United Kingdom). The surface porosity was determined using light microscopy (LM, Olympus PME-3, Tokyo, Japan), obtaining at least 20 micrographs with a total area of 0.55 mm^2, which were then analyzed by the threshold method.

For the compressive tests, cuboid samples with a height length corresponding to 1.5 times the length of the bottom side were used. The compressive tests were done on a universal testing device (LabTest SP 250.1-VM, Labortech s.r.o., Opava, Czech Republic) with a strain speed of 0.001 s^{-1}. Prepared alloys were also tested for thermal stability, which was determined by the hardness change during long-term annealing and by compressive tests, which were done either at laboratory temperature after annealing or at an elevated temperature of 800 °C.

Furthermore, the samples were also investigated for the kinetics of cyclic oxidation at 800 °C. For this purpose, the samples were placed into the electric resistance furnace for time segments composed of 4, 9, 25, 50, 75, and 100 h, and then cooled down outside the furnace. The weight gain due to a formation of oxidic products was measured on an analytical balance (Pioneer PA224, Ohaus, Parsippany, NJ, US).

To fully describe the mechanical properties, tribological tests were done using a pin-on-disc setup (TRIBOtechnic, Clichy, France). The tests were done on polished samples at laboratory temperature in an oscillating regime with an Al_2O_3 ball 6 mm in diameter that was moving with a speed of 10 $mm \cdot s^{-1}$ until reaching a total distance of 15 m. The ball was loaded with 5 N, and the wear track profile has been measured with a profilometer. The temperature and humidity during the tests were constant during all the tests corresponding to 22.2 °C and 35.5%. Obtained results were compared with the results of tool steels 1.2379 (AISI D2) and 1.3343 (AISI M 2) supplied from an external company. Both the steels were heat-treated by the supplier accordingly to the conditions specified in the relevant standards.

3. Results and Discussion

3.1. Phase Composition and Microstructure

The phase composition of all the MA + SPS alloys has been determined by the Rietveld X-ray diffraction analysis, and the patterns are shown in Figure 1. Accordingly to the results, all the prepared alloys were composed of two binary FeSi and Fe_3Si phases and of two ternary $Fe_3Al_2Si_3$ and AlMoSi phases, and the lattice parameters are shown in Table 1. Compared to the work of others [34–36], the short-term MA formed only intermetallic phases instead of solid solutions, which are created during much longer process durations. It was discovered that the different amount of the Mo and Ni addition did not change the phase compositions within the tested range of chemical compositions. All the phases were showing the presence of crystallites with average dimensions around 50 nm (Table 2). Only the FeSi phase in the MA + SPS FeAl20Si20Mo5Ni15 alloy contained larger crystallites with average dimensions of 100 nm. The volume fraction of the phases varied with the increasing content of Mo, favoring the formation of a ternary AlMoSi phase. Thus, the content of the AlMoSi phase increased from 8.5 wt % up to 23.0 wt %, mostly at the expanse of the FeSi phase. All of the mentioned phases were saturated with other elements, which slightly changed the lattice parameters when compared to the known values.

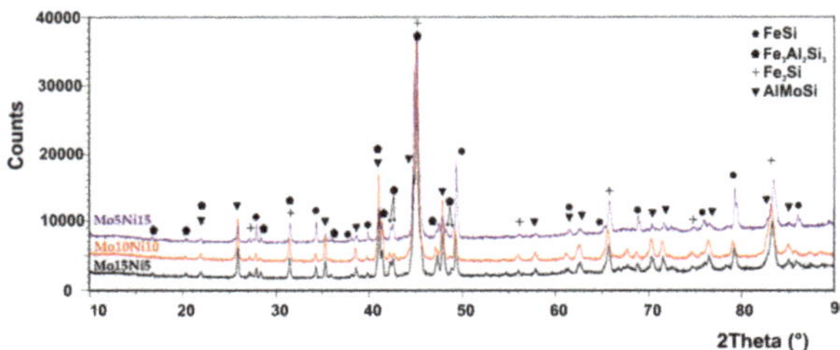

Figure 1. XRD diffraction patterns of the mechanical alloying and spark plasma sintering (MA + SPS) FeAl20Si20-Mo-Ni (wt %) alloys containing Mo and Ni in a range from 5–15 wt %.

Table 1. Lattice parameters determined by the XRD analysis of present phases identified in the MA + SPS FeAl20Si20-Mo-Ni alloys.

Phases	Lattice Parameters					
	a (nm)	*b* (nm)	*c* (nm)	α (°)	β (°)	γ (°)
FeSi	0.4519 ± 0.0001	–	–	–	–	–
$Fe_3Al_2Si_3$	0.4668 ± 0.0003	0.6325 ± 0.0004	0.7486 ± 0.0004	101.12 ± 0.04	105.88 ± 0.04	101.32 ± 0.04
Fe_3Si	0.5637 ± 0.0001	–	–	–	–	–
AlMoSi	0.4655 ± 0.0002	–	0.6532 ± 0.0004	–	–	–
FeSi	0.4530 ± 0.0001	–	–	–	–	–
$Fe_3Al_2Si_3$	0.4707 ± 0.0003	0.6222 ± 0.0004	0.7472 ± 0.0004	100.60 ± 0.04	105.36 ± 0.05	101.87 ± 0.05
Fe_3Si	0.5688 ± 0.0001	–	–	–	–	–
AlMoSi	0.4668 ± 0.0001	–	0.6538 ± 0.0001	–	–	–
FeSi	0.4530 ± 0.0001	–	–	–	–	–
$Fe_3Al_2Si_3$	0.4686 ± 0.0002	0.6320 ± 0.0002	0.7504 ± 0.0003	100.95 ± 0.01	105.70 ± 0.02	101.54 ± 0.03
Fe_3Si	0.5684 ± 0.0001	–	–	–	–	–
AlMoSi	0.4665 ± 0.0001	–	0.6539 ± 0.0002	–	–	–

In comparison to our previous work [37], the addition of Mo and Ni resulted in the formation of an AlMoSi phase, which depleted the content of the elements within the remaining phases. As a result, the lattice parameters of the FeSi, Fe_3Si, and $Al_2Fe_3Si_3$ phases were in the majority of the cases lower compared to the lattice parameters obtained for identical phases in the FeAl20Si20 alloy. These results also differed from results observed by others, which, e.g., either calculated the values of the Fe_3Si phase as $a = 0.5650$ nm [38] or determined the values of $Fe_3Al_2Si_3$ by an XRD measurement [39].

The observed change in the lattice parameters was caused by the partial substitution of elements that caused stress–strains in the lattice, changing its parameters. Such observations have been already mentioned in our previous work [37] and the works of others [16,40].

Table 2. The phase parameters, phase fractions, and crystallite sizes determined by Rietveld analysis.

Alloy	Phases	Space Group	Wt %	Crystallite Size (nm)
FeAl20Si20Mo5Ni15	FeSi	$P2_13$	32.5 ± 0.1	100
	$Fe_3Al_2Si_3$	$P\bar{1}$	24.5 ± 0.1	≈ 50
	Fe_3Si	$Fm\bar{3}m$	34.5 ± 0.1	43
	AlMoSi	$P6_222$	8.5 ± 0.1	63
FeAl20Si20Mo10Ni10	FeSi	$P2_13$	30.0 ± 0.1	65
	$Fe_3Al_2Si_3$	$P\bar{1}$	18.5 ± 0.2	≈ 50
	Fe_3Si	$Fm\bar{3}m$	39.0 ± 0.1	35
	AlMoSi	$P6_222$	12.5 ± 0.1	≈ 60
FeAl20Si20Mo15Ni5	FeSi	$P2_13$	17.0 ± 0.1	50
	$Fe_3Al_2Si_3$	$P\bar{1}$	27.0 ± 0.2	≈ 50
	Fe_3Si	$Fm\bar{3}m$	33.0 ± 0.1	40
	AlMoSi	$P6_222$	23.0 ± 0.1	55

It should be noted that the increasing content of the AlMoSi phase was followed by a decrease of the FeSi phase, reducing from 32.5 wt % to 17.0 wt %, which corresponded to FeAl20Si20Mo15Ni5 alloy. Additionally, the peak width was almost the same in all the alloys, which coincided with microstructural observations that confirmed almost identical dimensions of phases, regardless of the chemical composition.

The surface porosity of the prepared MA + SPS alloys (Figure 2) has been determined by a threshold method for which the light micrographs prior etching were used. All of the prepared alloys were showing almost comparable porosity around 1.6%, which is almost three times higher than of the FeAl20Si20 alloys prepared by the same conditions [37]. The reason for the higher porosity might be found in the increased lattice stress–strains in present FeSi, Fe_3Si, and $Al_2Fe_3Si_3$ phases due to their enrichment by alloying elements. Besides, the formation of fine-grained AlMoSi phases also contributed to the overall strengthening of the material and further decreasing the plasticity during SPS compaction. Thus, the porosity could be only decreased using higher compaction temperatures, which would, on the other hand, promote a higher rate of microstructural coarsening that would deteriorate the overall mechanical properties.

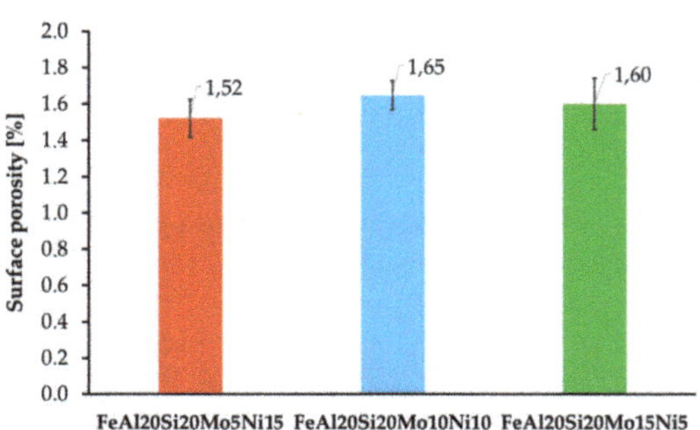

Figure 2. Surface porosity of the prepared MA + SPS FeAl20Si20-Mo-Ni alloys depending on the actual content of Mo and Ni within the alloy.

The SEM micrographs of all the MA + SPS alloys are shown in Figure 3. As can be seen, the MA + SPS alloys showed uniform microstructure with homogeneously distributed particles of intermetallic phases. These phases were mostly polyhedral in shape with various dimensions, as is shown. The present phases were roughly distinguished based on the physical background of the used backscattered electron detector, which displays elements with higher atomic number as bright areas, while lighter atoms manifest themselves as darker objects. Thus, based on the observations, three evident areas with different chemical content were discovered. The brightest rounded particles with an average diameter below 200 nm were containing Mo and thus corresponded to the AlMoSi phase. The bright gray and middle gray phases were showing sufficient brightness, suggesting the presence of elements such as Fe and Si, identifying themselves as FeSi or Fe_3Si phases. The dark gray phases were showing the presence of light elements such as Al, and thus were initially identified as $Fe_3Al_2Si_3$ phases. Among these phases, small dark and rounded objects were also observed. The particles could be either pores, which formed during etching in a reagent containing fluoride ions, or oxide particles. The origin of oxides particles might be found in the pre-oxidized powders, since the process of mechanical alloying was done in a protective Ar atmosphere. Comparing all the alloys used for MA, the lowest standard Gibbs energy of $\Delta G_{298,16} = -1584.0\ \mathrm{kJ \cdot mol^{-1}}$ corresponded to the formation of Al_2O_3, followed by the value of Fe_3O_4 ($\Delta G_{298,16} = -1015.3\ \mathrm{kJ \cdot mol^{-1}}$) [41]. When compared to other alloys, these oxides exhibit the highest affinity to oxygen and thus are the primary sources of contamination via oxygen. Among the oxidic particles, the presence of small dimples caused by fluorine ions, which were present in the etching solution, was observed.

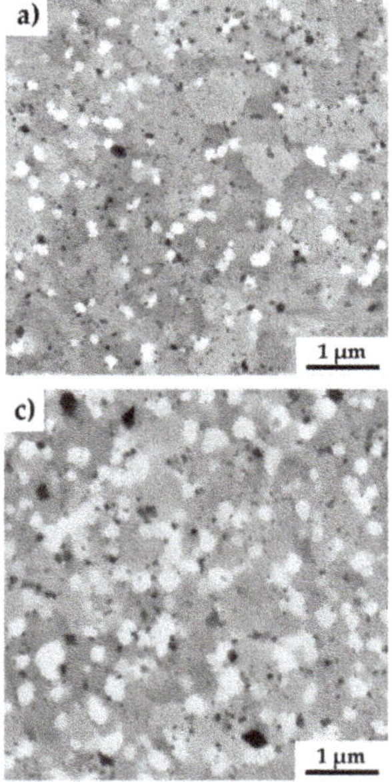

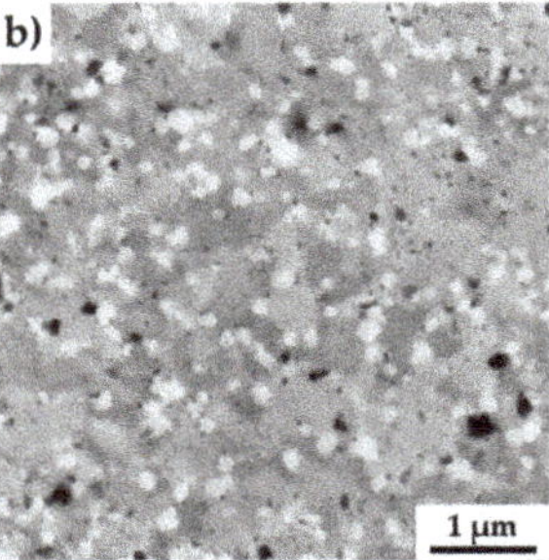

Figure 3. SEM micrographs of the MA + SPS FeAl20Si20-Mo-Ni alloys showing (**a**) FeAl20Si20Mo5Ni15; (**b**) FeAl20Si20Mo10Ni10; and (**c**) FeAl20Si20Mo15Ni5 alloys (a combination of BSE+SE detectors was used).

The present phases were distinguished by SEM+EDS element distribution maps, as shown in Figure 4. The maps show large areas, which were enriched mainly in Fe and Si. On the other hand, the areas enriched in Al were also containing Ni. Nevertheless, the present phases were hardly distinguishable by the appearance of the element distribution map, since the elements often supersaturate the phases, exceeding the expected concentrations. Such observations are nothing unusual, considering that the preparation via MA can be briefly described as a non-equilibria process allowing the creation of phases that are enriched of other elements. Thus, these areas were analyzed by the SEM+EDS point analysis to determine the average chemical composition of present phases, whose results are shown in Table 3.

Table 3. Results of the SEM+EDS analysis of the points marked in Figure 4.

Alloy	Points	Average Chemical Compositon (at %)					Corresponding Phase
		Al	Si	Fe	Ni	Mo	
FeAl20Si20Mo5Ni15	1–3	14.4 ± 3.9	38.1 ± 3.9	43.6 ± 1.0	3.3 ± 1.4	0.6 ± 0.3	FeSi
	4–6	31.7 ± 2.1	28.3 ± 2.4	34.1 ± 1.2	5.1 ± 3.9	0.9 ± 0.7	$Fe_3Al_2Si_3$
	7–9	34.9 ± 0.8	20.9 ± 0,2	23.2 ± 0.2	19.3 ± 0.1	1.3 ± 0.1	Fe_3Si
	10–12	31.5 ± 1.6	36.2 ± 4.2	14.7 ± 0.5	4.5 ± 3.8	13.0 ± 1.5	AlMoSi
FeAl20Si20Mo10Ni10	1–3	14.2 ± 3.4	37.1 ± 2.1	45.0 ± 1.2	2.1 ± 0.7	1.5 ± 0.6	FeSi
	4–6	33.8 ± 1.9	26.0 ± 1.0	34.8 ± 0.1	3.1 ± 0.9	2.2 ± 0.2	$Fe_3Al_2Si_3$
	7–9	34.0 ± 0.3	20.9 ± 0.3	30.9 ± 0.8	12.7 ± 0.3	1.6 ± 0.2	Fe_3Si
	10–12	27.0 ± 0.3	38.5 ± 2.5	15.5 ± 5.4	3.8 ± 0.1	15.2 ± 3.3	AlMoSi
FeAl20Si20Mo15Ni5	1–3	16.4 ± 2.3	36.3 ± 2.3	43.0 ± 2.5	1.3 ± 0.7	3.0 ± 1.8	FeSi
	4–6	34.0 ± 0.8	30.8 ± 5.0	27.2 ± 10.1	0.8 ± 1.2	7.8 ± 7.1	$Fe_3Al_2Si_3$
	7–9	35.5 ± 0.1	22.0 ± 1.2	33.2 ± 0.4	8.5 ± 0.7	1.9 ± 0.2	Fe_3Si
	10–12	29.8 ± 0.8	37.4 ± 1.2	15.3 ± 1.2	1.3 ± 1.1	16.2 ± 1.7	AlMoSi

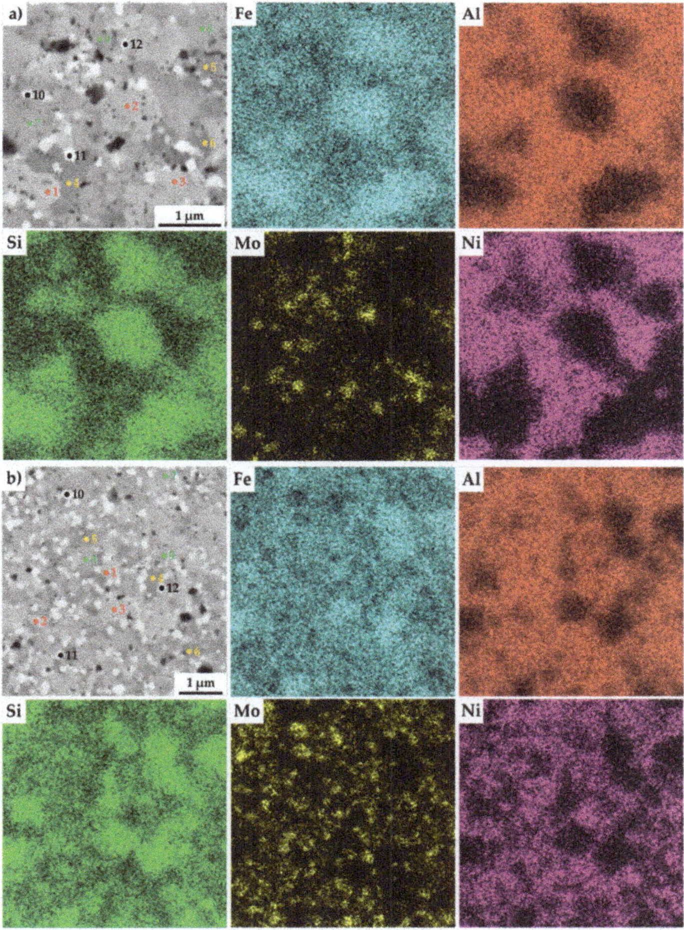

Figure 4. *Cont.*

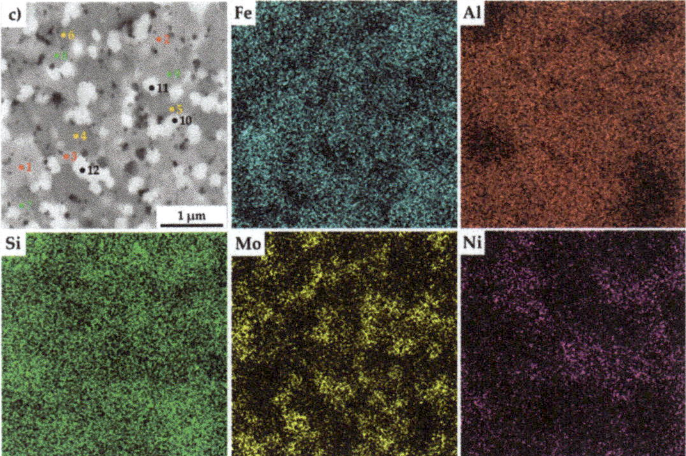

Figure 4. SEM+EDS element distribution maps of the MA+SPS: (**a**) FeAl20Si20Mo5Ni15; (**b**) FeAl20Si20Mo10Ni10; (**c**) FeAl20Si20Mo15Ni5 alloys with the marked places of chemical analysis, whose results are shown in Table 3.

The results of the point analysis show the already mentioned enrichment of binary FeSi and Fe_3Si phases by other alloying elements, namely of Al. Its concentration ranged in the FeSi phase, regardless of the chemical composition of the MA + SPS alloy, from 14.2 up to 16.4 at %. The same phase contained a significantly lower amount of Ni whose concentration decreased from 3.3 at % to 1.3 at %, reflecting the decreasing content of Ni in the MA + SPS alloy. On the other hand, the Fe_3Si phase showed the supersaturation of Ni, whose lowest concentration of 8.5 at % further increased up to 19.3 at.% in the FeAl20Si20Mo5Ni15 alloy. Both the binary silicides contained only small amounts of Mo below 3.0 at %. The ternary $Fe_3Al_2Si_3$ phase was showing a competing substitution of Ni and Mo, which had content that was changing concerning the chemical composition of the investigated alloy. Despite this, the AlMoSi phase showed, as the only one, almost constant chemical composition across all the investigated alloys. All of the present phases were showing the enrichment of other elements, while the chemical composition was shifted out of the equilibria state.

3.2. Mechanical Properties

The prepared alloys were after compaction via SPS tested for Vickers hardness, whose results are shown in Figure 5. As is shown, all the prepared MA + SPS alloys exhibited high hardnesses, which exceed those observed in only the ternary FeAl20Si20 alloy prepared in our previous research [37]. In direct comparison, the FeAl20Si20Mo5Ni15 exceeded the hardness of FeAl20Si20 alloy by almost more than 300 HV 0.1. Such an increase in hardness can be attributed to a higher content of especially FeSi and $Fe_3Al_2Si_3$ phases which reached in total up to 58.5 wt % for the first alloy (see Table 2). These phases, respecting the order of their appearance, are known to exhibit hardnesses up to 958 HV and 1553 HV, respectively. The presence of the Fe3Si phase more than surely softened the alloy since exhibiting a maximal hardness of 514 HV [16,19].

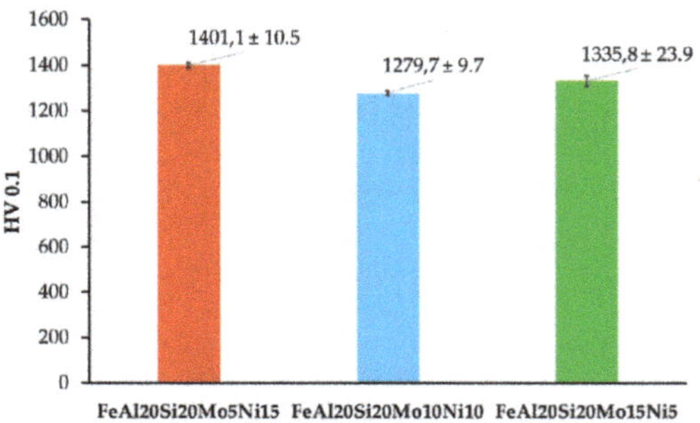

Figure 5. Vickers hardness of the prepared MA + SPS FeAl20Si20-Mo-Ni alloys.

However, this presumption corresponds to the results of the FeAl20Si20Mo15Ni5 alloy that showed the second-highest hardness, containing only 54 wt % of the previously mentioned phases. Thus, considering the phase fractions of each present phases (see Table 4), the highest contribution toward the hardness is caused by the ternary $Fe_3Al_2Si_3$ phase. This presumption seems to be correct, since the content of this phase was lowest in the case of FeAl20Si20Mo10Ni alloy, which also showed the lowest hardness 1279.7 ± 9.7 HV 0.1 of all the alloys. Besides, the deformation strengthening of the present phases needs to be also taken into account, among which at least the FeSi has been reported to achieve plastic deformation under extreme conditions [42,43]. Besides, the presence of the oxidic particles within the grains of present phases might contribute to the overall strengthening of the prepared MA + SPS alloys. Thus, the ultrahigh hardness of 1401 HV 0.1 of the FeAl20Si20Mo5Ni15 alloy has been achieved by a synergic contribution of all the above-mentioned effects, exceeding the hardness of laser-cladding Fe–Al–Si layers (560 HV 0.1) [44] almost three times over and almost two times over compared to those prepared by SHS reaction (860 HV 5) [15].

Table 4. Phase fractions and their correlation with the measured hardness of the MA + SPS alloys. *(differences from FeAl20Si20Mo5Ni15 are shown in brackets).*

Phases	Phase Fractions in the FeAl20Si20-Mo-Ni alloy [%]		
	Mo5Ni15	**Mo10Ni10**	**Mo15Ni5**
FeSi	32.5	30.0 (−2.5)	17.0 (−15.5)
Fe_3Si	34.5	39.0 (+4.5)	33.0 (−1.5)
$Fe_3Al_2Si_3$	24.5	18.5 (−6.0)	27.0 (+2.5)
AlMoSi	8.5	12.5 (+4.0)	23.0 (+14.5)
HV 0.1	1401.1 ± 10.5	1279.7 ± 9.7	1335.8 ± 23.9

The MA + SPS alloys have also been tested for thermal stability, which was expressed as hardness change during long-term annealing at 800 °C, as shown in Figure 6. All the alloys showed an initial increase in hardness by approximately 60 HV 0.1, followed by a slow decrease in hardness as the duration of annealing prolonged up to a total of 100 h. In the end, all the alloys showed, considering the confidence intervals, almost identical values of hardness reaching over 1100 HV 0.1. The highest hardness prior and after the tests was obtained by the FeAl20Si20Mo5Ni15 alloy, which contained the highest volume fraction of FeSi, Fe_3Si, and $Fe_3Al_2Si_3$ phases reaching up to 91.5 wt %. During the first 4 h of annealing, the hardness increased probably due to a formation of precipitates within the material, which further either dissolved or coarsened, reducing its strengthening contribution toward the alloy. Such a presumption might be supported by the already-mentioned enrichments of present phases by other elements due to a non-equilibria preparation process. A further decrease in hardness was caused by the microstructural coarsening of each constituent.

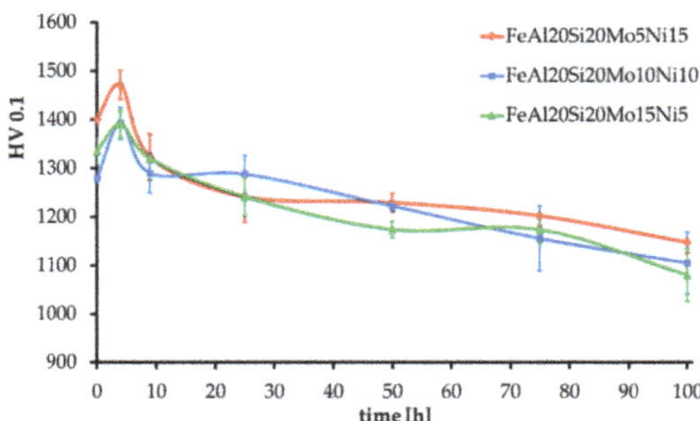

Figure 6. Thermal stability of the MA + SPS alloys expressed as HV 0.1 change during long-term annealing at 800 °C.

Besides, the MA + SPS alloys have been compressively tested either at laboratory temperature (Figure 7a), at a laboratory temperature after 100 h of annealing at 800 °C (Figure 7b), or at an elevated temperature of 800 °C (Figure 7c). As is shown, all the alloys showed ultrahigh ultimate compressive strengths (UCS) at laboratory temperature, among which the FeAl20Si20Mo10Ni10 alloy reached the highest UCS of approximately 2200 MPa, outperforming the second-best FeAl20Si20Mo5Ni15 alloy by more than 200 MPa.

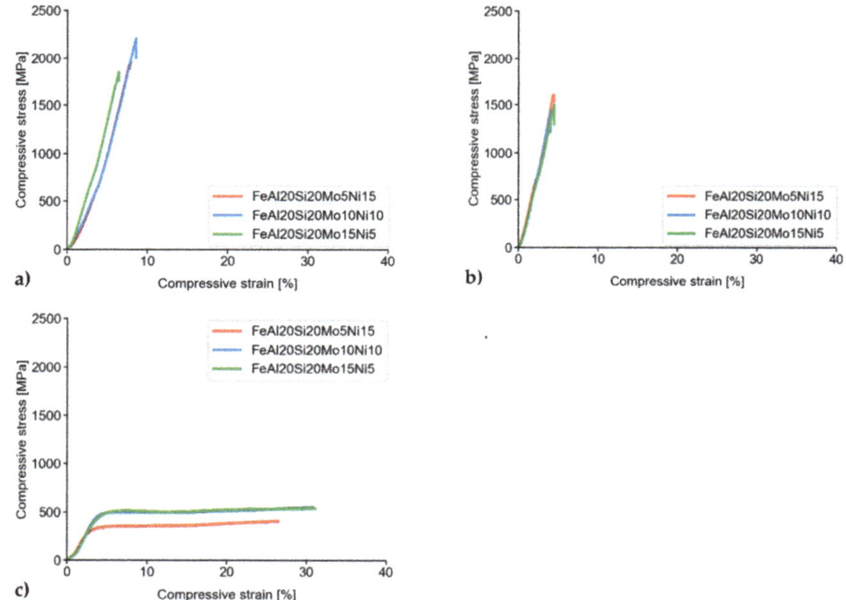

Figure 7. Compressive stress–strain curves of MA + SPS alloys at (**a**) laboratory temperature; (**b**) laboratory temperature after 100 h of annealing at 800 °C; and (**c**) a temperature of 800 °C.

When tested after 100 h annealing at 800 °C, all of the MA + SPS alloys softened, reducing its UCS down to approximately 1600 MPa. The observed decrease in the UCS value was in good agreement with the already observed hardness decrease due to a coarsening of present phases. The observed differences between each alloy were almost negligible, although the highest UCS of 1600 MPa was achieved in the case of the FeAl20Si20Mo5Ni15 alloy, which also showed the highest hardness after annealing.

The MA + SPS alloys were also tested at an elevated temperature of 800 °C (Figure 7c). During these tests, all of the previously brittle alloys changed their behavior, exhibiting significant plasticity due to the activation of non-discrete dislocation movements. Although the tests were done at elevated temperature, the FeAl20Si20Mo10Ni10 and FeAl20Si20Mo15Ni5 showed almost identical values of compressive yield strength (CYS), which were 428 and 437 MPa, respectively.

To fully describe the mechanical properties, the alloys were also tested by the pin-on-disc method to determine wear-related characteristics. After the tests, the morphology of wear tracks was observed with SEM, as is shown in Figure 8. The wear tracks show the presence of thermally induced microcracks whose origins were randomly distributed across all the present phases. Among that, some areas were showing the presence of wrinkles, which were pointed at localized plastic deformation (PD) within the wear track. Such behavior comes along with the already observed plastic deformation during compressive tests at elevated temperature. However, since the wear tests are highly localized, the heat dissipation in the bulk of the material is enormous, allowing the creation of thermally induced cracking of the surface. Besides, the wear debris (WD) at the ends of the wear tracks were composed of oxides, which confirms the presumption. These particles, which were mostly composed of different oxides, were also randomly present in the wear track as is shown, especially in Figure 8a.

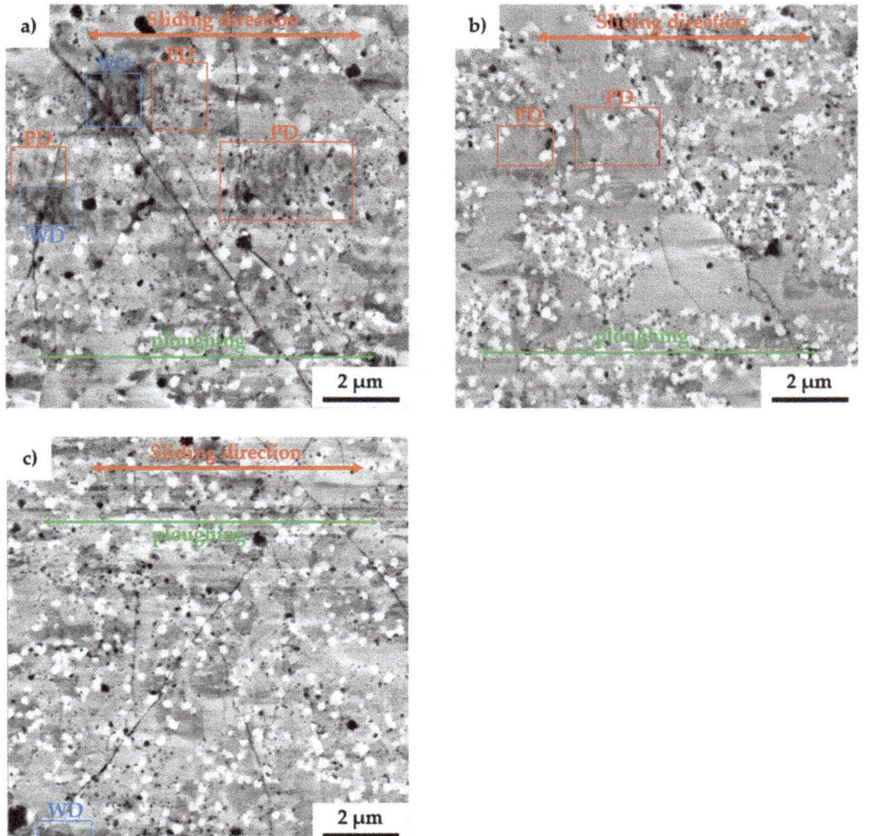

Figure 8. SEM micrographs of the wear tracks of the MA + SPS: (**a**) FeAl20Si20Mo5Ni15; (**b**) FeAl20Si20Mo10Ni10; and (**c**) FeAl20Si20Mo15Ni5 alloys. (PD – plastic deformation; WD – wear debris).

All the MA + SPS alloys were showing exceptional wear resistance with a wear rate that was almost one magnitude lower than that of the reference tool steel 1.3343 (Table 5). Among that, the friction coefficients of these materials were significantly lower and much steadier than those of the reference tool steels. The lowest friction coefficient of 0.446 was achieved in the case of FeAl20Si20Mo10Ni10 alloy, which also exhibited the lowest wear rate of 3.14 10^{-6} $mm^3 \cdot N^{-1} \cdot m^{-1}$. On the other hand, the reference tool steels were showing significantly higher friction coefficients as well as higher wear rates reaching up to 1.46 10^{-5} $mm^3 \cdot N^{-1} \cdot m^{-1}$. Such a high friction coefficient indicated high tangential forces between the ball and tested materials, which need to be overcome to maintain the movement during the wear test. As a result, a lot of energy dissipates during the intensive plastic deformation of sublayers beneath the sliding ball. As a result, the wear track of the tool steels contained deep and wide grooves from ploughing the released particles, enhancing the three-body abrasion.

Table 5. Results of the wear tests done at laboratory temperatures of all the MA + SPS alloys and two references *(Ra – surface roughness)*.

Alloy	Ra (μm)	Wear ($mm^3 \cdot N^{-1} \cdot m^{-1}$)	RSD (±)	Friction Coefficient (–)
FeAl20Si20Mo5Ni15	0.0166	5.97×10^{-6}	3.33×10^{-7}	0.495
FeAl20Si20Mo10Ni10	0.0062	3.14×10^{-6}	3.00×10^{-7}	0.446
FeAl20Si20Mo15Ni5	0.0139	5.68×10^{-6}	4.48×10^{-7}	0.498
Steel 1.2379	0.0096	1.46×10^{-5}	1.51×10^{-6}	0.732
Steel 1.3343	0.0080	2.84×10^{-6}	1.93×10^{-7}	0.669

Compared to that, the MA + SPS alloys showed a rather shallow profile of the wear track with only minor traces of ploughing, as is shown in Figure 8. The grooves were present in all the observed phases, implying that the cohesion between the phases was sufficient, and none of them chipped off, acting as a powerful abrasive medium that would significantly increase the wear rate. Such behavior of the MA + SPS alloys was responsible for achieving a low friction coefficient, as was discovered.

These excellent results of wear resistance were a direct consequence of the phase composition of all the MA + SPS alloys containing binary and ternary silicides, which comes along the ultrafine-grained microstructure and good cohesion of powder particles as well as of the present phases, which did not tend to chip. Considering the high hardness of all the MA + SPS alloys, the primary wear mechanism seems to be oxidation wear together with a minor contribution of abrasive wear. This presumption is supported by the presence of wrinkles pointing at the plastic deformation that are these materials capable of only at elevated temperatures as well as by the presence of oxides found within the wear track or at the end of the wear track.

3.3. Oxidation Resistance

The MA + SPS alloys have also been investigated for cyclic oxidation resistance at 800 °C during the early beginnings in standard atmosphere. All the alloys formed a layer made of oxidic products without any traces of delamination. The layer growth during the first hours of annealing manifested as a steep weight increase followed by a decrease in weight gain speed, since the layer has been effectively shielding the material, slowing the kinetics of oxidation. The initial steps of the oxide layer formation were reaction controlled, while a steep decrease in a weight gain suggested the formation and growth of a protective oxidic membrane whose presence further slowed oxidation. Thus, as the time of oxidation prolonged, the kinetics became controlled by oxygen diffusion through the developed oxidic layer. As is shown in Figure 9, all of the MA + SPS alloys showed exceptional oxidation resistance as the time of the test prolonged, which corresponded to a formation of the protective oxidic barrier. The highest oxidation resistance was observed in the case of FeAl20Si20Mo5Ni15 alloy, which during the first 4 h of cyclic oxidation did not create any traces of an oxidic layer. As the duration of the oxidation test prolonged, the same alloy developed a compact layer of oxides which effectively shielded the material, resulting in the lowest specific weight gain of approximately 2.5 $g \cdot m^{-2}$ among all tested alloys. However, the other MA + SPS alloys showed somewhat higher weight gains reaching up to

10 g·m^{-2}, which are still good enough. It should be noted that due to the limited dimensions of the samples, the weight gains were typical in the order of mg.

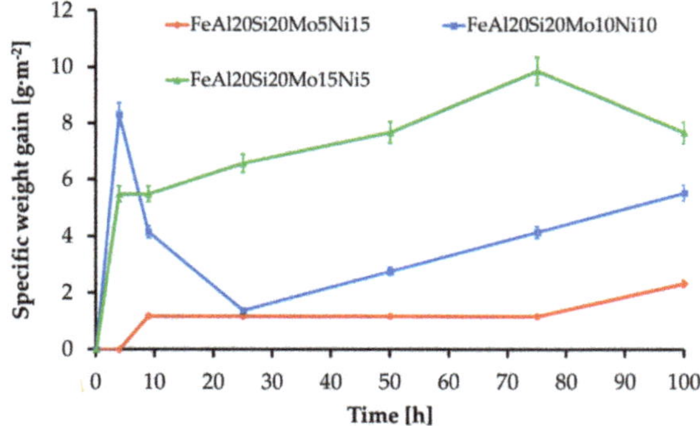

Figure 9. Oxidation kinetics of the MA + SPS FeAl20Si20-Mo-Ni alloys during annealing at 800 °C.

An oblique cross-section has been prepared to display the oxidic layer sufficiently. The thickness of the present oxidic layer (Table 6) obtained via the oblique cross-section has been calculated using Equation (1):

$$d_r = d_m \cdot \sin\left(\tan^{-1}\left(\frac{r_s}{l}\right)\right) \quad (1)$$

where d_r is the real thickness of the oxidic layer, d_m is the measured thickness on the oblique cross-section; r_s is the diameter of used support, and l is the distance between the support and the oxidic layer. As is shown in Table 6, the real thickness of the oxidic layers after 100 h of oxidation done at 800 °C was around 1 μm.

Table 6. Real thickness of an oxidic layer after 100 h at 800 °C observed on an oblique cross-section calculated using Equation (1).

Alloy	Thickness of Oxidic Layer (μm)
FeAl20Si20Mo5Ni15	0.99
FeAl20Si20Mo10Ni10	1.13
FeAl20Si20Mo15Ni5	0.83

The thicknesses of the present oxidic layers observed on oblique cross-sections were after 100 h of cyclic oxidation almost identical, reaching approximately 1 μm.

The SEM + EDS line profiles across the present oxidic layer for all the MA + SPS alloys are shown in Figure 10. From the EDS line profiles, it is visible that the oxidic layer is on the outside containing only Al and O, whose atomic ratios corresponds to Al_2O_3. Its presence is visible throughout the entire oxidic layer. However, the content of Al_2O_3 changes near the oxide–alloy interface, while the content of Si and Fe increases. The ratio of present oxides is changing, favoring the presence of SiO_2 and FeO-based oxides. The thickness of this sublayer, where the content of SiO_2 and FeO-based oxides increased, was approximately 0.5 μm for the first two alloys.

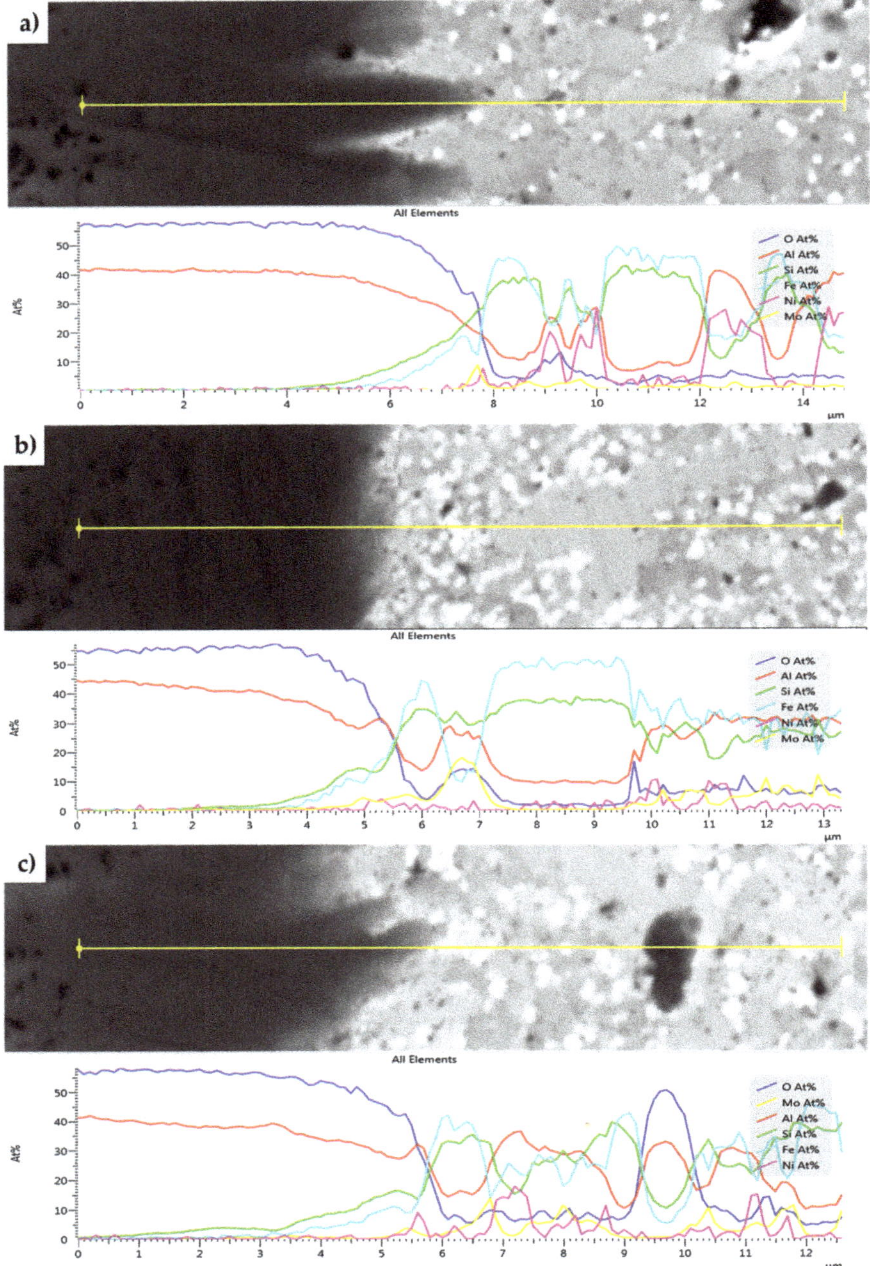

Figure 10. SEM+EDS linescans across the oxide layers formed on the MA + SPS: (**a**) FeAl20Si20Mo5Ni15; (**b**) FeAl20Si20Mo10Ni10; and (**c**) FeAl20Si20Mo15Ni5 alloys after 100 h of cyclic oxidation at 800 °C.

On the other hand, the FeAl20Si20Mo15Ni15 alloy showed an increased concentration of Si already in a distance of 1 µm from the interface of the environment oxidic layer. The increasing concentration of Si was later followed by an increasing concentration of Fe as the distance increased. The presence of

Si in the deeper parts of the oxidic layer significantly improved the oxidation resistance, outperforming those of the other tested MA + SPS alloys.

The formation of Al_2O_3 layers depleted the sub-areas beneath the layer, showing the presence of phases whose concentration of elements corresponded to the FeSi phase, still saturated, among other elements, with Al reaching up to 15 at.%. Such a finding was already discussed in the work of [45], who proposed and also verified that the increased diffusion of Al through the newly developed barrier depleted the present intermetallic phases from Al, resulting in a formation of silicides. Additionally, some of the previously mentioned black areas were showing an increased concentration of Al and O, whose ratios corresponded to Al_2O_3. The origin of these particles can be found in partially pre-oxidized powder particles of Al and others, which during the MA formed the most thermodynamically stable Al_2O_3 product.

4. Conclusions

A combination of mechanical alloying and compaction via spark plasma sintering successfully prepared the FeAl20Si20-Mo-Ni alloys containing from 5–15 wt % of the alloying elements. Prepared alloys showed a uniform microstructure composed of four phases, namely of FeSi, Fe_3Si, $Fe_3Al_2Si_3$, and AlMoSi phases. Formation of the AlMoSi phase increased the hardness and compressive strength of the alloys. It was found that increasing the amount of Mo in the alloy reduced the fraction of FeSi phase at the expanse of the newly created AlMoSi phase. All of the present phases were enriched with alloying elements, increasing the lattice stress–strains due to their deformation. The main contribution toward the hardness of the alloys was, among the MA process itself, caused by $Fe_3Al_2Si_3$, whose content together with the second-hardest FeSi phase reached in the FeAl20Si20Mo5Ni15 up to 58.5 wt %.

As a direct correlation with the hardness, the investigated alloys outperformed the reference tool steels by more than one order of magnitude regarding the wear rate while exhibiting lower friction coefficients. The primary wear mechanism was found to be oxidation wear, which subsequently allowed abrasive wear due to the presence of oxidic debris. During the cyclic oxidation tests, all the alloys showed exceptional oxidation resistance while creating a compact layer of oxidic products that was mainly composed of Al_2O_3 without any traces of delamination. Furthermore, it was found that the highest oxidation resistance of the FeAl20Si20Mo5Ni15 was caused by slightly different ratios of oxides present in the layer. The oxidic layer was initially composed of Al_2O_3 at the environment–layer interface and slightly changed, revealing an increase in the content of Si followed by Fe.

Author Contributions: Conceptualization, F.P.; methodology, F.P.; investigation, O.P., A.Š, V.K., and F.L.; data curation, A.Š. and V.K.; writing—original draft preparation, F.P.; writing—review and editing, V.K. and A.Š.; funding acquisition, F.P. All authors have read and agreed to the published version of the manuscript.

Funding: This research was carried out in the frame of the project 17-07559S, financed by Czech Science Foundation.

Conflicts of Interest: The authors declare no conflict of interest.

References

1. Xu, C.H.; Gao, W.; He, Y.D. High temperature oxidation behaviour of FeAl intermetallics—Oxide scales formed in ambient atmosphere. *Scr. Mater.* **2000**, *42*, 975–980. [CrossRef]
2. Novák, P.; Zelinková, M.; Šerák, J.; Michalcová, A.; Novák, M.; Vojtěch, D. Oxidation resistance of SHS Fe–Al–Si alloys at 800 °C in air. *Intermetallics* **2011**, *19*, 1306–1312. [CrossRef]
3. Senčekova, L.; Palm, M.; Pešička, J.; Veselý, J. Microstructures, mechanical properties and oxidation behaviour of single-phase Fe3Al (D03) and two-phase α-Fe,Al (A2)+Fe3Al (D03) FeAlV alloys. *Intermetallics* **2016**, *73*, 58–66. [CrossRef]
4. Sina, H.; Corneliusson, J.; Turba, K.; Iyengar, S. A study on the formation of iron aluminide (FeAl) from elemental powders. *J. Alloy. Compd.* **2015**, *636*, 261–269. [CrossRef]

5. Li, X.; Prokopčáková, P.; Palm, M. Microstructure and mechanical properties of Fe–Al–Ti–B alloys with additions of Mo and W. *Mater. Sci. Eng. A* **2014**, *611*, 234–241. [CrossRef]
6. Xu, C.H.; Gao, W.; Li, S. Oxidation behaviour of FeAl intermetallics – the effect of Y on the scale spallation resistance. *Corros. Sci.* **2001**, *43*, 671–688. [CrossRef]
7. Haušild, P.; Siegl, J.; Málek, P.; Šíma, V. Effect of C, Ti, Zr and B alloying on fracture mechanisms in hot-rolled Fe–40 (at.%)Al. *Intermetallics* **2009**, *17*, 680–687. [CrossRef]
8. Palm, M. Concepts derived from phase diagram studies for the strengthening of Fe–Al-based alloys. *Intermetallics* **2005**, *13*, 1286–1295. [CrossRef]
9. Schmitt, A.; Kumar, K.S.; Kauffmann, A.; Li, X.; Stein, F.; Heilmaier, M. Creep of binary Fe-Al alloys with ultrafine lamellar microstructures. *Intermetallics* **2017**, *90*, 180–187. [CrossRef]
10. Hadef, F. Solid-state reactions during mechanical alloying of ternary Fe–Al–X (X=Ni, Mn, Cu, Ti, Cr, B, Si) systems: A review. *J. Magn. Magn. Mater.* **2016**, *419*, 105–118. [CrossRef]
11. Fei, W.; Kuiry, S.C.; Seal, S. Inhibition of metastable alumina formation on Fe-Cr-Al-Y alloy fibers at high temperature using titania coating. *Oxid. Met.* **2004**, *62*, 29–44. [CrossRef]
12. Kadiri, H.E.; Molins, R.; Bienvenu, Y.; Horstemeyer, M.F. Abnormal High Growth Rates of Metastable Aluminas on FeCrAl Alloys. *Oxid. Met.* **2005**, *64*, 63–97. [CrossRef]
13. Novák, P.; Knotek, V.; Šerák, J.; Michalcová, A.; Vojtěch, D. Synthesis of Fe-Al-Si intermediary phases by reactive sintering. *Powder Metall.* **2011**, *54*, 167–171. [CrossRef]
14. Gupta, S.P. Intermetallic compound formation in Fe–Al–Si ternary system: Part I. *Mater. Charact.* **2002**, *49*, 269–291. [CrossRef]
15. Novák, P.; Knotek, V.; Voděrová, M.; Kubásek, J.; Šerák, J.; Michalcová, A.; Vojtěch, D. Intermediary phases formation in Fe–Al–Si alloys during reactive sintering. *J. Alloy. Compd.* **2010**, *497*, 90–94. [CrossRef]
16. Novák, P.; Michalcová, A.; Voděrová, M.; Šíma, M.; Šerák, J.; Vojtěch, D.; Wienerová, K. Effect of reactive sintering conditions on microstructure of Fe–Al–Si alloys. *J. Alloy. Compd.* **2010**, *493*, 81–86. [CrossRef]
17. Han, Y.; Ban, C.; Zhang, H.; Nagaumi, H.; Ba, Q.; Cui, J. Investigations on the Solidification Behavior of Al-Fe-Si Alloy in an Alternating Magnetic Field. *Mater. Trans.* **2006**, *47*, 2092–2098. [CrossRef]
18. Golovin, I.S.; Strahl, A.; Neuhäuser, H. Anelastic relaxation and structure of ternary Fe–Al–Me alloys with Me=Co, Cr, Ge, Mn, Nb, Si, Ta, Ti, Zr. *Int. J. Mater. Res.* **2006**, *97*, 1078–1092. [CrossRef]
19. Milekhine, V.; Onsøien, M.; Solberg, J.K.; Skaland, T. Mechanical properties of FeSi (ε), FeSi2 ($\zeta\alpha$) and Mg2Si. *Intermetallics* **2002**, *10*, 743–750. [CrossRef]
20. Galano, M.; Rubiolo, G.H. Creep behaviour of a FeSi-base metallic glass containing nanocrystals. *Scr. Mater.* **2003**, *48*, 617–622. [CrossRef]
21. Jóźwiak, S.; Karczewski, K.; Bojar, Z. The effect of loading mode changes during the sintering process on the mechanical properties of FeAl intermetallic sinters. *Intermetallics* **2013**, *33*, 99–104. [CrossRef]
22. Sundar, R.S.; Deevi, S.C. High-temperature strength and creep resistance of FeAl. *Mater. Sci. Eng. A* **2003**, *357*, 124–133. [CrossRef]
23. Wu, J.; Chong, X.; Jiang, Y.; Feng, J. Stability, electronic structure, mechanical and thermodynamic properties of Fe-Si binary compounds. *J. Alloy. Compd.* **2017**, *693*, 859–870. [CrossRef]
24. Novák, P.; Mejzlíková, L.; Hošek, V.; Martínek, M.; Marek, I.; Michalcova, A. Structure and Properties of Fe-Ni-Al-Si Alloys Produced by Powder Metallurgy. *Acta Phys. Pol. A* **2012**, *122*, 524–527. [CrossRef]
25. Nová, K.; Novák, P.; Arzel, A.; Průša, F. Alloying of Fe-Al-Si Alloys by Nickel and Titanium. *Manuf. Technol.* **2018**, *18*, 645–649. [CrossRef]
26. Jóźwiak, S.; Karczewski, K.; Bojar, Z. Kinetics of reactions in FeAl synthesis studied by the DTA technique and JMA model. *Intermetallics* **2010**, *18*, 1332–1337. [CrossRef]
27. Canakci, A.; Erdemir, F.; Varol, T.; Ozkaya, S. Formation of Fe–Al intermetallic coating on low-carbon steel by a novel mechanical alloying technique. *Powder Technol.* **2013**, *247*, 24–29. [CrossRef]
28. Dobromyslov, A.V.; Taluts, N.I.; Pilyugin, V.P.; Tolmachev, T.P. Mechanical alloying of Al–Fe alloys using severe deformation by high-pressure torsion. *Phys. Met. Metallogr.* **2015**, *116*, 942–950. [CrossRef]
29. Neikov, O.D. *Handbook of Non-Ferrous Metal Powders*; Neikov, O.D., Naboychenko, S.S., Murashova, I.V., Gopienko, V.G., Frishberg, I.V., Lotsko, D.V., Eds.; Elsevier: Oxford, UK; pp. 63–79. [CrossRef]
30. Čech, J.; Haušild, P.; Karlík, M.; Kadlecová, V.; Čapek, J.; Průša, F.; Novák, P. Mechanical Properties of FeAlSi Powders Prepared by Mechanical Alloying from Different Initial Feedstock Materials. *Matériaux Tech.* **2019**, *107*, 6. [CrossRef]

31. Azzaza, S.; Alleg, S.; Sunol, J.J. Phase Transformation in the Ball Milled Fe31Co31Nb8B30 Powders. *Adv. Mater. Phys. Chem.* **2013**, *3*, 90–100. [CrossRef]
32. Izadi, S.; Akbari, G.H.; Janghorban, K. Sintering and mechanical properties of mechanically alloyed Fe–Al–(B) nanostructures. *J. Alloy. Compd.* **2010**, *496*, 699–702. [CrossRef]
33. Jiang, T. Investigation of Phase Composition and Microstructure of the FeAl Intermetallics Compounds Bulks Fabricated by Mechanical Alloying Process and Hot-Pressing Process. *Adv. Mater. Res.* **2011**, *228*, 899–904. [CrossRef]
34. Kalita, M.P.C.; Perumal, A.; Srinivasan, A. Structural analysis of mechanically alloyed nanocrystalline Fe75Si15Al10 powders. *Mater. Lett.* **2007**, *61*, 824–826. [CrossRef]
35. Kalita, M.P.C.; Perumal, A.; Srinivasan, A. Structure and magnetic properties of nanocrystalline Fe75Si25 powders prepared by mechanical alloying. *J. Magn. Magn. Mater.* **2008**, *320*, 2780–2783. [CrossRef]
36. Boukherroub, N.; Guittoum, A.; Laggoun, A.; Hemmous, M.; Martínez-Blanco, D.; Blanco, J.A.; Souami, N.; Gorria, P.; Bourzami, A.; Lenoble, O. Microstructure and magnetic properties of nanostructured (Fe0.8Al0.2)100–xSix alloy produced by mechanical alloying. *J. Magn. Magn. Mater.* **2015**, *385*, 151–159. [CrossRef]
37. Průša, F.; Šesták, J.; Školáková, A.; Novák, P.; Haušild, P.; Karlík, M.; Minárik, P.; Kopeček, J.; Laufek, F. Application of SPS consolidation and its influence on the properties of the FeAl20Si20 alloys prepared by mechanical alloying. *Mater. Sci. Eng. A* **2019**, *761*, 138020. [CrossRef]
38. Ma, R.P.; Wan, M.; Huang, J.; Xie, Q.; Suzuki, T. Calculation of electronic structure and mechanical properties of DO 3 – Fe 75-x Si 25 Ni x intermetallic compounds by first principles. *Int. J. Mod. Phys. B* **2015**. [CrossRef]
39. Yanson, T.I.; Manyako, M.B.; Bodak, O.I.; German, N.V.; Zarechnyuk, O.S.; Cerný, R.; Yvon, K. Triclinic Fe3Al2Si3 and Orthorhombic Fe3Al2Si4 with New Structure Types. *Acta Crystallogr. Sect. C Cryst. Struct. Commun.* **1996**, *52*, 2964–2967. [CrossRef]
40. Apiñaniz, E.; Legarra, E.; Plazaola, F.; Garitaonandia, J.S. Influence of addition of Si in FeAl alloys: Theory. *J. Magn. Magn. Mater.* **2008**, *320*, e692–e695. [CrossRef]
41. Gale, W.F.; Totemeier, T.C. *Smithells Metals Reference Book*; Elsevier Science: Amsterdam, The Netherlands, 2003.
42. Kupenko, I.; Merkel, S.; Achorner, M.; Plückthun, C.; Liermann, H.-P.; Sanchez-Valle, C. Plastic deformation of FeSi at high pressures: Implications for planetary cores. In *EGU General Assembly Conference Abstracts*; EGU2017: Vienna, Austria, 2017.
43. Ehlers, S.K.; Mendiratta, M.G. Tensile behaviour of two DO3-ordered alloys: Fe3Si and Fe-20 at % Al-5 at % Si. *J. Mater. Sci.* **1984**, *19*, 2203–2210. [CrossRef]
44. Zhao, L.-Z.; Zhao, M.-J.; Li, D.-Y.; Zhang, J.; Xiong, G.-Y. Study on Fe–Al–Si in situ composite coating fabricated by laser cladding. *Appl. Surf. Sci.* **2012**, *258*, 3368–3372. [CrossRef]
45. Novak, P.; Nova, K. Oxidation Behavior of Fe-Al, Fe-Si and Fe-Al-Si Intermetallics. *Materials* **2019**, *12*, 1748. [CrossRef] [PubMed]

Article

Effect of Initial Powders on Properties of FeAlSi Intermetallics

Jaroslav Čech [1,*], Petr Haušild [1], Miroslav Karlík [1], Václav Bouček [1], Kateřina Nová [2], Filip Průša [2], Pavel Novák [2] and Jaromír Kopeček [3]

[1] Department of Materials, Faculty of Nuclear Sciences and Physical Engineering, Czech Technical University in Prague, 12000 Prague, Czech Republic
[2] Department of Metals and Corrosion Engineering, University of Chemistry and Technology, 16628 Prague, Czech Republic
[3] Institute of Physics of the Czech Academy of Sciences, 18221 Prague, Czech Republic
* Correspondence: jaroslav.cech@fjfi.cvut.cz; Tel.: +420-224-358-518

Received: 16 August 2019; Accepted: 31 August 2019; Published: 4 September 2019

Abstract: FeAlSi intermetallics are materials with promising high-temperature mechanical properties and oxidation resistance. Nevertheless, their production by standard metallurgical processes is complicated. In this study, preparation of powders by mechanical alloying and properties of the samples compacted by spark plasma sintering was studied. Various initial feedstock materials were mixed to prepare the material with the same chemical composition. Time of mechanical alloying leading to complete homogenization of powders was estimated based on the microstructure observations, results of XRD and indentation tests. Microstructure, phase composition, hardness and fracture toughness of sintered samples was studied and compared with the properties of powders before the sintering process. It was found that independently of initial feedstock powder, the resulting phase composition was the same (Fe_3Si + FeSi). The combination of hard initial powders required the longest milling time, but it led to the highest values of fracture toughness.

Keywords: FeAlSi; intermetallic alloys; mechanical alloying; spark plasma sintering; microstructure; nanoindentation; mechanical properties

---

1. Introduction

Intermetallic materials are used in a wide range of applications for some of their irreplaceable physical and mechanical properties [1]. Iron aluminides were one of the most studied materials especially due to their low weight and good oxidation resistance compared to austenitic stainless steels or nickel-based superalloys [2]. Improvement of Fe_3Al and FeAl based iron aluminides was achieved by alloying by Cr, Nb, Zr, Co, Ce, Ti, etc. (e.g., [3–7]). Much less attention was focused on the ternary system Fe-Al-Si because of the brittle behavior of these alloys. In the previous studies, it was found that the system FeAl20Si20 (wt.%) has excellent high temperature oxidation and sulphidation resistance [8,9]. Therefore, they are the potential candidates for replacement of stainless steels and nickel superalloys, which can reduce the amount of the necessary critical raw materials used for their production [10].

The production of these alloys by standard metallurgical processes such as casting followed by hot/cold rolling is practically not possible because of their brittleness. For this reason, mechanical alloying (MA) was used to prepare the powders suitable for further sintering. Mechanical alloying is the process which starts from blended powder mixtures which are exposed to severe deformation in a high-energy ball charge. During mechanical alloying, consisting of repeated cold welding, fracturing and rewelding, the powder particles become homogeneous [11,12]. Mechanical alloying can lead to formation of very fine-grained microstructure with metastable phase composition which cannot be reached by other methods. The conditions of mechanical alloying process, especially the ball-to-powder ratio, rotational speed and lubrication, have a crucial effect on the structure and phase composition of the product. The use of a lubricant minimizes the adhesion of the powder to the walls of the milling jar and balls and hence it reduces the contamination caused by the adhesion wear of these parts. On the other hand, it decreases friction force and lowers the temperature of the milled powder [13]. The ball-to-powder ratio affects the required milling duration strongly. The higher is the ratio, the higher energy is supplied to the powder and thus the process is more efficient and faster [14]. Growing ball-to-powder ratio also increases the micro-strain and decreases the grain size of the product [14]. The rotational velocity also affects the required duration of mechanical alloying and also the morphology of the particles. In the case of brittle materials, higher rotational velocity produces finer powders. On the other hand, in the case of plastic materials it promotes flattening and cold welding of particles [15,16].

Ultra-high energy mechanical alloying was developed in the previous works [13,17]. It is based on high ball-to-powder ratio with large balls leading to collisions with high kinetic energy. Milling without any lubricant increases the friction forces and consequently the local temperature, which leads to or accelerates thermally activated reactions. As a consequence, the process of powder homogenization is faster than in the case of commonly used conditions of mechanical alloying.

In order to prevent grain coarsening or undesirable phase changes, the powder consolidation must be fast. For this reason, spark plasma sintering (SPS) was chosen as a sintering method. During SPS [18] the powder is pressed between two electrodes and high amperage electric current is applied. Joule heat and electric discharges rapidly heat the powder and consolidate it into the final sample. As it uses high sintering rates, it prevents the growth of fine-grained microstructure, which can be beneficial for mechanical properties of the material (especially the hardness and the fracture toughness which is very important in the case of brittle intermetallic alloys).

Kinetics of the mechanical alloying depends on several factors, such as batch size, powder to ball mass ratio, rotational speed, lubrication, milling time or initial powder composition [11]. This paper focuses on the ternary alloy FeAlSi with the (optimal) composition FeAl20Si20. The effect of the other compositions (various ratio of Al and Si) was studied in previous papers (e.g., [8,9,17,19]). It was found that changes in the ratio of Al and Si can affect final phase composition (phases FeAl, Fe_3Si, Fe_2Al_5, FeSi, $Fe_3Al_2Si_3$, $FeAl_2Si$ in different amounts were identified for different compositions), oxidation resistance and/or mechanical properties (hardness). In this study, the effect of the initial powders on the process of mechanical alloying of FeAl20Si20 alloy was investigated. The aim of this paper is to improve the processing method of this alloy. Structural and mechanical properties of the powders and compacted materials prepared by SPS technique are examined and compared.

2. Materials and Methods

2.1. Materials Preparation

Compacts of the final composition FeAl20Si20 were prepared from different initial feedstock materials. The initial composition FeAl20Si20 was chosen on the basis of the previous works, where the excellent high-temperature oxidation resistance and promising mechanical properties were identified for this alloy [9]. A given number of prerequisites were mixed in required ratios to obtain desired final composition of 60 wt.% of Fe, 20 wt.% of Al and 20 wt.% of Si. As an initial input powders pure Fe (Strem Chemicals, Newburyport, MA, USA, ~7 μm, purity 99.9%), Al (Alfa Aesar, Haverhill, MA, USA, ~44 μm, purity 99.5%), Si (Strem Chemicals, ~44 μm, purity 99.7%) and pre-alloyed AlSi30 (phase composition: Fe, Al and Si, particle size < 400 μm), FeAl25 (phase composition: Fe_3Al and FeAl, particle size < 150 μm), FeSi25 (phase composition: Fe_3Si and FeSi, particle size < 200 μm) were used. Scanning electron microscope (SEM) backscattered electron (BSE) images of pre-alloyed powders are presented in Figure 1. Four mixtures were prepared:

- Fe + Al + Si (denoted Fe_Al_Si)

- FeAl25 + Si (denoted FeAl_Si)
- Fe + AlSi30 + Si (denoted Fe_AlSi)
- FeSi25 + Al (denoted FeSi_Al)

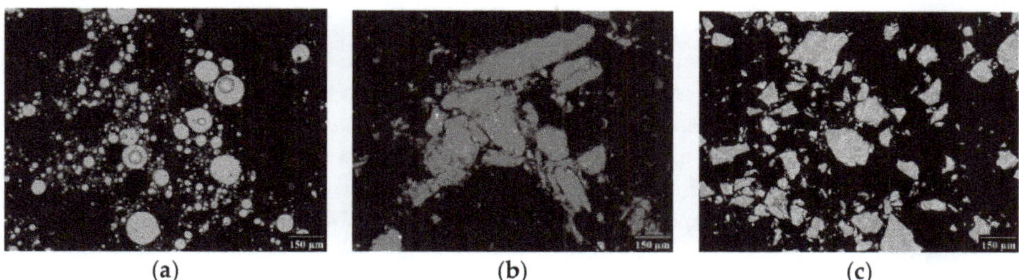

Figure 1. Microstructure of pre-alloyed powders (SEM BSE micrographs): (**a**) FeAl25, (**b**) AlSi30, (**c**) FeSi25.

Powders were mixed and mechanically alloyed in planetary ball mill PM 100 CM (Retsch, Haan, Germany) in argon atmosphere. Five grams of powder were milled by balls made of the same material as the mill-AISI 420 steel, the powder to ball mass ratio was 1:60 and rotational speed 400 rpm. Powders were milled for 1, 2, 4, 6 and 8 h and the evolution of their microstructure, phase composition and mechanical properties were studied.

Powders were compacted into discs with diameter of 20 mm by spark plasma sintering using HP D10 device (FCT Systeme GmbH, Rauenstein, Germany). Powders were heated to 1000 °C (heating rate 300 °C/min up to 900 °C and then 100 °C/min to 1000 °C to avoid overshoot of the temperature) and pressed to 48 MPa. The maximum temperature was held for 10 min followed by slow cooling rate 50 °C/min to avoid sample cracking due to thermal shocks.

2.2. Characterization Techniques

Microstructure, phase composition, hardness and Young's modulus of the powders after different milling times and compacted samples after sintering were evaluated. The powders and sintered samples were embedded into carbon filler containing phenolic resin and metallographic cuts were prepared by standard procedures. The surface was finished by polishing in 0.04 µm colloidal silica. Microstructure was examined in backscattered electron (BSE) signal in scanning electron microscope (SEM) JEOL JSM 5510LV (JEOL, Tokyo, Japan) with iXRF 500 energy dispersive X-ray spectrometer (EDS) and transmission electron microscope (TEM) JEOL 2200FS equipped by JEOL large angle EDS. TEM characterization was first carried out on loose mechanically alloyed powders deposited on copper grid with holey carbon film. The sintered materials were very brittle and thus the first TEM samples were prepared by crushing of the SPS compacts in an agate mortar. Later, focused ion beam (FIB) milling (in Zeiss Auriga device) was used to prepare thin lamellae about 15 µm by 15 µm in size. Phase composition was studied by PANalytical X'Pert Pro X-Ray diffractometer (PANalytical, Almelo, The Netherlands) in Bragg-Brentano geometry with Cu cathode.

Instrumented indentation tests were performed to determine mechanical properties of the studied materials. Indentation force-penetration depth curves were evaluated by Oliver-Pharr method [20,21] and the hardness and Young's modulus were calculated according to standard ISO 14,577 [22] on NHT^2 nanoindentation tester (Anton Paar, Graz, Austria). At least 15 particles from every powder mixture were tested to obtain statistically representative value of hardness and Young's modulus. As the hard powder particles are relatively small and embedded in the matrix with much lower hardness and Young's modulus, nanoindentation tests with partial unloadings at 1, 2, 5 and 10 mN were performed to determine the effect of the surrounding matrix on the measured values of hardness and Young's modulus of powder particles. It was found [23] that for the size of particles in the order of tens of micrometers, the values of hardness measured at the maximum load 2 mN (corresponding to the penetration depth of approximately 80 nm) were not affected by the surrounding resin and the scatter of the data was reasonably small. On the other hand, the measurements of Young's modulus of hard particles in such softer surrounding matrix were affected even for the penetration depths lower than 0.25% of the size of the particles. This means that in the case of the studied materials, penetration depths should be less than 50 nm, which is not possible because it will not meet the conditions of the ISO standard [22]. As a consequence, measured values of Young's modulus for powder particles were underestimated and they are not presented. Moreover, to avoid the effect of the particle size, which is changing with milling time [17], only the particles larger than 10 μm were tested to avoid the effect of the particle size in soft surrounding matrix on hardness measurements.

Based on this research, sintered samples were also tested at the maximum load of 2 mN in order to obtain the data for the same experimental conditions and to enable direct comparison of results.

To determine fracture toughness of the sintered materials, Vickers indentations with maximum load of 5 N were performed on MHT microindentation tester (Anton Paar, Graz, Austria) and the dimensions of radial cracks were measured (Figure 2). To evaluate the fracture toughness, model of Niihara [24] for Palmqvist cracks was employed. This model is based on the equation.

$$K_{IC} = 0.035\left(\frac{l}{a}\right)^{-\frac{1}{2}} Ha^{\frac{1}{2}}\frac{1}{\phi}\left(\frac{E\phi}{H}\right)^{\frac{2}{5}}, \tag{1}$$

where l is the crack length, a is a half-diagonal of the indent, H is hardness, E stands for Young's modulus and ϕ is the constraint factor (ratio of hardness and yield stress reaching the value of approximately 3).

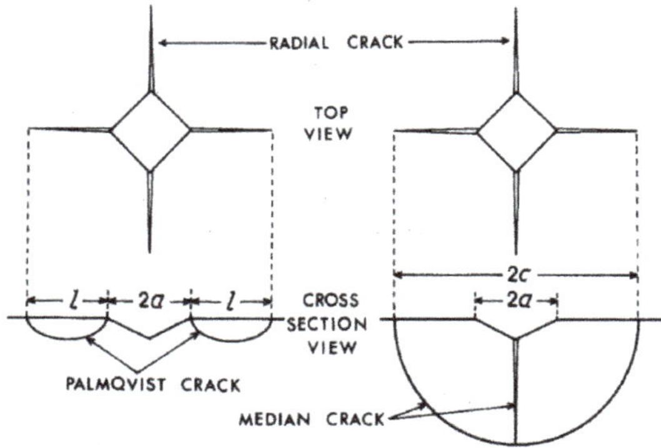

Figure 2. Geometry of the cracks created by the Vickers indentation [24].

3. Results

3.1. Powders

Typical evolution of the microstructure of the powders is presented in Figure 3. At the beginning of the milling process, coarse lamellas of the initial powders in the individual powder particles and the unchanged particles of the initial powders were present. With increasing milling time (2 h), the lamellas got thinner, the occurrence of particles unaffected by the milling process was still rarer and first homogeneous particles with required chemical composition were observed. After four hours of milling (Figure 4), the homogeneous particles were present for all four powder mixtures. For the mixture of pure elements Fe_Al_Si, the homogeneous particles completely replaced heterogeneous particles. For the mixtures prepared from pre-alloyed powders, particles with lamellar structure could still be observed. Complete homogenization of Fe_AlSi and Fe_SiAl powders was observed after 6 h of milling. Only the powder Fe_AlSi was not completely homogenous and the particles with lamellar structure were observed. The complete homogenization of FeAl_Si powder occurred after 8 h of milling.

Figure 3. Typical example of the microstructure evolution with increasing time of mechanical alloying (Fe_AlSi powder mixture, SEM BSE micrographs).

From the low magnification TEM micrographs of the Fe_Al_Si powder after MA for 8 h (not presented here) it was obvious, that only particles under 5 µm in size had stuck on the carbon film. The chemical composition of these small particles obtained from TEM-EDS was in the range (wt.%) 55% to 66% of Fe, 10% to 15% of Al, 13% to 16% of Si, and 0.7% to 4% of O. Two typical Fe-rich particles, agglomerates of very small grains, 20 to 100 nm in size, are presented in Figure 5a. According to simulations performed using JEMS software (version 4.7830U2019b11) by Stadelmann [25], the diameters of the rings in the related electron diffraction pattern (inset in Figure 5a) correspond to the Fe_3Si phase ($Fm\overline{3}m$, space group 225, a = 0.5655 nm).

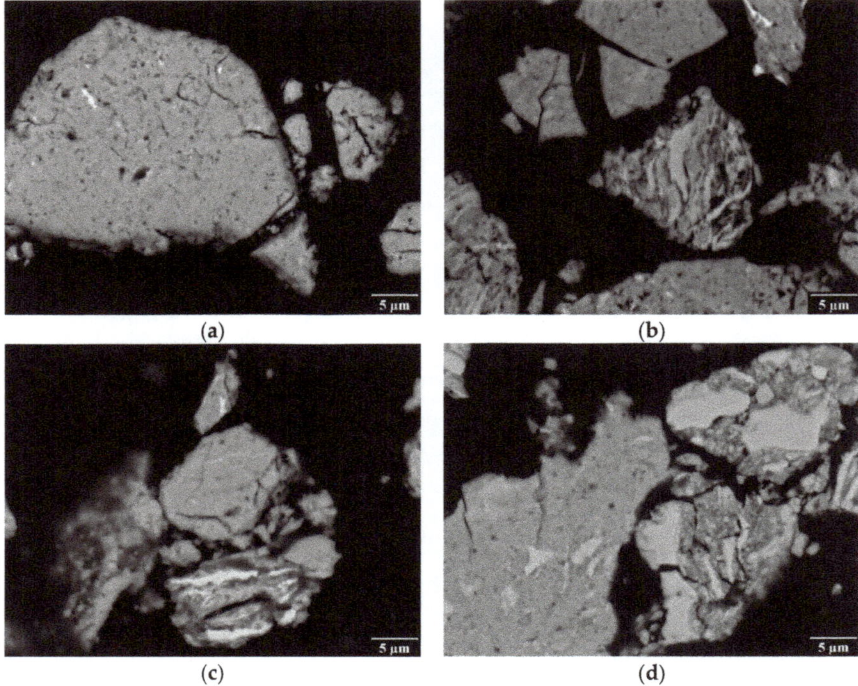

Figure 4. SEM BSE micrographs of powders after 4 h of mechanical alloying: (**a**) Fe_Al_Si, (**b**) FeAl_Si, (**c**) Fe_AlSi, (**d**) FeSi_Al.

Characteristic change of phase composition during mechanical alloying is shown in Figure 6 in the example of the Fe_AlSi powder mixture. At the beginning of the alloying process, the initial powders of the respective mixtures (i.e., Fe, Al, Si, and phases FeAl, Fe_3Al, FeSi, Fe_3Si in pre-alloyed powders) could be identified. With increasing milling time, the peaks of pure elements (Fe, Al and Si) disappeared and the peaks of the phases FeSi, Fe_3Si and FeAl occurred and started to prevail for higher milling times. The change rate was different for different powder mixtures (Figure 7). At the end of the mechanical alloying process, only the phases FeSi, Fe_3Si (supersaturated by Al) and small fraction of FeAl were present. The supersaturation of the phases FeSi and Fe_3Si by Al atoms caused the increase of lattice parameters and shift of the peaks to the lower diffraction angles.

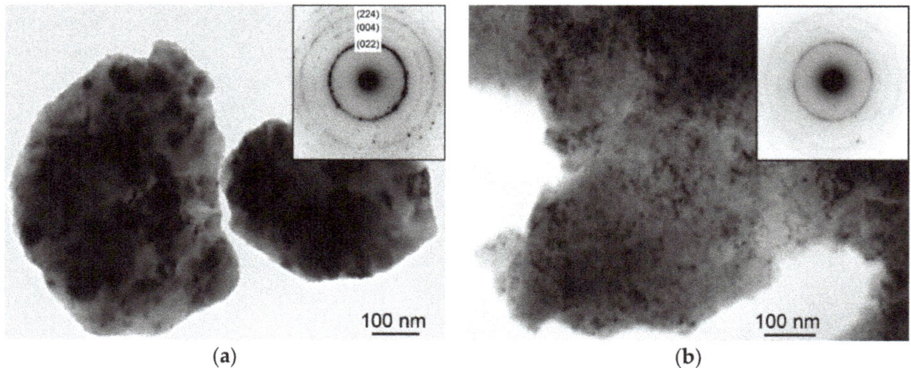

(**a**) (**b**)

Figure 5. *Cont.*

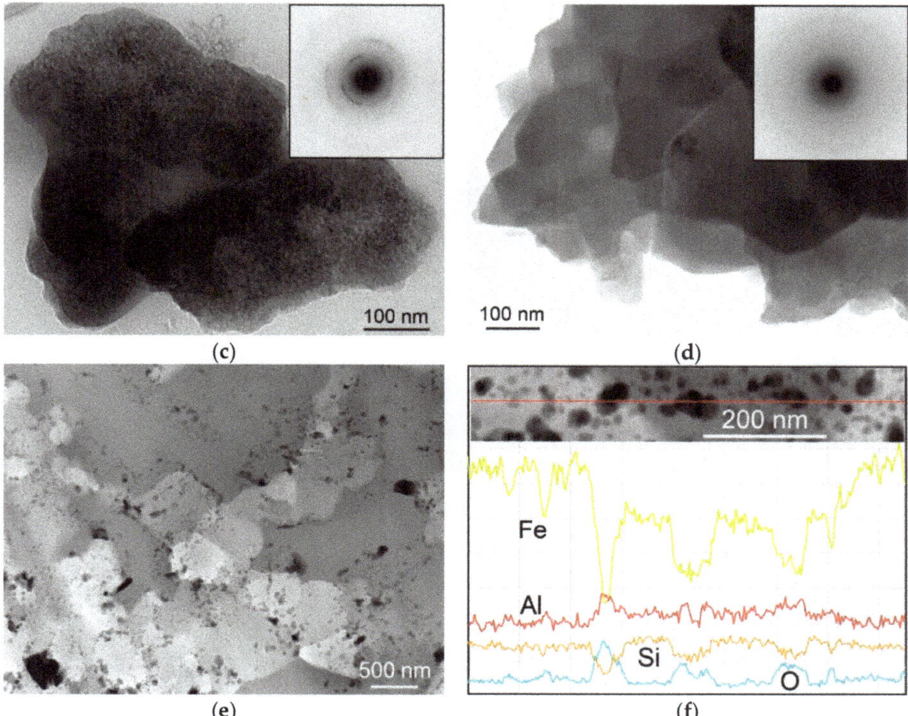

Figure 5. (**a–d**) Bright field TEM micrographs of: (**a**) nanocrystalline Fe_3Si particles in the loose powder mixture from pure elements (Fe_Al_Si) after 8 h of mechanical alloying (related ring diffraction pattern is in the inset), (**b–d**) corresponding spark plasma sintering (SPS) compact crushed in agate mortar: (**b**) refined crystallites of the Fe_3Si phase, (**c**) another (unidentified) phase with nanometric grains (**d**) amorphous phase, (**e**) STEM HAADF micrograph of a focused ion beam (FIB) lamella prepared from the same SPS compact, (**f**) STEM EDS line analysis across dark alumina particles in the micrograph (**e**).

For Fe_Al_Si powder mixture from pure elements, the phases FeSi, Fe_3Si and FeAl occurred after 2 h of milling and the peaks of original pure powders vanished after 4 h when the homogenization process was completed. The phases FeSi and Fe_3Si, which formed the final powder mixture, started to occur after 4 h of milling for the mixture FeAl_Si. The phases of the initial powders (i.e., FeAl, Fe_3Al and Si) were present even after 6 h of mechanical alloying and only after 8 h were not observed. The final phases (FeSi, Fe_3Si and FeAl) for the mixture Fe_AlSi were firstly observed after 2 h of the mechanical alloying and the pure elements were lastly observed after 4 h of milling. The phases FeSi and Fe_3Si in powder mixture FeSi_Al were present already from the beginning of milling because they formed the pre-alloyed powder; other phases disappeared after 4 h of milling.

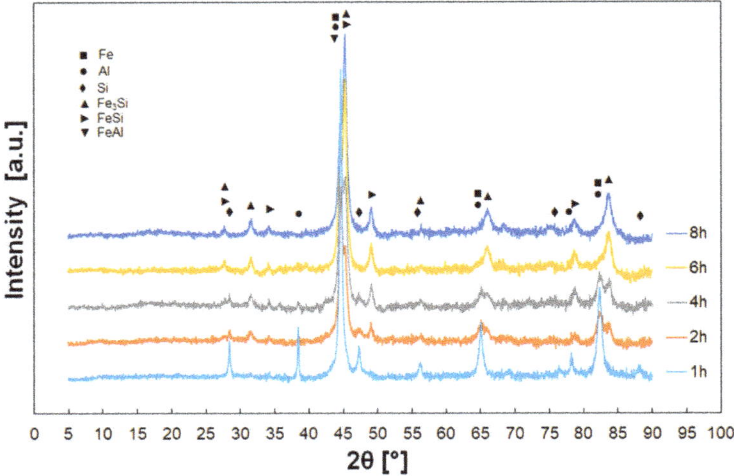

Figure 6. Typical example of the phase composition evolution with increasing time of mechanical alloying (Fe_AlSi powder mixture).

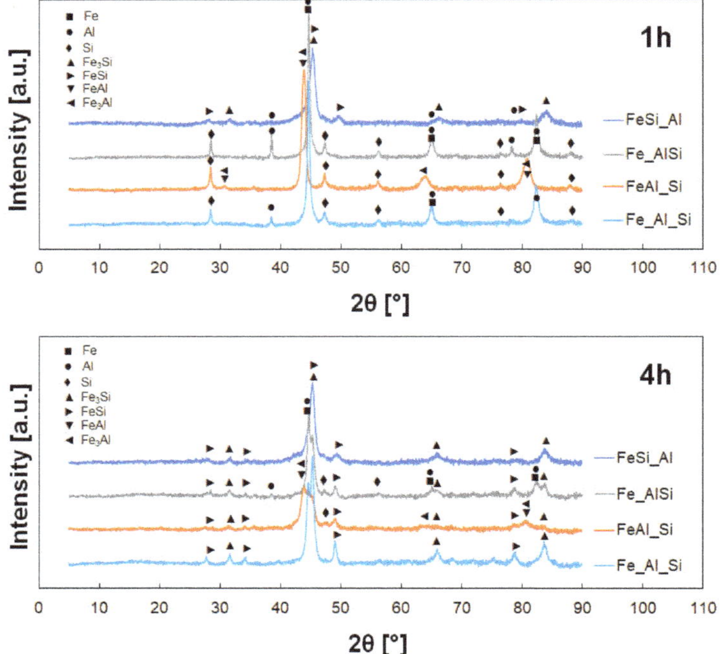

Figure 7. *Cont.*

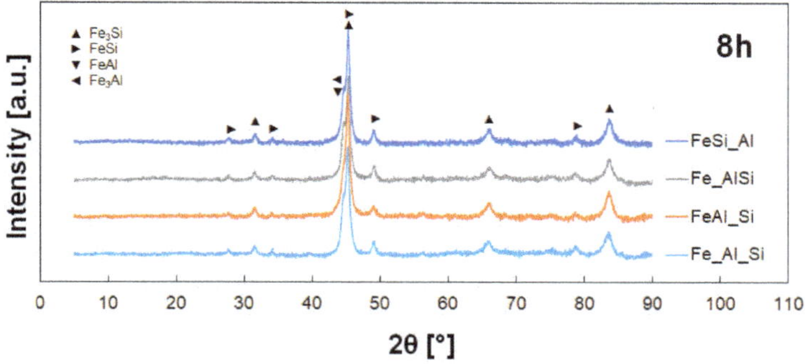

Figure 7. Phase composition of the powder mixtures in different stages of mechanical alloying.

Hardness values evaluated for maximum load 2 mN are presented in Figure 8. Values presented for 0 h correspond to initial pre-alloyed powders (i.e., FeAl25, AlSi30, FeSi25). Except FeSi_Al powder mixture, hardness increased at the beginning of the milling process. For Fe_Al_Si mixture, the significant increase ended after 2 h and the hardness stayed constant. For FeAl-Si mixture the increase was slower and more progressive up to the end of mechanical alloying at 8 h. Hardness of Fe_AlSi mixture started from the lowest values, but it increased very rapidly and it reaches the stable value after 2 h. On the other hand, mixture FeSi_Al started at very high hardness values (hard pre-alloyed powder FeSi25), hardness decreased to the minimum after 2 h after which it increased and it reached the constant value after 4 h. All the powder mixtures had hardness of approximately 15 GPa at the end of the mechanical alloying.

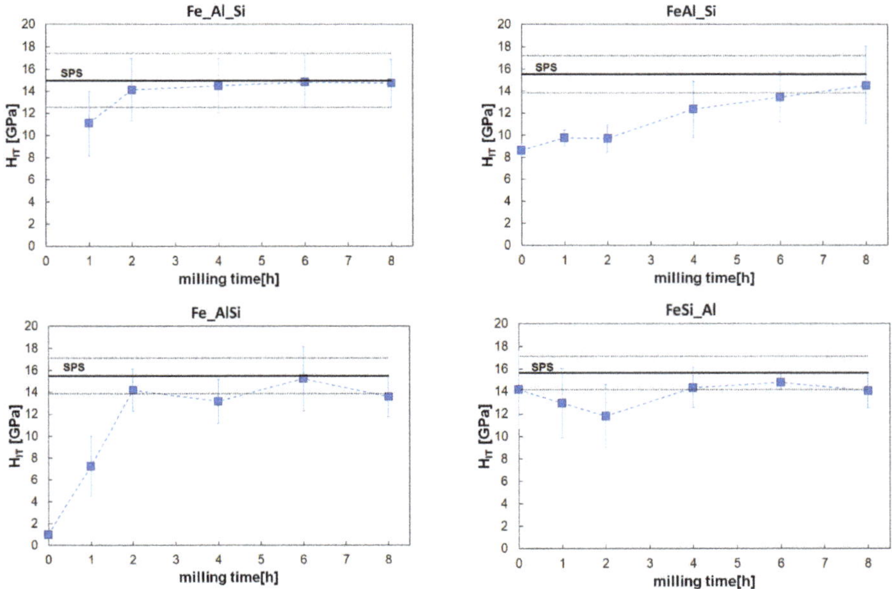

Figure 8. Hardness of the powder mixtures in different stages of mechanical alloying (solid horizontal line shows the value of the compacted sample measured at the maximum load of 2 mN, dotted lines represent the error bars).

3.2. Compacted Materials

An example of the microstructure of the sintered sample by the SPS method is shown in Figure 9. It was homogeneous with a very low porosity (less than 0.1% for Fe_Al_Si sample [19,26], relative density nearly 100%, and fine microstructure (the grain size was around 1 μm for all four sintered samples—Table 1). There were not observed any significant differences between different initial powder mixtures as the powder mixtures used for SPS were of the same chemical and phase composition after MA.

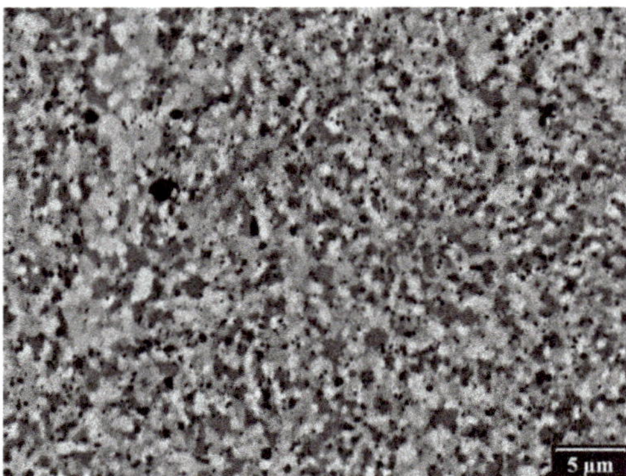

Figure 9. Microstructure of the compacted powder Fe_AlSi.

Table 1. Average grain size and fracture toughness of the samples prepared by SPS (measured at the maximum load of 5 N).

Powder Mixture	Grain Size [μm]	H [GPa]	E [GPa]	a [μm]	l [μm]	l/a [-]	K_{IC} [MPa·m$^{1/2}$]
Fe_Al_Si	1.04	13.48	233.55	14.29	18.22	1.28	2.57 ± 0.18
FeAl_Si	1.07	13.46	220.45	14.33	7.77	0.54	3.61 ± 0.32
Fe_AlSi	0.94	13.37	231.03	14.50	16.53	1.14	2.73 ± 0.28
FeSi_Al	0.88	12.61	220.66	14.26	9.95	0.70	3.21 ± 0.40

From TEM analysis of the corresponding powder from the crushed Fe_Al_Si SPS compact it follows that SPS led to refining of the crystallites of the Fe_3Si phase to nanometric size (Figure 5b), probably due to recrystallization (micrographs and diffraction patterns in Figure 5a,b are to scale). Isolated particles of other phases were also observed (Figure 5c). However, the crystallographic identification of these phases was not successful. Furthermore, amorphous nanoparticles were found (Figure 5d), formed possibly due to the local liquid phase sintering and rapid cooling of the compact. These amorphous phases may be at the origin of the high brittleness of the compact. According to EDS analysis, the chemical composition of the particles from the crushed compact was practically the same as of the loose MA powder.

From the scanning transmission electron microscopy (STEM) imaging of the FIB lamella prepared also from the Fe_Al_Si SPS compact using high-angle annular dark field detector (HAADF) it follows, that the grain size ranges from 100 nm up to 3 μm (Figure 5e). Many small particles were observed inside the grains and also on grain boundaries. The STEM HAADF signal is proportional to the scattering power (Z-contrast, where Z is the proton number); hence heavy elements appear bright, and holes and light particles appear dark. Therefore, the differences in the gray level of the grains in Figure 5e indicate a quite heterogeneous Fe concentration—grains appearing bright contain more iron, grains appearing dark contain more Al and Si. It can also be seen that the SPS compact includes a very high density of dark oxide particles, mostly 5 to 100 nm in size. These oxides are distributed inside of the majority of the grains and also on grain boundaries (Figure 5e). Figure 5f presents results of line

STEM-EDS analysis across the particles, the red line in the inset indicating the path of the electron beam. From the plot it is obvious that majority of particles are oxides of aluminum. Therefore, argon flushing atmosphere during mechanical alloying did not prevent alumina formation. TEM analyses of the powders and compacts of the other three materials (FeAl_Si, Fe_AlSi, and FeSi_Al gave almost the same results, so they are not presented here).

XRD patterns are in Figure 10 and they show only small changes compared to the mechanically alloyed powders. The XRD revealed that the sintered materials were predominantly composed of FeSi and Fe_3Si phases (supersaturated by Al). Contrary to the MA powders the peaks of ternary phase $Fe_3Al_2Si_3$ were identified in the sintered samples (compare Figures 7 and 10).

Hardness measured for maximum load of 2 mN is represented in Figure 8 by solid horizontal line (dotted horizontal lines show the scatter of the data). The measured hardness was about 15–16 GPa and it was in good agreement with the values reached by the mechanically alloyed powders after 8 h of milling.

The fracture toughness was determined by measuring hardness, Young's modulus and crack dimensions formed by the Vickers indentation to the maximum load of 5 N. The results are summarized in Table 1 and observed crack systems are shown in Figure 11. The measured hardness was about 13 GPa and Young's modulus 225 GPa. Fracture toughness calculated using Equation (1) from measured radial (Palmqvist) cracks was about 3 MPa·m$^{1/2}$. It was lowest for the Fe_Al_Si sample reaching only 2.57 MPa·m$^{1/2}$. The highest fracture toughness was found for the sample FeAl_Si (3.61 MPa·m$^{1/2}$). The SEM images confirmed that the cracks were not connected under the indent, which means that they were radial (not median) cracks. The system of the radial cracks was well developed for the samples Fe_Al_Si and Fe_AlSi. For the samples FeSi_Al and FeAl_Si, the systems with only three cracks were also sometimes observed. These indents were excluded from the analysis and only the systems with four cracks were analyzed. On the other hand, lateral cracks were occasionally observed for the samples FeSi_Al and FeAl_Si with higher fracture toughness, which could be the reason for the increase in K_{IC} (change of the mechanism of cracking associated with plastic deformation).

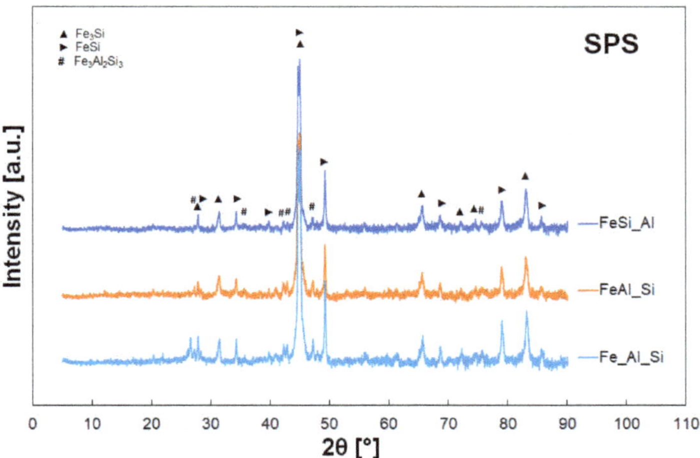

Figure 10. Phase composition of the sintered samples.

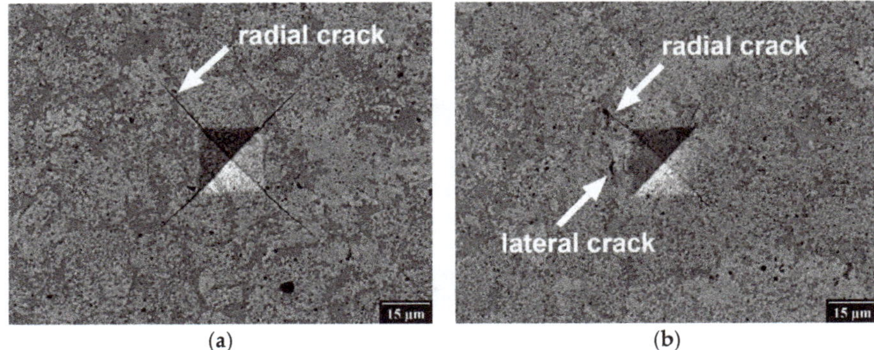

(**a**) (**b**)

Figure 11. Cracks propagating from the Vickers indentation: (**a**) Fe_Al_Si, (**b**) FeAl_Si.

4. Discussion

The results presented in the previous section show that the kinetics of the mechanical alloying process depend on the initial powders which were mixed to obtain final homogeneous product. It can be stated that the microstructure, phase composition, hardness and Young's modulus at the end of the mechanical alloying and after the spark plasma sintering are very similar. However, the kinetics of the formation of the homogeneous powder is different. All the measurements show that the fastest homogenization is for the Fe_Al_Si mixture made of pure elements. Full homogenization is reached after 4 h of mechanical alloying. 6 h are sufficient to homogenize the powders Fe_AlSi and FeSi_Al. The longest time (8 h) is required for the preparation of FeAl_Si mixture.

These differences are caused by the properties of the initial powders. For fast homogenization, it is advantageous to combine hard brittle and soft ductile powders. Ductile particles are easily plastically deformed and create cold welds. Brittle particles fracture, they get smaller and they can easily get inside soft plastic particles [11]. Only a small amount of the brittle hard particles causes fast increase in the hardness and accelerates the homogenization. This is the case for Fe_Al_Si and Fe_AlSi mixtures, in which Si has the role of the hard and brittle material in soft ductile Fe, Al, or AlSi30 pre-alloyed powder, respectively. Very fast increase in hardness of Fe_AlSi alloy during the first two hours of mechanical alloying can be caused by Si fine dispersion in the Al-Si eutectic. Increased temperature during mechanical alloying together with the fine Si eutectic particles in Al matrix accelerates the reactions with Fe and the formation of solid solution at the beginning of mechanical alloying. At the later stages of mechanical alloying, the primary Si acts similarly as the pure Si powder and the kinetics of the MA is equivalent with the case of the mixture of pure elements.

For the mixture FeSi_Al, the dominant component is hard FeSi25 intermetallic alloy. During the first two hours, the predominant hard particles are mixed with soft aluminum which causes the decrease in the hardness. Hardening processes are not fast enough to compensate this decrease caused by initial mixing. After 2 h of milling, homogenization occurs, hardness increases and the process continues in the same way as for Fe_Al_Si and Fe_AlSi mixtures.

The case of FeAl_Si mixture is fundamentally different because two hard brittle components are mixed. The brittle particles are preferentially crushed during the collisions and the formation of cold welds is more complicated. The diffusion between the components and the plastic deformation is lower and the homogenization takes a longer time.

As it was mentioned above, the particle size is slightly changing with milling time (from 30 to about 10 µm on Fe_Al_Si [17]). Only the particles larger than 10 µm were tested to avoid the effect of the particle size in soft surrounding matrix on hardness measurements. No substantial effect of particle size on hardness values was however noted, which is in agreement with the values measured on sintered samples (where the whole particle size distribution intervenes). Thus, the particles size distribution should not have a significant effect on the presented values of the hardness. As a result, the changes in hardness should correspond only to the changes in microstructure and homogenization of the powder particles caused by mechanical alloying.

Increase in hardness and the saturation on the constant value during mechanical alloying is generally observed [27,28]. The hardening rate depends on the kinetic energy of the collisions which can be driven by the milling parameters such as rotational speed, powder to ball ratio (material of the balls), etc. The lower rotational speed or lighter balls cause slower increase in hardness and require more time for the completion of the MA process [27]. The increase in hardness of the MA powders can be caused by different mechanisms. The first is work hardening when the particles are plastically deformed during the collisions with balls causing the increase of dislocation density [29]. The second mechanism is the grain refinement (Hall-Petch hardening) [28,30]. The next strengthening mechanism is dispersion hardening where the nanoparticles blocking the dislocation movement are homogenously distributed in the matrix. This type of hardening is typical for oxide dispersion strengthened steels [12]. The last type is the solid solution hardening where the atoms from the crystal structure are replaced by other atoms with not exactly the same atomic radius, causing the distortion of the lattice (the case of Al and Si atoms in the studied material). All these mechanisms can cause the increase in hardness observed in this study. When the equilibrium with the recovery mechanisms is established, the hardness does not increase more and saturates at the final level.

As it was discussed in previous paragraphs, the kinetics of the mechanical alloying and thus the strengthening mechanisms depend on the initial combination of the powders. If at least one component is ductile, the hardening mechanisms based on the plastic deformation and dislocation movement are easier and the hardening is faster. Increasing the amount of hard brittle components causes cracking and the particles cannot be easily plastically deformed. Thus, the hardening and the final homogenization is more time-demanding.

After the MA, FeSi and Fe_3Si are predominant phases in homogeneous powders instead of the stable ternary phases [31]. It is caused by the easy contact of two particles during the milling and increased solubility of the elements caused by mechanical alloying [32]. Other phases found for the alloys with different Al/Si ratio (e.g., Fe_2Al_5 [17]) were not identified. Spark plasma sintering was chosen for the consolidation of the mechanically alloyed powders because it is a fast method which prevents significant changes in microstructure, grain growth and mechanical properties. Nevertheless, the presence of stable ternary phase $Fe_3Al_2Si_3$ was observed after the sintering in all the samples. Exact phase ratio for Fe_Al_Si mixture was determined in the previous study [19] (33.0 wt.% of Fe_3Si, 37.5 wt.% of FeSi, 29.5 wt.% of $Fe_3Al_2Si_3$) and according to the diffractograms in Figure 10, there is no significant difference between various mixtures. This ternary phase usually complicates the processing of this type of materials by conventional methods because of its low plasticity. It would therefore be desirable to prevent this phase as well as the amorphous brittle phases before final processing. Compared to other initial compositions or processing ways (self-propagating high-temperature synthesis), the presence of the phases Fe_2Al_5 [8] or $FeAl_2Si$ [9], respectively, was not observed.

As can be seen from Figure 8, hardness after mechanical alloying was not affected by the sintering process and it stays very high (15 GPa). Hardness of compacted samples measured for maximum load of 5 N is slightly lower than the value measured for the load of 2 mN. This can be caused by several reasons which could include porosity of the sample (even if it was very low), cracking at higher loads or indentation size effect that could already be present at used low loads [33].

A wide range of formulas for different crack systems and materials exists for the determination of the fracture toughness by indentation methods (e.g., [34–39]). It was confirmed by SEM observations that the system of cracks in the studied material is radial (Palmqvist cracks). For this reason, Equation (1) which is valid in the range of crack lengths $0.25 \leq l/a \leq 2.5$ was employed. The fracture toughness is quite low, in the range of ceramics [40], which could be caused by the high amount of stored plastic deformation and/or the presence of brittle amorphous phases. On the other hand, the manufacturing process performed in this study (i.e., mechanical alloying and spark plasma sintering) leads to significant fracture toughness improvement compared to standard casting of these materials (about 10 times [19]).

The lowest fracture toughness was found for the mixture Fe_Al_Si which was homogenized in the shortest time. The highest fracture toughness was measured for FeAl_Si mixture, for which the homogenization took the longest time of 8 h. This means that the fast homogenization leads to a decrease of the fracture toughness. It can be caused by the stored plastic deformation in the powder particles. When the homogenization process is finished soon (e.g., after 4 h), no significant changes happen during further milling and only plastic deformation is stored in the particles. This means that from the point of the fracture toughness, the reduction of the milling time to the shortest time necessary for the homogenization of the powder would be optimal.

Other reasons for the differences in fracture toughness could be different levels of supersaturation of Fe_3Si and/or FeSi phases by aluminum or different volume fraction of amorphous phase. All aluminum in solid solution in FeAl pre-alloyed powders could suppress the local liquid phase sintering. However, the TEM observation does not allow the exact quantification of the amorphous phase content in different pre-alloyed powders. On the other hand, no correlation between fracture toughness and grain size could be found as the grain size is practically the same for all pre-alloyed powders, which is probably closely connected with the saturation of the plastic deformation during milling.

5. Conclusions

The presented results show the differences in the kinetics of the mechanical alloying process depending on the initial feedstock materials used for the production of FeAl20Si20 alloy. The results can be summarized as follows:

- The final fine microstructure and measured mechanical properties of the powder mixtures are not significantly different. They are composed mainly by Fe_3Si and FeSi phases supersaturated by Al.
- The mixture of hard (brittle) and soft (ductile) powders leads to shorter milling times, the mixture of hard initial powders requires longer milling times for complete homogenization.
- For the optimization of the production process (time for complete homogenization), the most efficient is the mixture Fe_Al_Si, followed by Fe_AlSi and FeSi_Al. Longest time to prepare homogenous powder is needed for FeAl_Si mixture.
- Fracture toughness is improved when the milling process is stopped immediately after the complete homogenization (highest fracture toughness was observed for the mixture FeAl_Si with the longest homogenization time).

Author Contributions: Conceptualization, J.Č. and P.H.; methodology, J.Č., P.H., F.P. and P.N.; formal analysis, J.Č., M.K. and P.H.; investigation, J.Č., P.H., M.K., V.B. and K.N.; resources, K.N., J.K.; data curation, J.Č.; writing–original draft preparation, J.Č.; writing–review and editing, P.H.; funding acquisition, F.P.

Funding: The funding by Czech Science Foundation (grant No. 17-07559S) and European Regional Development Fund in the frame of the project Centre of Advanced Applied Sciences (No. CZ.02.1.01/0.0/0.0/16_019/0000778) is gratefully acknowledged.

Acknowledgments: This research was financially supported by Czech Science Foundation (grant No. 17-07559S) and European Regional Development Fund in the frame of the project Centre of Advanced Applied Sciences (No. CZ.02.1.01/0.0/0.0/16_019/0000778).

Conflicts of Interest: The authors declare no conflict of interest. The funders had no role in the design of the study; in the collection, analyses, or interpretation of data; in the writing of the manuscript, or in the decision to publish the results.

References

1. Deevi, S.C.; Sikka, V.K. Nickel and iron aluminides: An overview on properties, processing, and applications. *Intermetallics* **1996**, *4*, 357–375. [CrossRef]
2. Zhu, X.; Yao, Z.; Gu, X.; Cong, W.; Zhang, P. Microstructure and corrosion resistance of Fe-Al intermetallic coating on 45 steel synthesized by double glow plasma surface alloying technology. *Trans. Nonferrous Met. Soc. China* **2009**, *19*, 143–148. [CrossRef]

3. Janda, D.; Fietzek, H.; Galetz, M.; Heilmaier, M. The effect of micro-alloying with Zr and Nb on the oxidation behavior of Fe3Al and FeAl alloys. *Intermetallics* **2013**, *41*, 51–57. [CrossRef]
4. Zamanzade, M.; Vehoff, H.; Barnoush, A. Effect of chromium on elastic and plastic deformation of Fe3Al intermetallics. *Intermetallics* **2013**, *41*, 28–34. [CrossRef]
5. Kratochvíl, P.; Karlík, M.; Haušild, P.; Cieslar, M. Influence of Annealing on Mechanical Properties of an Fe-28Al-4Cr-0.1Ce Alloy. *Intermetallics* **1999**, *7*, 847–853. [CrossRef]
6. Karlík, M.; Haušild, P.; Šíma, V.; Málek, P.; Vlasák, T. High Temperature Mechanical Properties of Fe-40-at% Al Based Intermetallic Alloys with C or Ti Addition. *Int. J. Mater. Res.* **2009**, *100*, 386–390. [CrossRef]
7. Prahl, J.; Haušild, P.; Karlík, M.; Crenn, J.-F. Fracture Behaviour of Fe3Al Alloy with Additions of Zr and C at Different Temperatures. *Kovové Materiály* **2005**, *43*, 134–144.
8. Novák, P.; Nová, K. Oxidation Behavior of Fe-Al, Fe-Si and Fe-Al-Si Intermetallics. *Materials* **2019**, *12*, 1748. [CrossRef]
9. Novák, P.; Zelinková, M.; Šerák, J.; Michalcová, A.; Novák, M.; Vojtěch, D. Oxidation resistance of SHS Fe–Al–Si alloys at 800 °C in air. *Intermetallics* **2011**, *19*, 1306–1312. [CrossRef]
10. Critical Raw Materials. Available online: https://ec.europa.eu/growth/sectors/raw-materials/specific-interest/critical_cs (accessed on 28 July 2019).
11. Suryanarayana, C. Mechanical alloying and milling. *Prog. Mater. Sci.* **2001**, *46*, 1–184. [CrossRef]
12. Bhadeshia, H.K.D.H. Mechanically alloyed metals. *Mater. Sci. Technol.* **2000**, *16*, 1404–1411. [CrossRef]
13. Novák, P.; Průša, F.; Nová, K.; Bernatiková, A.; Salvetr, P.; Kopeček, J.; Haušild, P. Application of Mechanical Alloying in Synthesis of Intermetallics. *Acta Phys. Pol. A* **2018**, *134*, 720–723. [CrossRef]
14. Zakeri, M.; Ramezanib, M.; Nazari, A. Effect of Ball to Powder Weight Ratio on the Mechanochemical Synthesis of MoSi2-TiC Nanocomposite Powder. *Mater. Res.* **2012**, *15*, 891–897. [CrossRef]
15. Baig, Z.; Mamat, O.; Mustapha, M.; Mumtaz, A.; Sarfraz, M.; Haider, S. An Efficient Approach to Address Issues of Graphene Nanoplatelets (GNPs) Incorporation in Aluminium Powders and Their Compaction Behaviour. *Metals* **2018**, *8*, 90. [CrossRef]
16. Hao, X.-N.; Zhang, H.-P.; Zheng, R.-X.; Zhang, Y.-T.; Ameyama, K.; Ma, C.-L. Effect of mechanical alloying time and rotation speed on evolution of CNTs/Al-2024 composite powders. *Trans. Nonferrous Met. Soc. China* **2014**, *24*, 2380–2386. [CrossRef]
17. Nová, K.; Novák, P.; Průša, F.; Kopeček, J.; Čech, J. Synthesis of Intermetallics in Fe-Al-Si System by Mechanical Alloying. *Metals* **2019**, *12*, 20. [CrossRef]
18. Orru, R.; Licheri, R.; Locci, A.M.; Cincotti, A.; Cao, G. Consolidation/synthesis of materials by electric current activated/assisted sintering. *Mater. Sci. Eng. R Rep.* **2009**, *63*, 127–287. [CrossRef]
19. Novák, P.; Vanka, T.; Nová, K.; Stoulil, J.; Průša, F.; Kopeček, J.; Haušild, P. Structure and properties of Fe-Al-Si alloy prepared by mechanical alloying. *Materials* **2019**, *12*, 2463. [CrossRef]
20. Oliver, W.C.; Pharr, G.M. An improved technique for determining hardness and elastic modulus using load and displacement sensing indentation experiments. *J. Mater. Res.* **1992**, *7*, 1564–1583. [CrossRef]
21. Oliver, W.C.; Pharr, G.M. Measurement of hardness and elastic modulus by instrumented indentation: Advances in understanding and refinements to methodology. *J. Mater. Res.* **2004**, *19*, 3–20. [CrossRef]
22. ISO 14577. *Metallic Materials—Instrumented Indentation Test for Hardness and Material Parameters*; ISO: Geneva, Switzerland, 2002.
23. Čech, J.; Haušild, P.; Karlík, M.; Kadlecová, V.; Čapek, J.; Průša, F.; Novák, P. Mechanical properties of FeAlSi powders prepared by mechanical alloying from different initial feedstock materials. *Matériaux Tech.* **2019**, *107*, 207. [CrossRef]
24. Niihara, K. A Fracture Mechanics Analysis of Indentation-Induced Palmqvist Crack in Ceramics. *J. Mater. Sci. Lett.* **1983**, *2*, 221–223. [CrossRef]
25. Jems Website. Available online: http://www.jems-saas.ch/Home/jemsWebSite/jems.html (accessed on 28 July 2019).
26. Nová, K.; Novák, P.; Vanka, T.; Průša, F. The effect of production process on properties of FeAl20Si20. *Manuf. Technol.* **2018**, *18*, 295–298. [CrossRef]
27. Kruger, M.; Schmelzer, J.; Helmecke, M. Similarities and Differences in Mechanical Alloying Processes of V-Si-B and Mo-Si-B Powders. *Metals* **2016**, *6*, 241. [CrossRef]

28. Hegde, M.R.; Surendranathan, A.O. Phase Transformation, Structural Evolution and Mechanical Property of Nanostructured FeAl as a Result of Mechanical Alloying. *Russ. J. Non-Ferr. Met.* **2009**, *50*, 474–484. [CrossRef]
29. Schmelzer, J.; Baumann, T.; Dieck, S.; Kruger, M. Hardening of V–Si alloys during high energy ball milling. *Powder Technol.* **2016**, *294*, 493–497. [CrossRef]
30. Lei, R.; Wang, M.; Xu, S.; Wang, H.; Chen, G. Microstructure, Hardness Evolution, and Thermal Stability Mechanism of Mechanical Alloyed Cu-Nb Alloy during Heat Treatment. *Metals* **2016**, *6*, 194. [CrossRef]
31. Marker, M.C.; Skolyszewska-Kühberger, B.; Effenberger, H.S.; Schmetterer, C.; Richter, K.W. Phase equilibria and structural investigations in the system Al–Fe–Si. *Intermetallics* **2011**, *19*, 1919–1929. [CrossRef]
32. Huang, B.L.; Perez, R.J.; Lavernia, E.J.; Luton, M.J. Formation of supersaturated solid solutions by mechanical alloying. *Nanostruct. Mater.* **1996**, *7*, 67–79. [CrossRef]
33. Nix, W.D.; Gao, H. Indentation size effects in crystalline materials: A law for strain gradient plasticity. *J. Mech. Phys. Solids* **1998**, *46*, 411–425. [CrossRef]
34. Ponton, C.B.; Rawlings, R.D. Vickers indentation fracture toughness test Part 1 Review of literature and formulation of standardised indentation toughness equations. *Mater. Sci. Technol.* **1989**, *5*, 865–872. [CrossRef]
35. Zhang, S.; Zhang, X. Toughness evaluation of hard coatings and thin films. *Thin Solid Films* **2012**, *520*, 2375–2389. [CrossRef]
36. Laugier, M.T. New formula for indentation toughness in ceramics. *J. Mater. Sci. Lett.* **1987**, *6*, 355–356. [CrossRef]
37. Lawn, B.R.; Evans, A.G.; Marshall, D.B. Elastic/plastic indentation damage in ceramics: The median/radial crack system. *J. Am. Ceram. Soc.* **1980**, *63*, 574–581. [CrossRef]
38. Chen, J. Indentation-based methods to assess fracture toughness for thin coatings. *J. Phys. D Appl. Phys.* **2012**, *45*, 203001. [CrossRef]
39. Feng, Y.; Zhang, T. Determination of Fracture toughness of Brittle Materials by Indentation. *Acta Mech. Solidica Sin.* **2015**, *28*, 221–234. [CrossRef]
40. Gogotsi, G.A. Fracture toughness of ceramics and ceramic composites. *Ceram. Int.* **2003**, *29*, 777–784. [CrossRef]

 *materials*

Article

Mechanism of the Intermediary Phase Formation in Ti-20 wt. % Al Mixture during Pressureless Reactive Sintering

Andrea Školáková [1,*], Pavel Salvetr [1], Pavel Novák [1], Jindřich Leitner [2] and Davy Deduytsche [3]

1. Department of Metals and Corrosion Engineering, University of Chemistry and Technology, Technická 5, 166 28 Prague 6, Czech Republic
2. Department of Solid State Engineering, University of Chemistry and Technology, Technická 5, 166 28 Prague 6, Czech Republic
3. Department of Solid State Sciences, Ghent University, Krijgslaan 281 S1 9000 Gent, Belgium

* Correspondence: skolakoa@vscht.cz; Tel.: +420-220-444-055

Received: 7 June 2019; Accepted: 4 July 2019; Published: 6 July 2019

Abstract: This work aims to describe the mechanism of intermediary phases formation in TiAl20 (wt. %) alloy composition during reactive sintering. The reaction between titanium and aluminum powders was studied by in situ diffraction and the results were confirmed by annealing at various temperatures. It was found that the Ti_2Al_5 phase formed preferentially and its formation was detected at 400 °C. So far, this phase has never been found in this alloy composition during reactive sintering processes. Subsequently, the Ti_2Al_5 phase reacted with the titanium, and the formation of the major phase, Ti_3Al, was accompanied by the minor phase, TiAl. Equations of the proposed reactions are presented in this paper and their thermodynamic and kinetic feasibility are supported by Gibbs energies of reaction and reaction enthalpies.

Keywords: in situ diffraction; aluminides; reactive sintering; mechanism; powder metallurgy

---

1. Introduction

A Ti-Al system consists of five important phases, including a Ti_3Al compound with a hexagonal close-packed superlattice (space group P63/mmc), an equiatomic TiAl compound with a tetragonal structure (space group P4/mmm), and aluminum-rich intermetallic compounds, namely $TiAl_2$ (space group I41/amd), Ti_2Al_5 (space group P4/mmm), and $TiAl_3$ (space group I4/mmm), also with a tetragonal structure. All intermetallic compounds are collectively called titanium aluminides. Titanium aluminides belong to the group of innovative materials that gradually replace nickel-based superalloys in highly demanding applications [1]. They possess a great combination of stable mechanical properties at high temperatures (500–900 °C), low density, and good oxidation resistance. For this reason, they are suitable candidates as structural materials for the aerospace and automotive industries. So far, they have been used in turbocharger wheels and turbine blades. The current research is mainly focused on the development of alloys with microstructures containing TiAl and Ti_3Al phases, which should ensure great creep resistance [1–5]. Despite this advantage, the application range of titanium aluminides is still limited because they suffer from room-temperature brittleness and have poor melt-metallurgic properties [5]. Moreover, the extreme reactivity of molten titanium usually causes contamination of the obtained products [6]. These compounds are produced by a melt-metallurgy process comprising vacuum induction melting (VIM), vacuum arc remelting (VAR), centrifugal casting, conventional melting, and hot isostatic pressing (HIP) [1]. Powder metallurgy techniques are much more suitable because melting is avoided. Thus, molten titanium does not occur, and a high purity of the products can be obtained [1]. One of the currently studied processes of these compounds is Self-propagating

High-temperature Synthesis (SHS) which uses solid state diffusion to obtain a mostly local, usually eutectic, composition and a subsequent formation of the liquid phase at temperatures lower than the melting points of the initial materials. This mechanism is, thus, different from reactive sintering using only solid-state diffusion. However, SHS is taken as a specific type of reactive sintering.

The SHS process is a modern method of sintering because time and energy are saved [7,8]. SHS lies in the heating of initial reactants, which are usually powders [9]. Highly exothermic reactions occur during heating [10]. Heat released by these reactions is supported other exothermic reactions, and, thus, the SHS reaction can propagate through the whole heated powder mixture spontaneously [9,11]. For this reason, this reaction is often called as "self-sustaining". Reactions are usually thermally activated. SHS reaction is divided into two modes, which differ in their heating initiation. A reaction can be initiated by heating of the whole sample, and then it is called Thermal Explosion mode or by heating of only one side of sample and this case of heating is called Ignition mode. The thermal explosion mode usually involves homogeneous heating, while the Ignition mode uses extremely fast heating in most cases [12]. Both modes have their own disadvantages. For the former mode, pre-combustion is necessary and can modify the reaction mechanism. On the other hand, this mode is more accurate in the determination of reaction temperatures. The latter mode is faster, but the results have poor reliability [12].

TiAl and $TiAl_3$ aluminides prepared by SHS reaction are the most studied ones [10,13–16]. Ti_3Al is not a commonly studied aluminide from the viewpoint of the SHS process, and, hence, its mechanism of phase formation has not yet been described. On the other hand, published research shows contradictory or incomplete results, and the description of the reaction mechanisms of phase formation is not united for TiAl and $TiAl_3$ alloy composition. It is known that the reaction between titanium and aluminum powders is initiated when the temperature reaches the melting point of aluminum during heating [17]. The $TiAl_3$ phase is concluded to be the first phase during the reaction between liquid aluminum and solid titanium, followed by the formation of the TiAl and Ti_3Al phases [13,16–19]. It is believed that the $TiAl_3$ phase forms preferentially due to its thermodynamic and kinetic system preferences [20]. The $TiAl_2$ phase should form after the formation of the TiAl and Ti_3Al phases by interdiffusion between the $TiAl_3$ phase and titanium [2]. Other works showed that the TiAl phase formed first after a reaction between the melted aluminum and titanium [21]. Further, the TiAl phase is considered to be the starting phase before the formation of the Ti_2Al_5 and $TiAl_2$ phases [20], followed by the formation of the $TiAl_3$ phase. The explanation for the relationship between phases can be found in their low free energies [19]. However, other works showed that a very thin layer of the Ti_3Al phase formed first on the titanium side surrounded by the TiAl phase and $TiAl_3$ formed as the last one [22]. This result shows a completely different phase formation sequence in comparison to the results mentioned before. For this reason, there are holes in the description of reaction mechanisms, and it is necessary to determine the accurate reaction conditions of phase formation.

Ti and Al foils, instead of powders, have also been intensively studied, mainly because of their better mechanical properties. The aim is to synthesize the composite as a sandwich. For this reason, foils are most often produced by electron beam deposition, which affects the phase composition of the interface between aluminum and titanium, and, subsequently, these foils are heated. It was found that diffusion welding formed. Authors usually observe the growth of the intermetallic layer while both reactants are in the solid state and obtained foundational results that they could use determine kinetics. However, they could study only the interface between aluminum and titanium whose chemical composition corresponds to 50:50 in at. % of Ti:Al [23–27]. For this reason, the reaction between solid aluminum and solid titanium is better to simulate the annealing of a compressed powder mixture at temperatures lower than the melting point of aluminum (used in this work). Advantage is possible to mix a different ratio of aluminum to titanium and study Ti_3Al, TiAl, or $TiAl_3$ phases. Moreover, the powder mixture is only compressed, and, thus, the interface between aluminum and titanium is not affected by pre-heating.

In this work, a Ti-20 wt. % Al powder mixture (corresponding to Ti_3Al compound) was studied. This chemical composition was deliberately chosen because this work is part of project in which the sintering of Ti_3Al, TiAl, and $TiAl_3$ compounds has been studied. The amount of aluminum (20 wt. %) lies in the center of the stability range of the Ti_3Al phase, according to diagram Ti-Al [28]. This mixture was subjected to the Thermal Explosion mode (TE) SHS, and the mechanism of phase formation was described. In situ diffraction was applied to determine the sequence of phase formation, and their formation was observed at the highest heating rate that the in situ XRD (X-ray diffraction) device allows. To determine whether the phase formed below the melting point of aluminum, powder mixtures were annealed at various temperatures and times. This enabled us to show and confirm the phases' formation sequence. This work offers the first detailed descriptions of the phase formation of this alloy composition.

2. Materials and Methods

All tested samples were prepared from blended powders of titanium (with a purity of 99.5% and a particle size of 44 μm, STREM CHEMICALS, Newburyport, MA, USA) and aluminum (99.62%, 44 μm, STREM CHEMICALS, Newburyport, MA, USA). Powder blends corresponding to the Ti_3Al compound, i.e., TiAl20 (in wt. %), with a weight of 3 g, were prepared. Subsequently, powder mixture was uniaxially cold pressed at an ambient temperature to cylindrical green bodies of 10 mm in diameter by a pressure of 450 MPa for 5 min using the LabTest 5.260SP1-VM universal loading machine (Labortech, Opava, Czech Republic). Reaction kinetics, especially the initiation of the reactions, strongly depends on the green density of the compacts. We have studied intermetallics for several years and we have experienced that reaction kinetics mainly depend on particle size and compaction pressure. For this reason, we used the finest powders of titanium and aluminum and a compaction pressure of 450 MPa to obtain the powder mixture with the best contacts of particles and the lowest possible porosity.

Afterwards, the sample was inserted into the induction furnace and heated under a protective Ar atmosphere. Heating was recorded by an optical pyrometer Optris OPTP20-2M (Optris, Portsmouth, NH, USA) to observe the emerging exothermic reaction in the Thermal Explosion (TE-SHS) mode. The heating rate was determined, according to the slope of the obtained curve, as 109 $°C \cdot min^{-1}$. In order to describe the phases' formation sequence during the SHS process, in situ XRD analysis (Department of Solid State Sciences, Ghent University, Ghent, Belgium) of the compressed powder mixture was performed under an He atmosphere during heating to 900 °C, with a heating rate of 60 $°C \cdot min^{-1}$. XRD source and linear Vantec detector were fixed at the positions of 21° (source) and 42° (detector).

Further, the formation of phases was observed at temperatures lower than the melting point of aluminium −400, 450, 500, and 600 °C. Samples were evacuated in silica ampoules and exposed at these temperatures for 8, 24, and 48 h.

The obtained samples were ground by sandpaper P80–P4000 with SiC abrasive particles (Hermes Schleifmittel GmbH, Hamburg, Germany), polished by suspension of colloidal silica Eposil F (ATM GmbH, Mammelzen, Germany) with hydrogen peroxide (volume 1:6), and etched by Kroll's reagent (5 mL HNO_3, 10 mL HF, 85 mL H_2O). The microstructure was examined by a scanning electron microscope TESCAN VEGA 3 LMU (Tescan, Brno, Czech Republic) equipped with an OXFORD Instruments X-max EDS SDD 20 mm^2 detector (Oxford Instruments, High Wycombe, UK) for identification of the chemical composition of individual phases (SEM-EDS). Phase composition was determined by X-ray diffraction using the PANalytical X'Pert Pro diffractometer with CuKα radiation (PANalytical, Almeo, The Netherlands).

3. Results and Discussion

Figure 1 shows the heating curve obtained during the reactive sintering of the Ti-20 wt. % Al powder mixture in an induction furnace. The exothermic peak is associated with the SHS reaction and formation of phases. The reaction started above the temperature of the melting point of aluminum because T_{onset} was 716 °C, indicating that the presence of a liquid phase triggered the reaction. This means that the reaction was not initiated at the melting point of aluminum at 660 °C, as is typical for aluminides [16]. The liquid phase was present, accelerated the sintering at higher temperatures, and was gradually consumed during the process [2]. The maximum SHS reaction, called the combustion temperature, was 760 °C. The microstructure was composed of unreacted titanium, which was surrounded by large areas of the Ti_3Al phase (Figure 2). TiAl was found between the layers of the Ti_3Al phase. These phases were also detected by XRD analysis (Figure 3), and, moreover, the Ti_2Al_5 phase was found. This phase is very fine and is probably dispersed in the matrix uniformly. According to the area fraction, it can be assumed that the Ti_3Al phase formed during an exothermic SHS reaction, which was observed by an optical pyrometer (Figure 1). This phase also formed as a major phase in Ti:Al = 1:1 system [16], and it was stated that its formation is associated with diffusion. On the other hand, the Ti_2Al_5 phase was not detected in this aluminum enriched system [16]. Results from the EDS analysis are shown in Table 1.

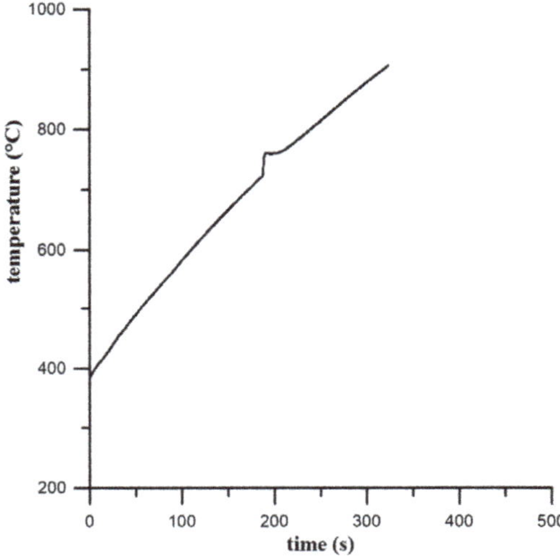

Figure 1. Heating curve obtained by the optical pyrometer at a heating rate of 109 $°C{\cdot}min^{-1}$.

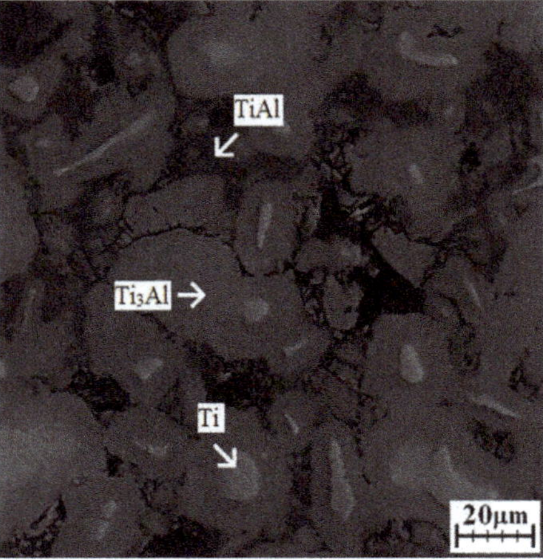

Figure 2. SEM image of microstructure obtained after induction at heating rate of 109 $°C{\cdot}min^{-1}$.

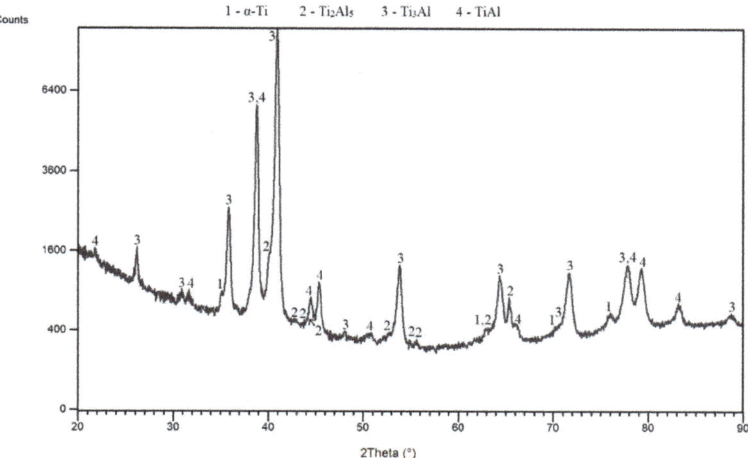

Figure 3. XRD patterns obtained after induction at a heating rate of 109 °C·min^{-1}.

Table 1. SEM + EDS point analysis results of the reactively sintered TiAl20.

Phases	Element Concentration (at. %)	
	Ti	Al
TiAl	50 ± 2	50 ± 2
Ti_3Al	72 ± 3	28 ± 3

The in situ XRD analysis shows the continuous formation of phases from the powder mixture during heating (Figure 4). This observation of phase composition during heating helps to describe the mechanisms of phase formation. It can be seen that titanium and aluminum were present up to approximately 500 °C. At this temperature, the Ti_2Al_5 phase started to form. Titanium and aluminum diffraction lines disappeared at approximately 700 °C. This disappearance is associated with the formation of other phases—Ti_3Al and TiAl. These phases were accompanied by the Ti_2Al_5 phase, which had formed earlier. It was impossible to determine which of the phases (Ti_3Al or TiAl) formed preferentially. All their diffraction lines appeared at the same temperature, i.e., approximately 700 °C. The intensity of the lines of the Ti_2Al_5 phase was distinctive from 500 °C to 700 °C. This result implies that the SHS reaction between the titanium and aluminum powders is initiated by the formation of an intermediary Ti_2Al_5 phase. Many works [2,16–20] claimed that the $TiAl_3$ phase formed before the formation of the TiAl, Ti_3Al, or $TiAl_2$ phases already at 580 °C. Their explanation as to why the $TiAl_3$ phase arose preferentially lies in its minimum free energy, suggesting that its formation is thermodynamically and kinetically favored over the formation of other aluminides. Only the presence of liquid aluminum is sufficient for the formation of the $TiAl_3$ phase [20]. On the other hand, in situ XRD analysis, and our other results, showed that the Ti_2Al_5 phase is formed preferentially in a TiAl20 powder blend. The reaction temperature is an important parameter that can support or induce synthesis of aluminides, as described in work [29]. Based on the presented results, the other part of this study focused on individual temperatures to clarify aluminide formation.

From the obtained results, it can be seen that the Ti_2Al_5 phase formed below the melting point of aluminum. For this reason, mixtures of powders were annealed at various temperatures, and times, microstructures, and phase compositions were observed. Figure 5a–c shows microstructures obtained after annealing at 400 °C. Unreacted particles of aluminum and titanium were found. However, XRD analysis (Figure 6) revealed that Ti_2Al_5 formed at this temperature, and this phase was observed at the interface of titanium and aluminum. In situ XRD analysis detected its formation at 500 °C, but analysis was performed at a heating rate of 60 °C·min^{-1} (the maximum heating rate that can be set), and it is known that with a decreasing heating rate, the initiation temperatures also decrease [30]. Moreover, the heating during the in situ XRD experiment was rather quick and continuous. Therefore, the amount in the Ti_2Al_5 phase, which is expected to form by a diffusion mechanism, is very low during continuous heating, and, therefore, it may not be detected. During annealing, the detectable amount of the Ti_2Al_5 phase was formed after 8 h at 400 °C (Figure 6). The EDS results are listed in Table 2. It can be seen that chemical composition does not correspond to Ti_2Al_5 but the analyzed area is small and, thus, strongly affected by titanium.

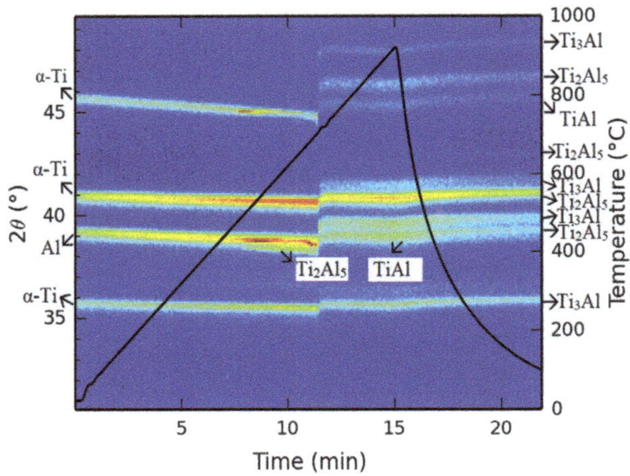

Figure 4. In situ diffraction obtained during heating (heating rate of 60 °C·min^{-1}).

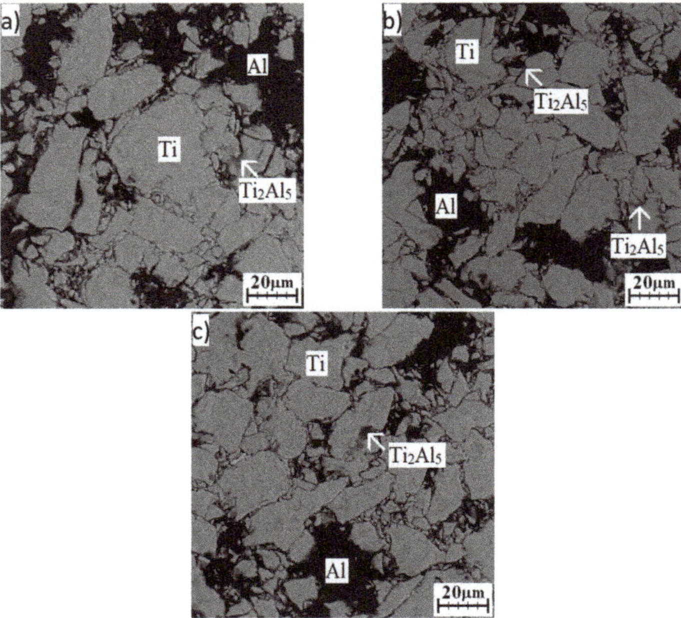

Figure 5. SEM images of a microstructure of a TiAl20 powder mixture annealed at 400 °C: (**a**) 8 h; (**b**) 24 h; (**c**) 48 h.

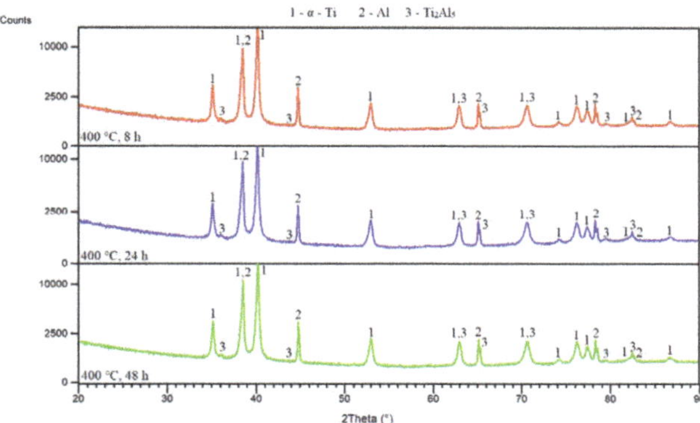

Figure 6. XRD patterns of TiAl20 powder mixture annealed at 400 °C.

Table 2. SEM + EDS point analysis results of TiAl20 annealed at 400 °C.

Phase	Element Concentration (at. %)	
	Ti	Al
Ti_2Al_5, 48 h	42 ± 3	58 ± 3

Annealing at a higher temperature (450 °C) for 8 h allowed the Ti_2Al_5 phase to be observed in the microstructure more markedly (Figure 7a). The microstructure was, again, composed of unreacted particles. One day annealing affected the microstructure significantly. The Ti_2Al_5 phase was detected by EDS analysis (see Table 3) on the interface between the titanium and aluminum powders (Figure 7b). With prolongation of annealing time, the areas of occurrence of the Ti_2Al_5 phase were larger (Figure 7c). XRD analysis confirmed the presence of hexagonal titanium, aluminum, and the Ti_2Al_5 phase (Figure 8).

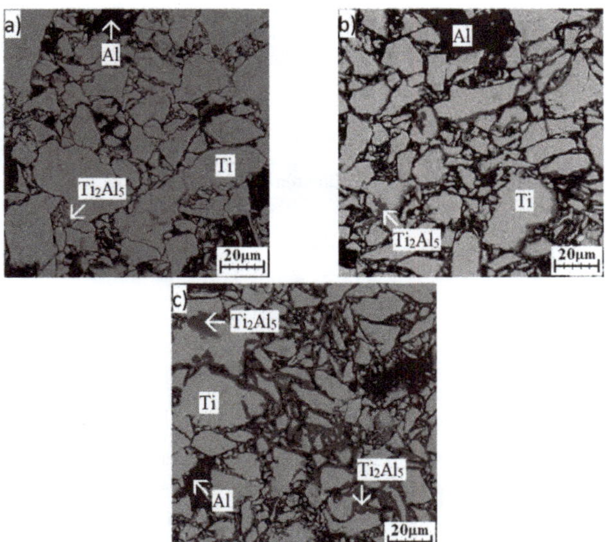

Figure 7. SEM images of the microstructure of the TiAl20 powder mixture annealed at 450 °C: (**a**) 8 h; (**b**) 24 h; (**c**) 48 h.

Table 3. SEM + EDS point analysis results of TiAl20 annealed at 450 °C.

Phase	Element Concentration (at. %)	
	Ti	Al
Ti_2Al_5, 8 h	38 ± 10	62 ± 10
Ti_2Al_5, 24 h	41 ± 4	59 ± 4
Ti_2Al_5, 48 h	30 ± 3	70 ± 3

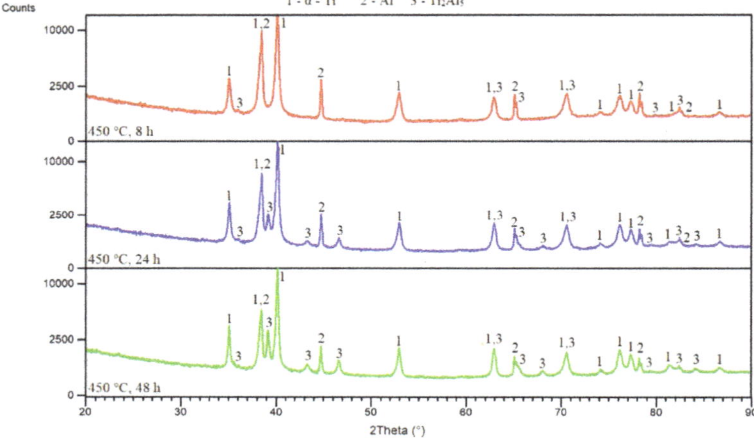

Figure 8. XRD patterns of TiAl20 powder mixture annealed at 450 °C.

The area of the Ti_2Al_5 phase was found during EDS analysis of the microstructure after annealing at 500 °C for 8 h (Figure 9a), but prolongation of times clearly revealed its location. The Ti_2Al_5 phase formed on the interfaces between aluminum and titanium particles, as well as at lower temperatures (Figure 9b,c). EDS analysis also confirmed its presence (Table 4). The phase composition was confirmed by XRD analysis (Figure 10).

Table 4. SEM + EDS point analysis results of TiAl20 annealed at 500 °C.

Phase	Element Concentration (at. %)	
	Ti	Al
Ti_2Al_5, 8 h	30 ± 10	70 ± 10
Ti_2Al_5, 24 h	32 ± 7	68 ± 7
Ti_2Al_5, 48 h	35 ± 3	65 ± 3

A temperature of 600 °C changed the microstructure significantly. Unreacted aluminum was not detected after all annealing times (Figure 11a–c). The Ti_3Al and TiAl phases were determined by a combination of EDS (Table 5) and XRD analysis (Figures 11a–c and 12). Ti_3Al replaced the Ti_2Al_5 phase and formed an interface between the original titanium particles. The TiAl phase was found only in some interfaces, and these interfaces were always darker (Figure 11a–c). In situ diffraction (Figure 4) and the results from microstructure observation showed that the Ti_3Al and TiAl phases formed simultaneously, but the Ti_3Al phase is the major phase. XRD analysis detected the Ti_2Al_5 phase, which was still present and did not disappear after formation of phases enriched by titanium (Figure 12). Ti_2Al_5 occurring in the microstructure was not detected by EDS analysis, likely due to its small dimensions and low volume fraction. This means that the Ti_3Al phase really formed as the main phase during the SHS reaction. Its formation was accompanied by the formation of the minor phase,

TiAl. The Ti_2Al_5 phase formed preferentially as a metastable reaction intermediate. The Ti_2Al_5 and TiAl phases both have a tetragonal structure [2], and this could be one of the reasons why the TiAl phase could be stabilized as a minor phase.

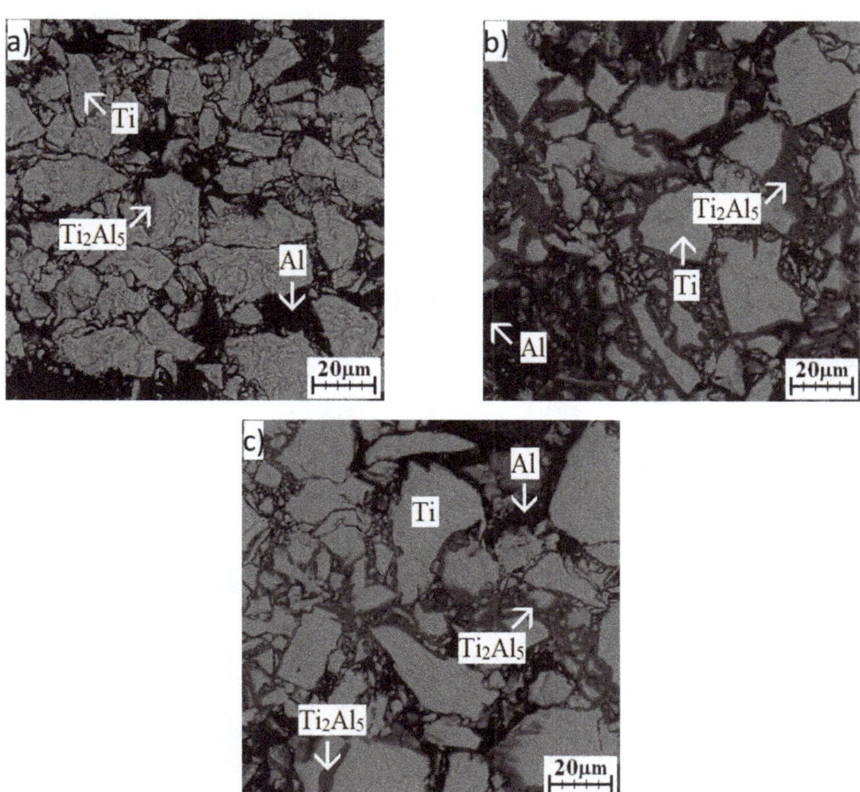

Figure 9. SEM images of the microstructure of TiAl20 powder mixture annealed at 500 °C: (**a**) 8 h; (**b**) 24 h; (**c**) 48 h.

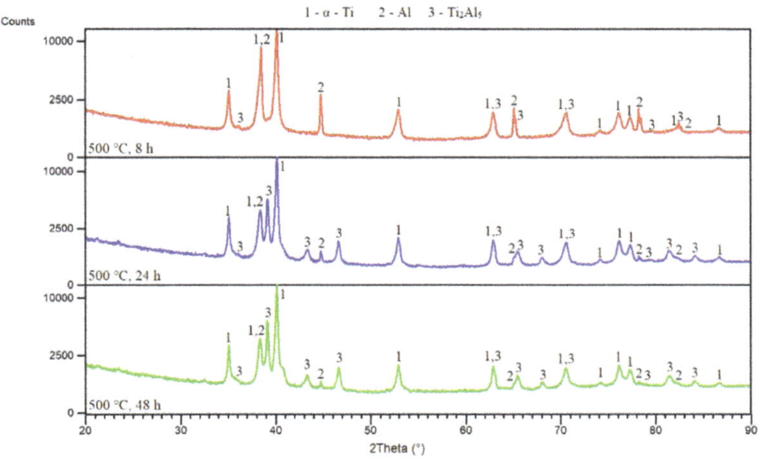

Figure 10. XRD patterns of TiAl20 powder mixture annealed at 500 °C.

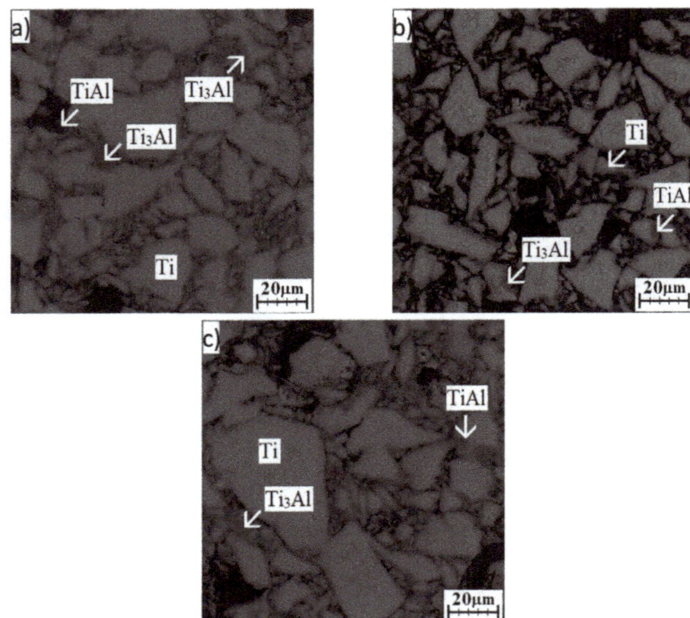

Figure 11. SEM images of the microstructure of TiAl20 powder mixture annealed at 600 °C: (**a**) 8 h; (**b**) 24 h; (**c**) 48 h.

Table 5. SEM + EDS point analysis results of TiAl20 annealed at 600 °C.

Phases	Element Concentration (at. %)	
	Ti	Al
TiAl, 8 h	51 ± 6	49 ± 6
Ti_3Al, 8 h	64 ± 4	36 ± 4
TiAl, 24 h	50 ± 6	50 ± 6
Ti_3Al, 24 h	65 ± 3	35 ± 3
TiAl, 48 h	48 ± 5	52 ± 5
Ti_3Al, 48 h	71 ± 7	29 ± 7

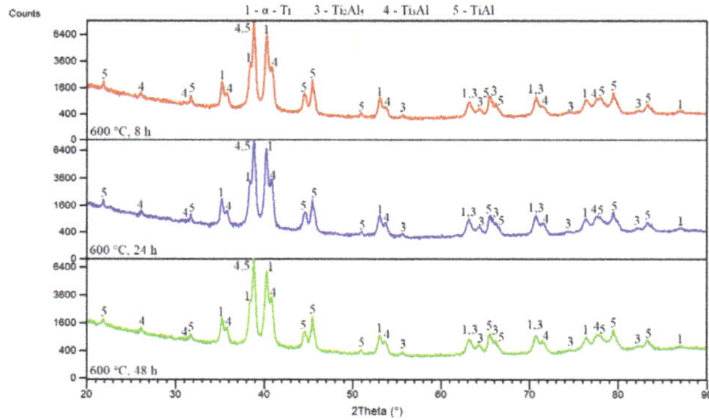

Figure 12. XRD patterns of TiAl20 powder mixture annealed at 600 °C.

The probable reaction scheme, with the Gibbs energies of reaction and reaction enthalpies, is as follows (Equations (1)–(3)):

$$2\ \mathrm{Ti} + 5\ \mathrm{Al} \rightarrow \mathrm{Ti_2Al_5},\quad \Delta H\ (400\ ^\circ\mathrm{C}) = -271.06\ \mathrm{kJ},\quad \Delta G\ (400\ ^\circ\mathrm{C}) = -226.29\ \mathrm{kJ} \tag{1}$$

$$\mathrm{Ti_2Al_5} + 13\ \mathrm{Ti} \rightarrow 5\ \mathrm{Ti_3Al},\quad \Delta H\ (600\ ^\circ\mathrm{C}) = -293.81\ \mathrm{kJ},\quad \Delta G\ (600\ ^\circ\mathrm{C}) = -242.84\ \mathrm{kJ} \tag{2}$$

$$\mathrm{Ti_2Al_5} + 3\ \mathrm{Ti} \rightarrow 5\ \mathrm{TiAl},\quad \Delta H\ (600\ ^\circ\mathrm{C}) = -72.38\ \mathrm{kJ},\quad \Delta G\ (600\ ^\circ\mathrm{C}) = 15.75\ \mathrm{kJ} \tag{3}$$

Our calculations are derived from the thermodynamic data published in [31,32].

4. Conclusions

According to the obtained results, the mechanism of phase formation in Ti-Al systems can be described. The SHS reaction started at temperatures higher than the melting point of aluminum. It was found that Ti_2Al_5 phase formed preferentially, even though a titanium-rich powder mixture was investigated. The Ti_2Al_5 phase formed at 400 °C by a diffusion-controlled reaction. This phase subsequently reacted with the titanium, and the Ti_3Al and TiAl phases formed simultaneously. These two phases could form only above 600 °C. The Ti_3Al phase was a major phase. The results were compared between the data obtained by in situ diffraction and the annealed samples, and the described mechanism was confirmed.

Author Contributions: A.Š. was responsible for the theoretical background, writing the article and data analysis. P.S. prepared samples for metallographic observations. P.N. was responsible for final correction. J.L. was responsible for calculations. D.D. provided in situ measurements.

Funding: This research was financially supported by specific university research (MSMT No 21-SVV/2019).

Conflicts of Interest: The authors declare no conflict of interest.

References

1. Agote, I.; Coleto, J.; Gutiérrez, M.; Sargsyan, A.; Garcia de Cortazar, M.; Lagos, M.A.; Sytschev, A. Production of gamma-TiAl based alloy by combustion synthesis+ compaction route, characterization and application. *Kov. Mater.* **2008**, *46*, 87.
2. Djanarthany, S.; Viala, J.C.; Bouix, J. An overview of monolithic titanium aluminides based on Ti_3Al and TiAl. *Mater. Chem. Phys.* **2001**, *72*, 301–319. [CrossRef]
3. Andreev, D.E.; Sanin, V.N.; Yukhvid, V.I. Cast alloy production on the basis of titanium aluminide with centrifugal SHS method. *Inorg. Mater.* **2009**, *45*, 867–872. [CrossRef]
4. Clemens, H.; Kestler, H. Processing and applications of intermetallic γ-TiAl-based alloys. *Adv. Eng. Mater.* **2000**, *2*, 551–570. [CrossRef]
5. Uenishi, K.; Kobayashi, K.F. Processing of intermetallic compounds for structural applications at high temperature. *Intermetallics* **1996**, *4*, S95–S101. [CrossRef]
6. Sung, S.Y.; Kim, Y.J. Alpha-case formation mechanism on titanium investment castings. *Mat. Sci Eng A* **2005**, *405*, 173–177. [CrossRef]
7. Jiang, Y.; He, Y.H.; Huang, B.Y.; Zou, J.; Huang, H.; Xu, N.P.; Liu, C.T. Criterion to control self-propagation high temperature synthesis for porous Ti–Al intermetallics. *Powder Metall.* **2011**, *54*, 404–407. [CrossRef]
8. Lagos, M.A.; Agote, I.; San Juan, J.M.; Hennicke, J. Fabrication of TiAl alloys by alternative powder methods. In *Gamma Titanium Aluminide Alloys A Collection of research on Innovation and Commercialization of Gamma Alloy Technology*; Wiley: Hoboken, NJ, USA, 2014; pp. 77–82. [CrossRef]
9. Agote, I.; Coleto, J.; Gutiérrez, M.; Sargsyan, A.; de Cortazar, M.G.; Lagos, M.A.; Vadchenko, S.G. Microstructure and mechanical properties of gamma TiAl based alloys produced by combustion synthesis+ compaction route. *Intermetallics* **2008**, *16*, 1310–1316. [CrossRef]
10. Kobashi, M.; Inoguchi, N.; Kanetake, N. Effect of elemental powder blending ratio on combustion foaming behavior of porous Al–Ti intermetallics and Al_3Ti/Al composites. *Intermetallics* **2010**, *18*, 1039–1045. [CrossRef]
11. Arakawa, Y.; Kobashi, M.; Kanetake, N. Foaming behavior of long-scale Al–Ti intermetallic foam by SHS mode combustion reaction. *Intermetallics* **2013**, *41*, 22–27. [CrossRef]

12. Bertolino, N.; Monagheddu, M.; Tacca, A.; Giuliani, P.; Zanotti, C.; Tamburini, U.A. Ignition mechanism in combustion synthesis of Ti–Al and Ti–Ni systems. *Intermetallics* **2003**, *11*, 41–49. [CrossRef]
13. Mirjalili, M.; Soltanieh, M.; Matsuura, K.; Ohno, M. On the kinetics of $TiAl_3$ intermetallic layer formation in the titanium and aluminum diffusion couple. *Intermetallics* **2013**, *32*, 297–302. [CrossRef]
14. Qin, J.; Chen, G.; Wang, B.; Hu, N.; Han, F.; Du, Z. Formation of in-situ Al_3Ti particles from globular Ti powders and Al alloy melt under ultrasonic vibration. *J. Alloys Compd.* **2015**, *653*, 32–38. [CrossRef]
15. Wei, N.; Han, X.; Zhang, X.; Cao, Y.; Guo, C.; Lu, Z.; Jiang, F. Characterization and properties of intermetallic Al_3Ti alloy synthesized by reactive foil sintering in vacuum. *J. Mater. Res.* **2016**, *31*, 2706–2713. [CrossRef]
16. Ma, Y.; Fan, Q.; Zhang, J.; Shi, J.; Xiao, G.; Gu, M. Microstructural evolution during self-propagating high-temperature synthesis of Ti-Al system. *J. Wuhan Univ. Technol.* **2008**, *23*, 381–385. [CrossRef]
17. Yi, H.C.; Petric, A.; Moore, J.J. Effect of heating rate on the combustion synthesis of Ti-Al intermetallic compounds. *J. Mater. Sci.* **1992**, *27*, 6797–6806. [CrossRef]
18. Yang, W.Y.; Weatherly, G.C. A study of combustion synthesis of Ti-Al intermetallic compounds. *J. Mater. Sci.* **1996**, *31*. [CrossRef]
19. Sujata, M.; Bhargava, S.; Sangal, S. On the formation of $TiAl_3$ during reaction between solid Ti and liquid Al. *J. Mater. Sci Lett.* **1997**, *16*, 1175–1178. [CrossRef]
20. Peng, L.M.; Wang, J.H.; Li, H.; Zhao, J.H.; He, L.H. Synthesis and microstructural characterization of Ti–Al_3Ti metal–intermetallic laminate (MIL) composites. *Scripta Mater.* **2005**, *52*, 243–248. [CrossRef]
21. Kamynina, O.K.; Vadchenko, S.G.; Sytschev, A.E.; Rogachev, A.S.; Umarov, L.M.; Sachkova, N.V. High-porosity TiAl foam by volume combustion synthesis. *Int. J. Self-Propagating High-Temp. Synth.* **2007**, *16*, 137–140. [CrossRef]
22. Novák, P.; Šerák, J.; Vojtěch, D.; Kubásek, J.; Michalcová, A. Where reactive sintering beats melt technology. *Metal Powder Rep.* **2008**, *63*, 20–23. [CrossRef]
23. Ustinov, A.I.; Falchenko, Y.V.; Ishchenko, A.Y.; Kharchenko, G.K.; Melnichenko, T.V.; Muraveynik, A.N. Diffusion welding of γ-TiAl based alloys through nano-layered foil of Ti/Al system. *Intermetallics* **2008**, *16*, 1043–1045. [CrossRef]
24. Xu, L.; Cui, Y.Y.; Hao, Y.L.; Yang, R. Growth of intermetallic layer in multi-laminated Ti/Al diffusion couples. *Mat. Sci Eng. A* **2006**, *435*, 638–647. [CrossRef]
25. Sun, Y.B.; Zhao, Y.Q.; Zhang, D.; Liu, C.Y.; Diao, H.Y. Multilayered Ti-Al intermetallic sheets fabricated by cold rolling and annealing of titanium and aluminum foils. *Trans. Nonferr Metal. Soc. China* **2011**, *21*, 1722–1727. [CrossRef]
26. Gachon, J.C.; Rogachev, A.S.; Grigoryan, H.E.; Illarionova, E.V.; Kuntz, J.J.; Kovalev, D.Y.; Nosyrev, A.N.; Sachkova, N.V.; Tsygankov, P.A. On the mechanism of heterogeneous reaction and phase formation in Ti/Al multilayer nanofilms. *Acta Mater.* **2005**, *53*, 1225–1231. [CrossRef]
27. Luo, J.G.; Acoff, V.L. Using cold roll bonding and annealing to process Ti/Al multi-layered composites from elemental foils. *Mat. Sci. Eng. A* **2004**, *379*, 164–172. [CrossRef]
28. Massalski, T.B. *Binary Alloy Phase Diagrams*, 2nd ed.; ASM International: Materials Park, OH, USA, 1990.
29. Xiong, X.; Huang, B. The process and mechanism of TiAl-Based alloy synthesized from Ti and Al powders. *J. Cent. South. Univ. Technol.* **1995**, *2*, 8–11. [CrossRef]
30. Školáková, A.; Salvetr, P.; Novák, P.; Vojtěch, D. Formation of Ti-Al phases during SHS process. *Acta Phys Pol. A* **2018**, *134*, 743–747. [CrossRef]
31. Kattner, U.R.; Lin, J.C.; Chang, Y.A. Thermodynamic assessment and calculations of the Ti-Al system. *Metall. Mater. Trans. A* **1992**, *23*, 2081–2090. [CrossRef]
32. Školáková, A.; Leitner, J.; Salvetr, P.; Novák, P.; Deduytsche, D.; Kopeček, J.; Detavernier, C.; Vojtěch, D. Kinetic and thermodynamic description of intermediary phases formation in Ti-Al system during reactive sintering. *Mater. Chem. Phys.* **2019**, *230*, 122–130. [CrossRef]

Article

Properties Comparison of Ti-Al-Si Alloys Produced by Various Metallurgy Methods

Anna Knaislová [1,*], Pavel Novák [1], Jaromír Kopeček [2] and Filip Průša [1]

1 Department of Metals and Corrosion Engineering, University of Chemistry and Technology, Technická 5, 166 28 Prague, Czech Republic; panovak@vscht.cz (P.N.); prusaf@vscht.cz (F.P.)

2 Institute of Physics of the Czech Academy of Sciences, Na Slovance 1999/2, 182 21 Prague, Czech Republic; kopecek@fzu.cz

* Correspondence: knaisloa@vscht.cz

Received: 9 September 2019; Accepted: 19 September 2019; Published: 21 September 2019

Abstract: Melting metallurgy is still the most frequently used and simplest method for the processing of metallic materials. Some of the materials (especially intermetallics) are very difficult to prepare by this method due to the high melting points, poor fluidity, or formation of cracks and pores after casting. This article describes the processing of Ti-Al-Si alloys by arc melting, and shows the microstructure, phase composition, hardness, fracture toughness, and compression tests of these alloys. These results are compared with the same alloys prepared by powder metallurgy by the means of a combination of mechanical alloying and spark plasma sintering. Ti-Al-Si alloys processed by melting metallurgy are characterized by a very coarse structure with central porosity. The phase composition is formed by titanium aluminides and titanium silicides, which are full of cracks. Ti-Al-Si alloys processed by the powder metallurgy route have a relatively homogeneous fine-grained structure with higher hardness. However, these alloys are very brittle. On the other hand, the fracture toughness of arc-melted samples is immeasurable using Palmqvist's method because the crack is stopped by a large area of titanium aluminide matrix.

Keywords: titanium aluminides and silicides; casting; powder metallurgy

---

1. Introduction

Titanium intermetallics with other light elements (for example aluminium or silicon) are prospective high-temperature alloys, applicable especially for construction components working at high temperatures under static loads [1–5]. They are considered for high-temperature service because they offer a balance of oxidation resistance and mechanical properties at higher temperatures superior to conventional titanium alloys [6,7]. Ti-Al(Si) alloys are characterized by low density, good resistance against oxidation at 600–800 °C [8,9], good thermal stability, high specific strength at high temperatures [10,11], and a favourable ratio of mechanical properties to density [12]. Alloys based on α_2-Ti_3Al and γ-TiAl can be used as high-temperature construction materials for aviation applications [13–15], but low room-temperature ductility, as well as insufficient high-temperature strength and toughness, are problems which limit their broader use [16].

The production of Ti-Al alloys (and generally intermetallic compounds) is focused on conventional casting techniques, for example, on melting in an electric arc furnace under an argon atmosphere or electroshock remelting in an inert atmosphere [17]. Despite the enormous efforts made over the last 40 years, no castings have been produced to meet the reliability and cost requirements of the aviation industry [18]. The problems are caused especially by the high reactivity of molten titanium alloys with the ceramic crucible, and hence it is needed to use cold wall crucibles. However, cold wall crucibles enable low superheating of only about 60 °C. Slow cooling rates after centrifugal casting lead to the production of castings with a coarse-grained structure and large porosity. In order to remove the internal defects, the castings are processed by the HIP (Hot Isostatic Pressing) method. This technology is used by GE for engines for Boeing, but it is very expensive [19].

The melting of Ti-Al alloys is a multi-step process. The process involves the melting of an alloy, casting into a mould, extended isothermal annealing of ingot and controlled rapid crystallization of the melted metal [17]. Casting is often followed by the isostatic pressing of hot castings, which has a positive effect on homogeneity, but the production is more expensive [12].

Ti-Al alloys for turbine wheels of turbochargers are now manufactured by lost-wax casting [20]. The model of the required object is made of the wax and coated by the ceramics, therefore the massive mould is made. Then the wax is melted and poured out the mould. The metallic melt is filled into the mould and the casting solidifies. The process finishes with the removal of the ceramic mould after the solidification of the alloy. Another technology is induction skull melting (ISM). This process combines the advantages of induction melting and cold crucible melting. ISM is used for melting highly reactive material (for example TiAl) with high purity. According to this method, γ-TiAl alloys are melted in the vacuum induction furnace in a water-cooled crucible. The contamination from the crucible into the alloy is minimal. The big advantage is the continuous stirring, which guarantees the composition's uniformity [21,22]. This process is very energy-intensive and highly expensive because of the high cost of melting furnace and its non-ecological operation, high consumption of cooling water and inert gas and low achievable melt overheating [20]. The problems of cast Ti-Al alloys also include the reactions of titanium and aluminium with the atmosphere of the melting facility, the reaction of titanium with melting crucibles or ceramic moulds, or the evaporation of aluminium. The γ-TiAl alloys prepared by this process have a polycrystalline lamellar microstructure. If the sample is heated to a temperature above 1150 °C, the polycrystalline lamellar structure is replaced by a fine homogeneous duplex one with uniform grains. The lamellar structure is less ductile than the duplex one, but it has better fracture toughness, fatigue resistance, and creep strength at high temperatures [20]. However, it is also very difficult to forge the Ti-Al alloys due to inherent poor deformability even at temperatures above 1000 °C [23,24]. Chesnutt et al. wrote that it is possible to form α_2-TiAl and γ-TiAl alloys (with difficulty). The α_2-TiAl alloys are produced in ingots of 3200 kg [25]. Lapin wrote about extrusion and forging for producing compressor blades for engine testing. Compressor blades were produced by Thyssen, GfE, Leistritz and GKSS for Rolls-Royce using Ti-Al-Nb-(B,C) alloy, but the microstructure was very heterogeneous, and it can be expected that the segregation effects would be even larger for the preparation of bigger components. The boron addition enabled the refinement of the grain size of the as-cast titanium aluminides [19].

The production of Ti-Al-Si alloys by melting metallurgy is very difficult. The disadvantages include the high melting points of intermetallics (titanium silicide Ti_5Si_3 melts at the temperature of 2130 °C [12]) and high melt reactivity with the melting crucibles [8], so it is necessary to use Y_2O_3 and ZrO_2 crucibles (which are more expensive than corundum or graphite crucibles). Other problems include the contamination of the melt from the reaction with the atmosphere in the furnace [12]. The Ti-Al-Si alloys have very poor fluidity, so the alloy has many casting defects after casting, for example, pores and cracks in the structure [26,27].

Hard and brittle sharp-edged titanium silicide (Ti_5Si_3) phases are formed by melting and casting of Ti-Al-Si alloy. These phases have a negative impact on mechanical properties, especially on the fracture toughness [28]. Hence, there is an effort to eliminate these big brittle phases of titanium silicides. In the previous study, the formation of in situ composites composed of elongated particles of Ti_5Si_3 phases in a tough matrix made of TiAl or Ti_3Al aluminides has been investigated [29]. In the structure, the cracks perpendicular to the direction of solidification were found because of the different coefficient of thermal expansion of Ti_5Si_3 in different crystal directions. Therefore, tensile stresses and crack initiation occur [12]. These problems limit the possibility of processing of Ti-Al-Si materials by directional crystallisation [30].

Applicability of forming of Ti-Al-Si alloys is also limited due to the low fracture toughness and ductility of the material, which persists even at temperatures above 1000 °C [12]. However, the preparation of Ti-Al alloys with the addition of Si by ingot metallurgy is limited to eutectic and hypoeutectic alloys based on α_2-Ti_3Al, since hypereutectic alloys are extremely brittle because of the coarse primary silicides formed upon solidification [10].

In this work, Ti-Al-Si alloys were processed by melting metallurgy in an arc melting furnace. The results were compared with the same alloys prepared by powder metallurgy by the means of a combination of mechanical alloying and spark plasma sintering.

2. Materials and Methods

The Ti-Al-Si alloys (TiAl10Si20, TiAl10Si30, TiAl15Si15, TiAl20Si20 (wt. %)) were prepared by melting metallurgy in the arc melting furnace Bühler MAM-5 in the Institute of Physics of the Czech Academy of Sciences. These compositions have been chosen according to our previous research [26]. The pieces of individual elements (purity 99.95%) were four times re-melted under the argon atmosphere.

The reference samples were prepared by powder metallurgy. The optimal preparation route was chosen as a combination of mechanical alloying and spark plasma sintering (MA + SPS). The mixture of pure titanium (with a purity of 99.5% and a particle size of 44 μm, STREM CHEMICALS, Newburyport, MA, USA), aluminium (99.62%, 44 μm, STREM CHEMICALS, Newburyport, MA, USA) and silicon (99.5%, 44 μm, Alfa Aesar, Haverhill, MA, USA) powders were filled in a steel vial with steel milling balls (ball-to-powder weight ratio was 60:1) and milled under the argon atmosphere. For the mechanical alloying, a planetary ball mill (PM 100 CM, Retsch, Haan, Germany) was applied. Milling was performed for four hours with rotation velocity of 400 min^{-1}. Mechanically alloyed powders were consolidated by spark plasma sintering (FCT Systeme GmbH, Rauenstein, Germany) under the pressure of 80 MPa and the temperature of 1100 °C for 15 min. The applied heating rate and cooling rates were 100 and 50 °C/min, respectively [31].

The phase composition was analyzed by the means of the X-ray diffraction analysis (XRD) using a X'Pert Pro (PANalytical, Almelo, Netherlands) X-ray diffractometer with CuKα radiation and a LynxEye XE detector (PANalytical, Almelo, The Netherlands). XRD patterns were evaluated qualitatively by X'Pert HighScore 3.0 software package (PANalytical, Almelo, Netherlands), exploiting PDF-2 2018 database. For the microstructure inspections, samples were ground by P80 to P4000 grinding papers (Hermes Schleifmittel GmbH, Hamburg, Germany) and polished by diamond paste with the particle size of 1–2 μm. The polished samples were etched by Kroll's reagent (it was prepared in our laboratory) (10 mL HF, 5 mL HNO_3 and 50 mL H_2O). The microstructure was investigated by the inverted optical microscope Olympus PME3 (Olympus, Prague, Czech Republic) and documented by the Carl Zeiss AxioCam ICc3 (Carl Zeiss, Jena, Germany) digital camera and AxioVision software package (version 4.8.2, Carl Zeiss, Jena, Germany). Porosity was evaluated by Lucia 4.8 image analyser (Laboratory Imaging, Prague, Czech Republic). TESCAN VEGA 3 LMU (TESCAN, Brno, Czech Republic) electron microscope with EDS analyser Oxford Instruments X-max 20 mm^2 (Oxford Instruments, High Wycombe, UK) (SEM-EDS) was used for the deeper microstructure investigation and the identification of present phases on the micrographs.

Compressive strength tests were performed with the use of the universal testing machine LabTest 5.250SP1-VM (LaborTech, Opava, Czech Republic). Values of ultimate compressive strength were determined from the measured stress-strain curves. Vickers hardness with a load of 5 kg (HV 5), 100 grams (HV 0.1) and 50 grams (HV 0.05) was measured on the polished samples (10 measurements on each specimen). Fracture toughness was measured by Vickers indentation method with the load of 1 kg on microhardness tester Future-Tech FM-700 (Future-Tech, Kawasaki, Japan). Indentions were observed by the means of the above-mentioned Olympus PME3 microscope. Fracture toughness was calculated using Palmqvist's Equation (1):

$$Kc = 0.016 \cdot \left(\frac{E}{HV}\right)^{\frac{1}{2}} \cdot \left(\frac{F}{c^{\frac{3}{2}}}\right) \quad (1)$$

where E is modulus of elasticity (GPa), HV is Vickers hardness (GPa), F is load (N), and c is half of the crack length after indention (mm).

3. Results

The phase composition of as-cast Ti-Al-Si alloys is displayed in Figure 1. The TiAl10Si20 and TiAl15Si15 alloys consists of titanium silicide (Ti_5Si_3) and titanium aluminide (TiAl). Titanium silicide TiSi, Ti_5Si_4 and pure silicon were, as expected, found in the TiAl10Si30 alloy due to the higher amount of silicon in the alloy. The TiAl20Si20 alloy is composed of Ti_5Si_3 and Ti_5Si_4 silicides in $TiAl_3$ aluminide matrix.

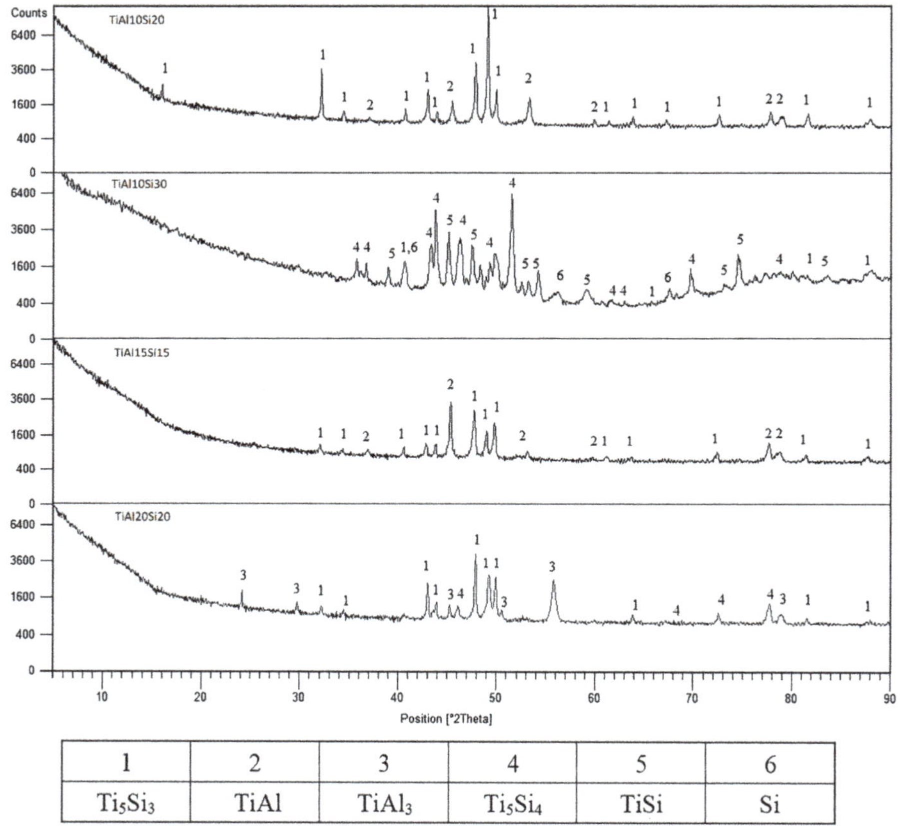

1	2	3	4	5	6
Ti_5Si_3	TiAl	$TiAl_3$	Ti_5Si_4	TiSi	Si

Figure 1. XRD patterns of Ti-Al-Si alloys processed by melting metallurgy.

Figure 2 shows the microstructure of Ti-Al-Si alloys prepared by arc melting. The as-cast alloys have coarse dendritic structures. As-cast TiAl10Si20 alloy is characterized by a very porous structure, as well as TiAl20Si20. The pores (black spots on the photos) are concentrated largely in the center of the sample and they are irregular in shape. Alloys TiAl10Si30 and TiAl15Si15 have much smaller pores. The particles of titanium silicides are very coarse in each alloy, and they have various local characters of morphology. The microstructure is characterized by primary silicides, silicides with fibrous and lamellar morphology and fine eutectic structure. The colonies of titanium silicides are differently oriented, depending on the local direction of heat transfer, which causes the elongation of silicides.

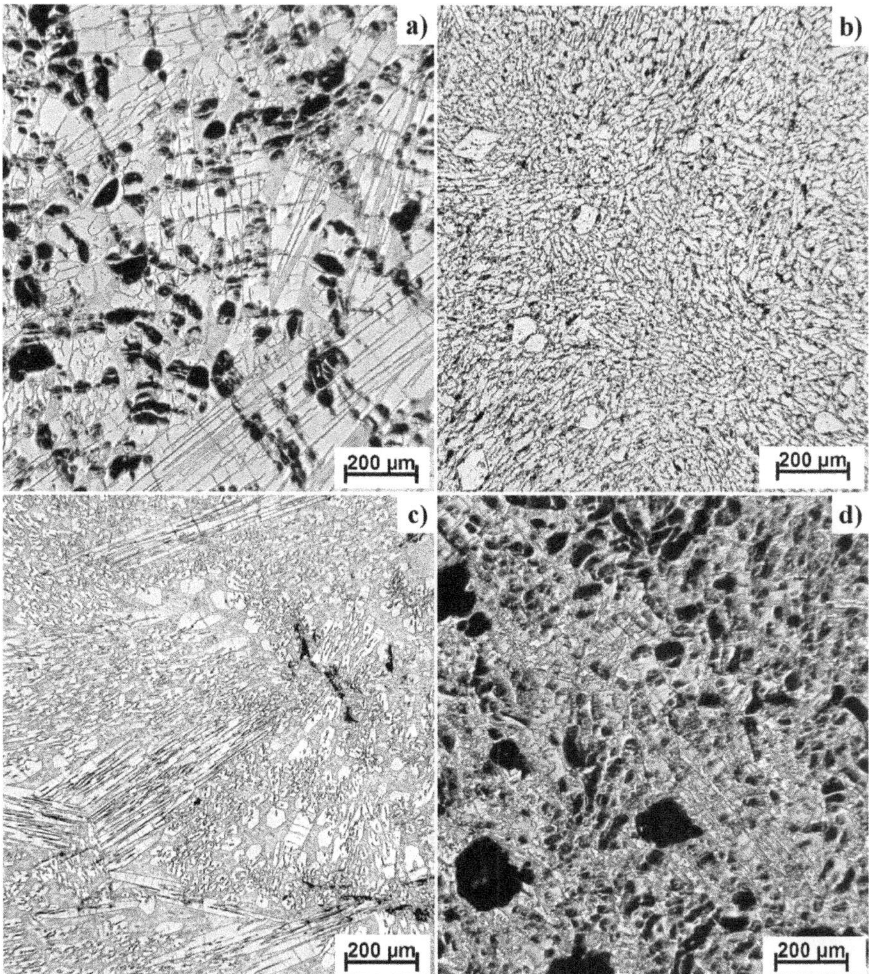

Figure 2. Microstructures of Ti-Al-Si alloys processed by melting metallurgy (light microscopy): (**a**) TiAl10Si20, (**b**) TiAl10Si30, (**c**) TiAl15Si15, (**d**) TiAl20Si20 (black areas are pores, lighter grey particles are titanium silicides, darker grey ones are titanium aluminides).

Figure 3 shows the porosity and average equivalent diameter of pores of Ti-Al-Si alloys prepared by melting metallurgy. The big differences between the values of porosity are given by preparation techniques. During the fast cooling, the gasses are closed in the middle of the sample and formed the middle porosity. At the edges of sample, the porosity is minimal.

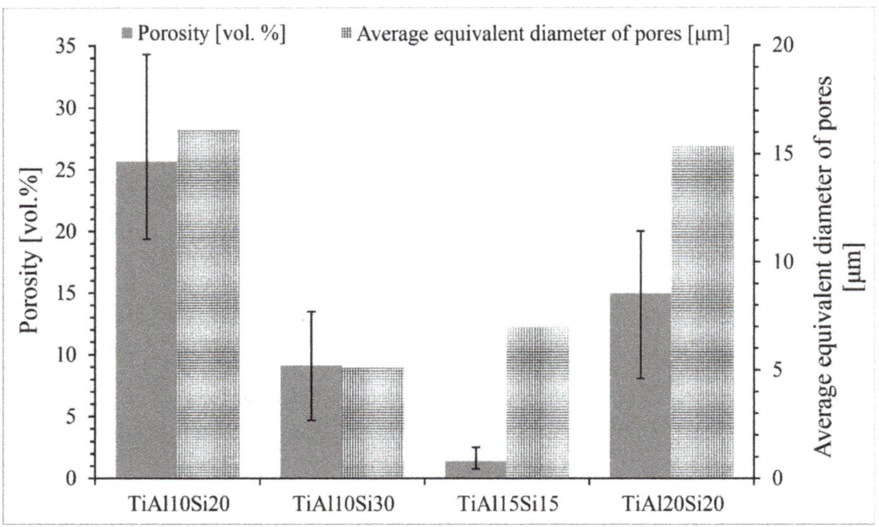

Figure 3. Porosity vol. and average equivalent diameter of pores of Ti-Al-Si alloys processed by melting metallurgy. no subscript

Figure 4 presents the SEM micrographs of Ti-Al-Si alloys acquired in a backscattered electrons (BSE) regime. Except for the TiAl10Si30, all alloys have multiple cracks in the titanium silicides. Composition of each phase (TiAl, Ti_5Si_3) in TiAl10Si20 alloy corresponds well with the phase diagram. Only spectrum four contains a local lower amount of silicon, which was detected near the crack. TiAl10Si30 alloy has the finest particles of titanium silicides. At the same time, TiAl10Si30 alloy has the lowest amount of aluminide matrix (Figure 5). It is due to the highest amount of silicon in the basic chemical composition and the higher ratio between aluminium and silicon in the alloy. The TiAl10Si30 alloy processed by casting exhibited the presence of areas of aluminides, containing higher degree of substitution of aluminium by silicon (Figure 4b). In addition, TiAl10Si30 alloy are contaminated by iron, copper and oxygen from the preparation route. In TiAl15Si15 alloy, the chemical composition of TiAl and Ti_5Si_3 phase corresponds well with the phase diagram, i.e., the phases are minimally substituted by other elements. According to the EDS (Energy Dispersive Spectroscopy) analysis (Spectrum 11) and Ti-Al phase diagram, TiAl20Si20 should contain aluminide phase ($TiAl_3$), which was confirmed by XRD diffraction.

The hardness of the Ti-Al-Si alloys (Figure 6) prepared by melting metallurgy varies between 416 and 549 HV 5. TiAl10Si30 alloy with the highest amount of silicon has a higher amount of hard silicides (Figure 5) and, hence, it reaches the highest values of hardness.

The microhardness of the Ti-Al-Si alloys (Figure 7) is in the range between 750 and 1066 HV 0.1 and TiAl10Si30 alloy achieves the highest hardness. It is caused by the fact that silicon in alloy is bonded in silicides (Ti_5Si_3, Ti_5Si_4 and TiSi), which are harder than the contained aluminides. The predicted hardness was calculated by Equation (2). The calculated hardness values of Ti-Al-Si alloys are compared with the measured ones in Figure 7. The values are comparable, only TiAl10Si30 alloy has lower values of calculated hardness than hardness, which was measured.

$$w_{(Ti-Si)} \cdot HV_{(Ti-Si)} + w_{(Ti-Al)} \cdot HV_{(Ti-Al)} = HV_{(Ti-Al-Si)} \tag{2}$$

where $w_{(Ti\text{-}Si)}$ is an area fraction of titanium silicides, $HV_{(Ti\text{-}Si)}$ is the Vickers hardness of titanium silicides, $w_{(Ti\text{-}Al)}$ is an area fraction of titanium aluminides, $HV_{(Ti\text{-}Al)}$ is Vickers hardness of titanium aluminides, and $HV_{(Ti\text{-}Al\text{-}Si)}$ is the hardness of the product.

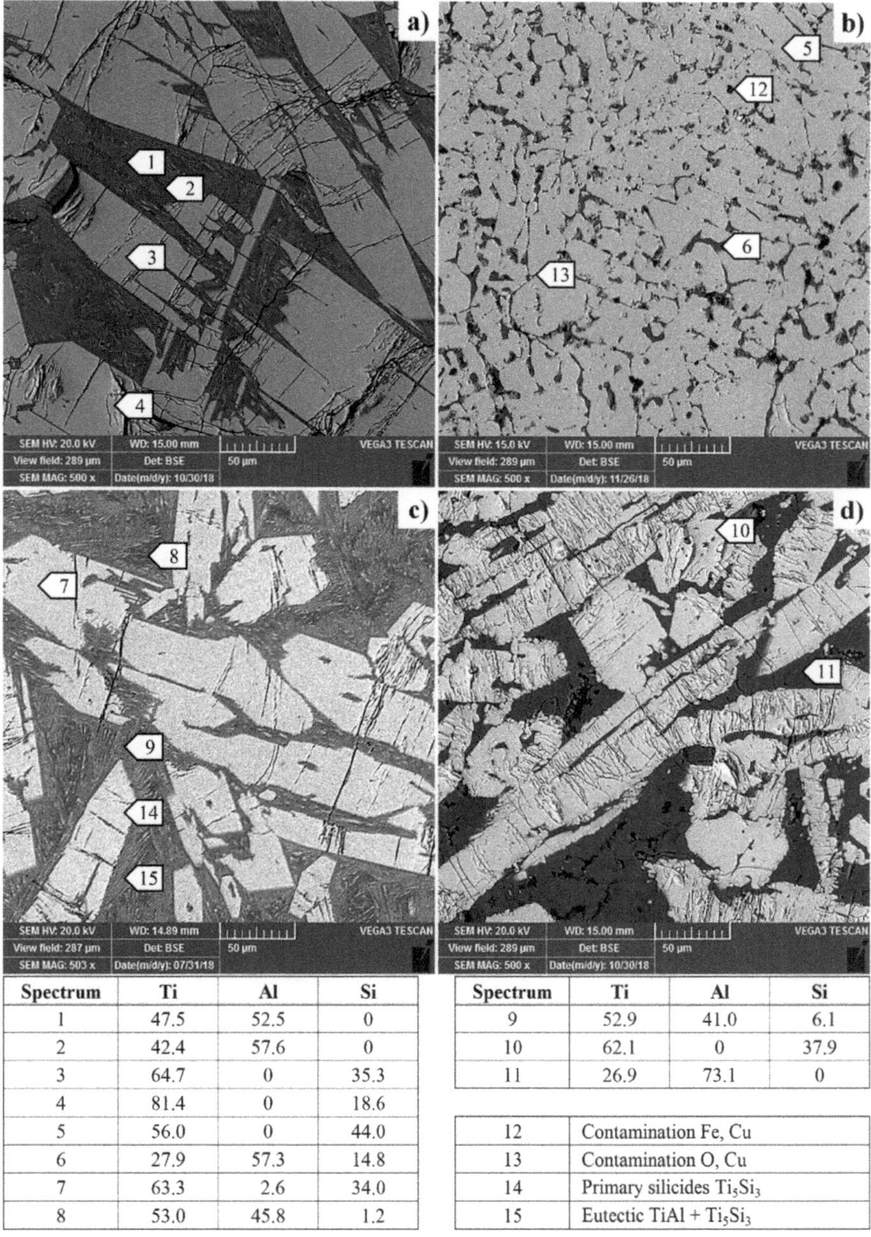

Spectrum	Ti	Al	Si
1	47.5	52.5	0
2	42.4	57.6	0
3	64.7	0	35.3
4	81.4	0	18.6
5	56.0	0	44.0
6	27.9	57.3	14.8
7	63.3	2.6	34.0
8	53.0	45.8	1.2

Spectrum	Ti	Al	Si
9	52.9	41.0	6.1
10	62.1	0	37.9
11	26.9	73.1	0

12	Contamination Fe, Cu
13	Contamination O, Cu
14	Primary silicides Ti_5Si_3
15	Eutectic TiAl + Ti_5Si_3

Figure 4. Microstructure of Ti-Al-Si alloys processed by melting metallurgy (light microscopy): (**a**) TiAl10Si20, (**b**) TiAl10Si30, (**c**) TiAl15Si15, (**d**) TiAl20Si20 (lighter particles are titanium silicides, darker ones are titanium aluminides) (composition in atomic %).

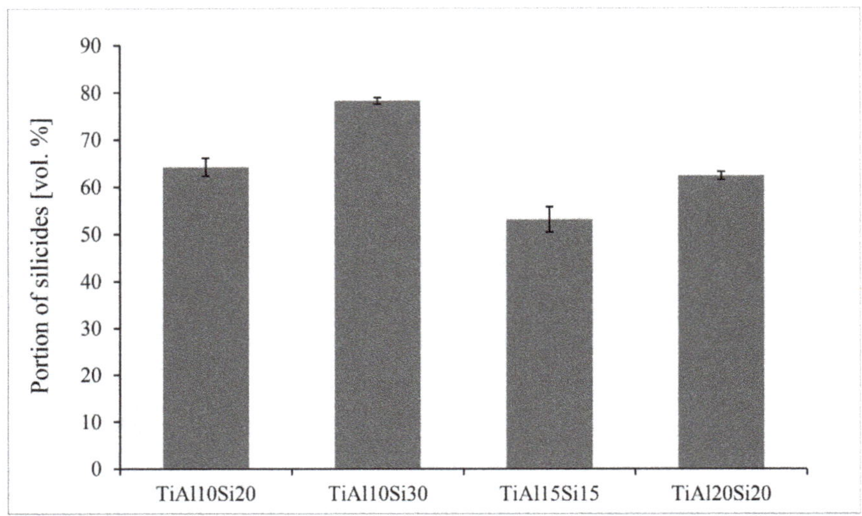

Figure 5. Amount of silicides in Ti-Al-Si alloys processed by melting metallurgy.

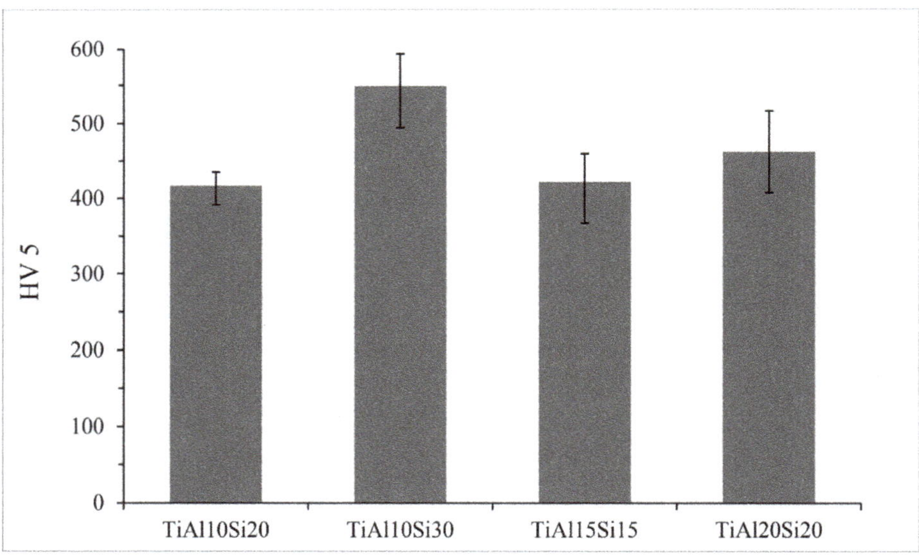

Figure 6. Hardness of Ti-Al-Si processed by melting metallurgy.

Figure 8 shows the comparison of microhardness of silicide and aluminide phase. Silicide phase hardness varies between 1070 and 1175 HV 0.05, the hardness of the aluminide phase varies between 200 and 530 HV 0.05. The big variations in the hardness of the aluminide phase are given by the substitution of silicon in titanium aluminide and the formation of small particles of titanium silicides in Ti-Al matrix. Titanium aluminide phase $TiAl_3$ present in TiAl20Si20 alloy has a higher hardness than TiAl phase. Values of the hardness of titanium silicides show only minor variations (Figure 8), so it is possible to say that silicides Ti_5Si_3 and Ti_5Si_4 have the same hardness, but titanium silicide TiSi present in TiAl10Si30 alloy has a slightly lower hardness.

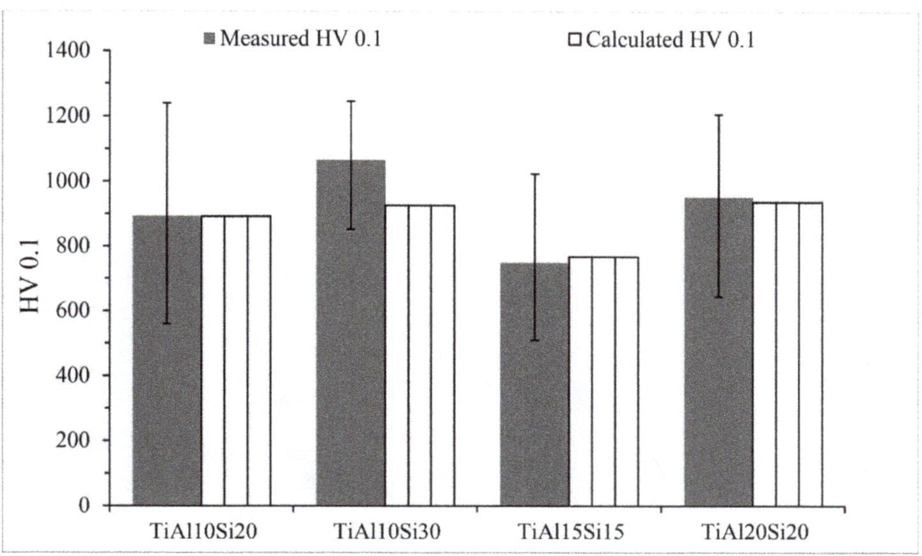

Figure 7. Microhardness of the Ti-Al-Si alloys processed by melting metallurgy.

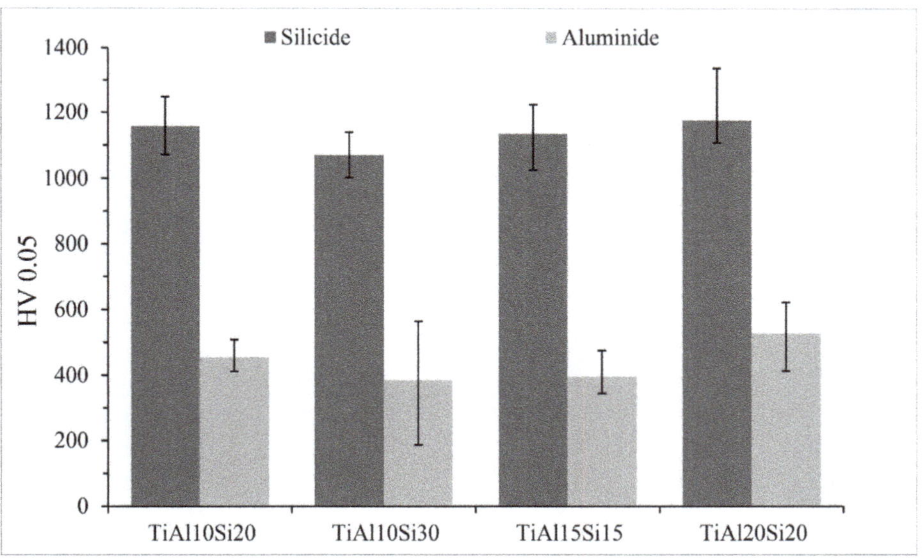

Figure 8. Comparison of microhardness of silicide and aluminide phase.

Intermetallics belong to the materials commonly referred as brittle. For these materials classes, the fracture toughness and compressive strength are the measures for potential use in the practice. Fracture toughness of Ti-Al-Si alloys processed by melting metallurgy was measured from the ten indentions into the silicide phase of each sample using Palmqvist's equation. Since titanium silicides are brittle, the cracks were formed during the indention. All of the alloys possess lower fracture toughness than the technical ceramics, such as alumina or silicon carbide having fracture toughness of around 4 MPa $m^{1/2}$ [32]. Crack propagation was stopped by the present aluminide phase at the TiAl15Si15 alloy prepared by arc melting. On the other hand, the crack spreads through the sample in case of the same alloy prepared by powder metallurgy (Figure 9). Therefore, the fracture toughness of the titanium silicides was measured (Figure 10). It is possible to say that a coarser structure is conducive to an increase on the toughness of the material and the samples are less brittle. The signs of the plasticity of the as-cast alloys are also shown in the deformation curves in compression (Figure 11). The TiAl15Si15 alloy achieves the higher ultimate tensile strength in compression—1700 MPa. This value of UTS is comparable with the same alloy processed by the combination of mechanical alloying and spark plasma sintering [31].

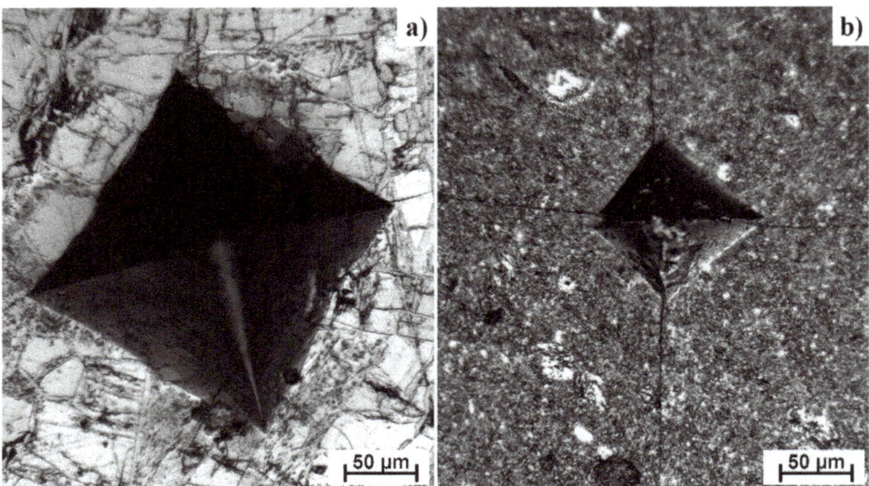

Figure 9. Cracks after indentation for calculating the fracture toughness of TiAl15Si15 alloy processed by: (**a**) melting metallurgy, (**b**) powder metallurgy.

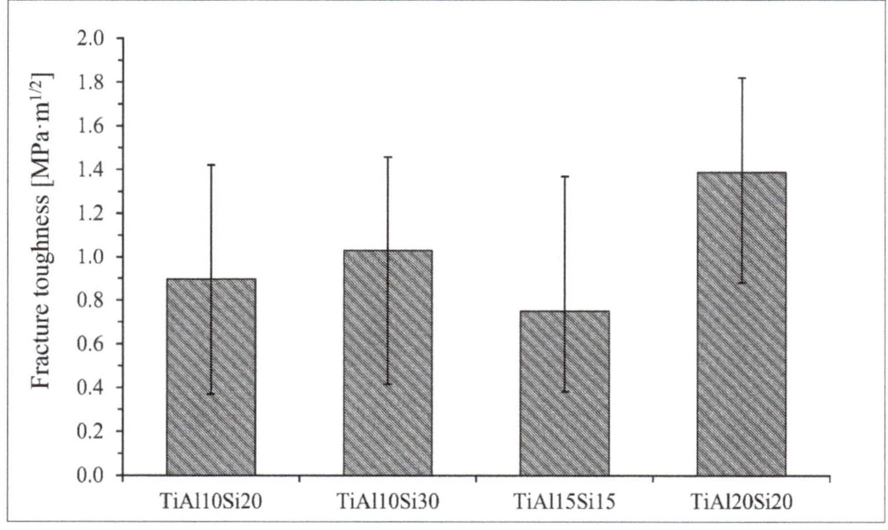

Figure 10. Fracture toughness of silicide phases in Ti-Al-Si alloys processed by melting metallurgy.

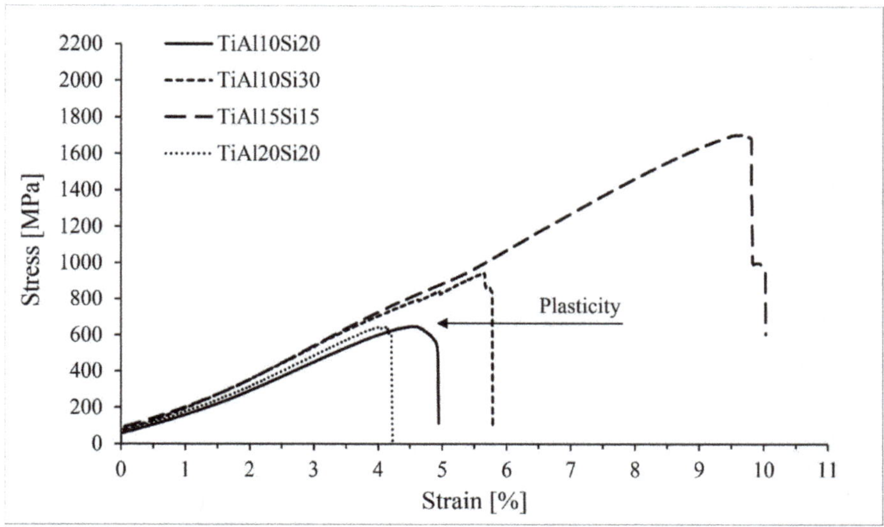

Figure 11. Compression tests of Ti-Al-Si alloys processed by melting metallurgy.

The reference samples of Ti-Al-Si alloys were prepared by powder metallurgy, by mechanical alloying and Spark Plasma Sintering (MA + SPS). SEM micrographs of the MA + SPS alloys are presented in Figure 12. The TiAl10Si20 alloy is characterized by a very homogeneous well-refined structure, and also the most fine-grained, with the Ti_5Si_3 strengthening phase in TiAl matrix. TiAl10Si30 is composed by Ti_5Si_3 and Ti_5Si_4 silicides in TiAl and $TiAl_2$ matrix. The TiAl15Si15 alloy contains a small amount of Ti_2Al phase. TiAl20Si20 alloy is composed of Ti_5Si_3, Ti_5Si_4, and $TiSi_2$ silicides, while the higher ratio between aluminium and titanium unbound in silicides leads to the single-phase $TiAl_3$ matrix. The sample also contains iron contamination originating from the milling steel vial.

Alloys processed by mechanical alloying and the spark plasma sintering method are characterized by lower porosity (Figure 13) than the alloys prepared by arc melting (Figure 3). The porosity of Ti-Al-Si alloys processed by powder metallurgy is lower than 1 vol. %. The pores are uniformly distributed (Figure 12).

The hardness of the Ti-Al-Si alloys (Figure 14) prepared by powder metallurgy varies between 800 and 1037 HV 5. These values of hardness are higher than values of Ti-Al-Si alloys processed by arc melting. The higher hardness corresponds to more homogeneous and fine-grained crackless microstructure. TiAl10Si30 and TiAl10Si20 reach the highest values of hardness.

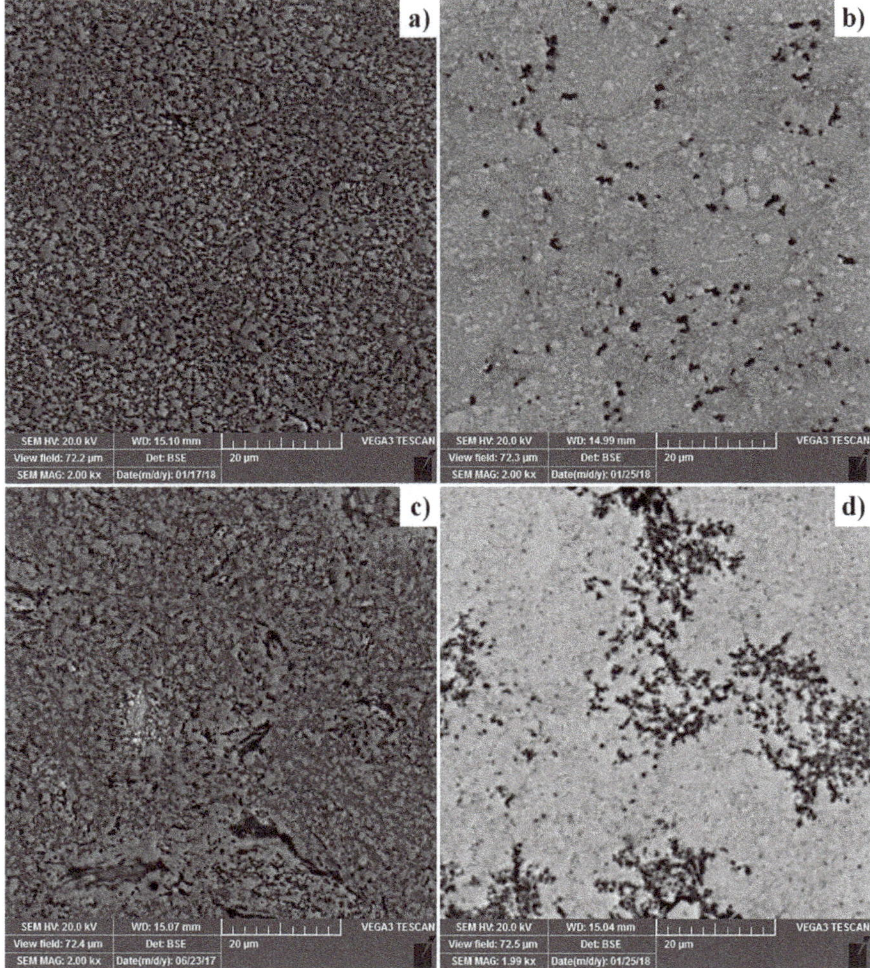

Figure 12. Microstructure of Ti-Al-Si alloys processed by powder metallurgy using mechanical alloying and Spark Plasma Sintering (scanning electron microscopy): (**a**) TiAl10Si20, (**b**) TiAl10Si30, (**c**) TiAl15Si15, (**d**) TiAl20Si20.

The microhardness of the Ti-Al-Si alloys prepared by powder metallurgy (Figure 15) varies between 1246 and 1430 HV 0.1. The measured microhardness is higher than the microhardness of Ti-Al-Si materials processed by melting metallurgy.

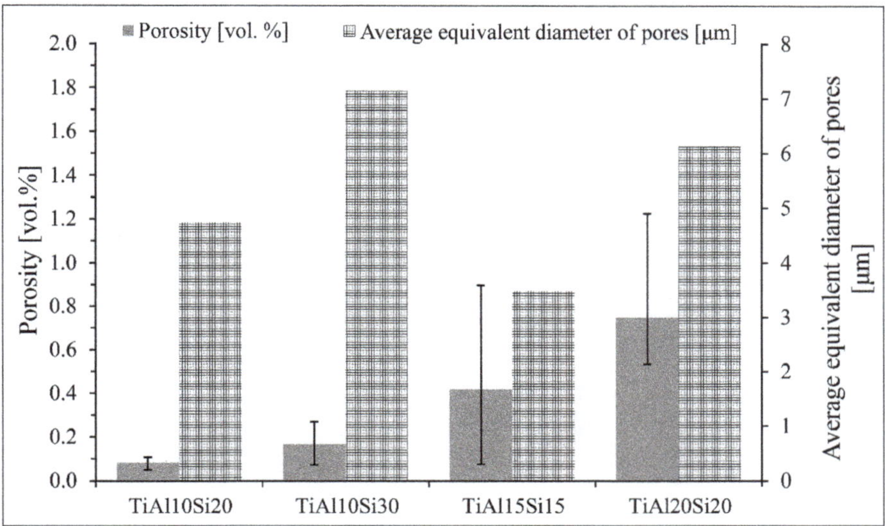

Figure 13. Porosity vol. and average equivalent diameter of pores of Ti-Al-Si alloys processed by powder metallurgy.

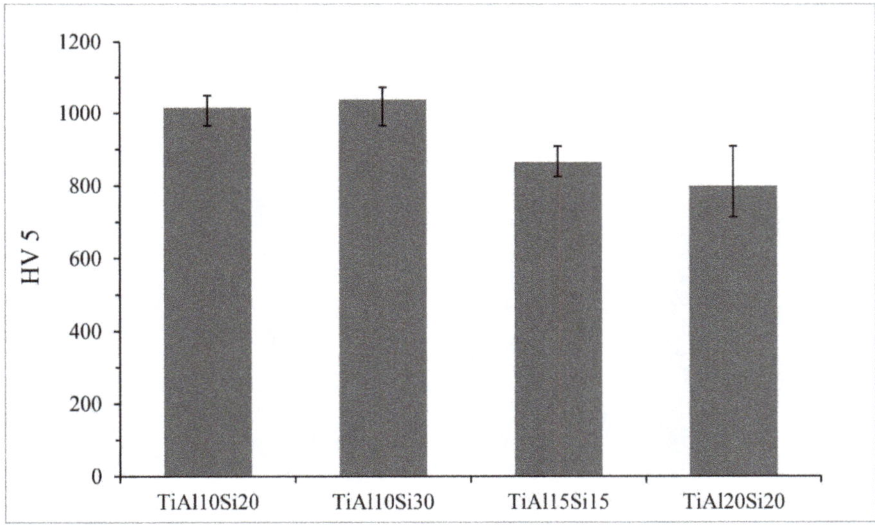

Figure 14. Hardness of Ti-Al-Si processed by powder metallurgy.

The fracture toughness of Ti-Al-Si alloys prepared by powder metallurgy was measured by the indentation and was calculated by Palmqvist's Equation. All of the alloys have low values of fracture toughness, because the intermetallic phases present in these alloys are very brittle (Figure 16). The alloys processed by powder metallurgy have lower values of fracture toughness than the same alloys prepared by melting metallurgy.

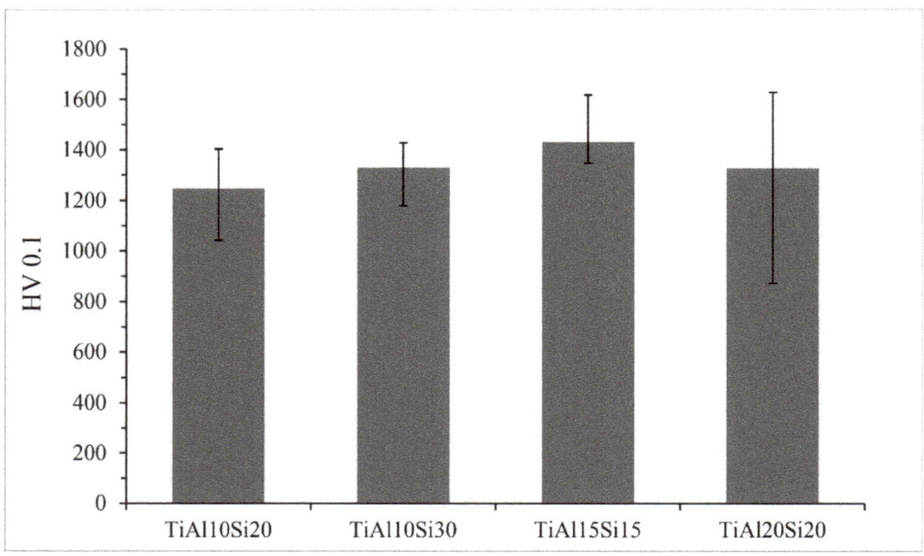

Figure 15. Microhardness HV 0.1 of Ti-Al-Si processed by powder metallurgy (mechanical alloying and Spark Plasma Sintering).

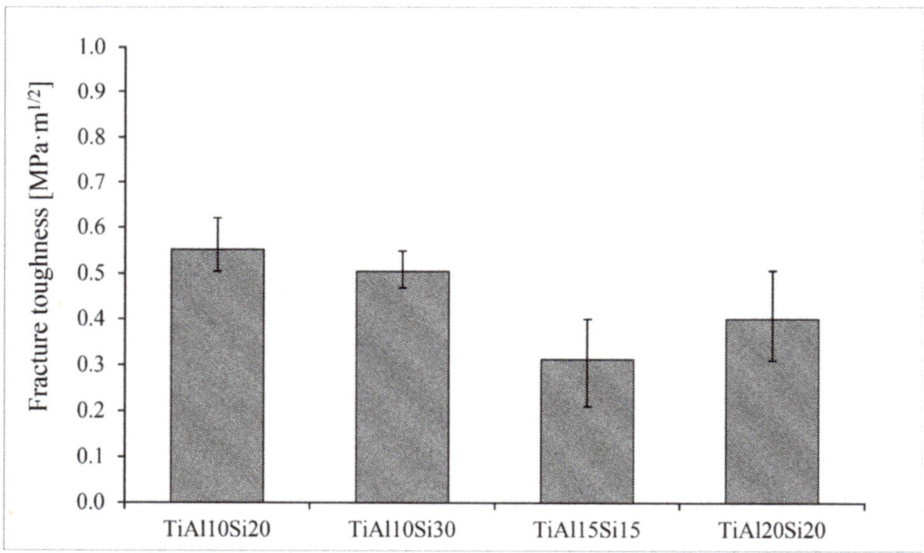

Figure 16. Fracture toughness of silicide phases in Ti-Al-Si alloys processed by melting metallurgy.

The Ti-Al-Si alloys processed by powder metallurgy achieve higher ultimate tensile strength in comparison with the same alloys processed by melting metallurgy (Figure 17). The alloys have better values due to the more homogeneous and fine-grained structure.

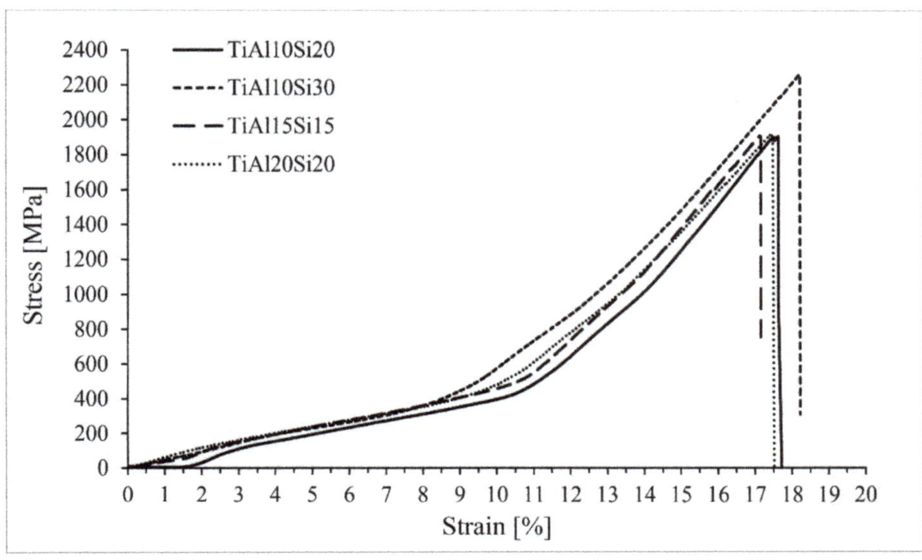

Figure 17. Compression tests of Ti-Al-Si alloys processed by melting metallurgy.

4. Discussion

Table 1 shows the comparison of the phase composition of the Ti-Al-Si materials processed by melting and powder metallurgy. The phase composition of TiAl10Si20 alloy processed by both the arc melting and the MA + SPS is in good agreement with the equilibrium phase diagram of Ti-Al-Si ternary system [31,33]. The same results were obtained by reactive sintering followed by spark plasma sintering (RS + SPS) described in our previous article [26]. Another titanium silicide, TiSi, was found in the TiAl10Si30 alloy due to the higher amount of silicon in the alloy. Formation of TiSi corresponds to the equilibrium phase diagram [33] and this titanium silicide was also found in the TiAl10Si30 alloy prepared by RS + SPS [26]. As-cast TiAl15Si15 alloy has the same phase composition as in the equilibrium phase diagram, but the alloy prepared by MA + SPS contains a small amount of Ti_2Al phase. The TiAl20Si20 alloy contains a combination of silicides Ti_5Si_3, Ti_5Si_4, and $TiSi_2$ (by type of preparation) and a single-phase $TiAl_3$ matrix caused by the higher ratio between aluminium and titanium unbound in silicides.

Table 1. Comparison of phase composition of Ti-Al-Si alloys (MA = mechanical alloying, RS = reactive sintering, SPS = Spark Plasma Sintering).

Alloy	Phase Diagram [27]	Melting	MA + SPS	RS + SPS [21]
TiAl10Si20	Ti_5Si_3, TiAl	Ti_5Si_3, TiAl	Ti_5Si_3, TiAl	Ti_5Si_3, TiAl
TiAl10Si30	TiSi, Ti_5Si_4, $TiAl_3$	TiSi, Ti_5Si_3, Ti_5Si_4, Si	Ti_5Si_3, TiAl, $TiAl_2$, Ti_5Si_4	TiSi, Ti_5Si_3, $TiAl_3$, Ti_2Al, Si
TiAl15Si15	Ti_5Si_3, TiAl	Ti_5Si_3, TiAl	Ti_5Si_3, TiAl, $TiAl_2$	Ti_5Si_3, TiAl
TiAl20Si20	Ti_5Si_3, $TiAl_3$	Ti_5Si_3, $TiAl_3$, Ti_5Si_4	Ti_5Si_3, $TiAl_3$, Ti_5Si_4, $TiSi_2$	Ti_5Si_3, $TiAl_3$, Ti_5Si_4, $TiSi_2$

Mechanical alloying causes that the aluminium in titanium aluminide and silicon in titanium silicides are strongly substituted by silicon and aluminium, respectively. Authors described that Ti(Al,Si) supersaturated solid solutions are formed after at least 10 or 20 h of milling [34,35], but it is shown in this work that intermetallic phases are formed at the shorter time of mechanical alloying (after 4 h).

The Ti-Al-Si alloys prepared by melting metallurgy (Figure 2) have a very coarse structure with pores, which are concentrated in the centre of the sample. The coarse particles of titanium silicides are oriented depending on the local direction of heat transfer, which causes the elongation of silicides. These silicides are cracked due to the high cooling rate after melting [36]. The central porosity is given by the processing technology. Gases are trapped in the sample as a result of evaporation of the elements (mostly aluminium) during the exothermic formation of intermetallics and relatively rapid cooling. The highest cooling rate is on the surface of the samples, so the samples solidified from the surface to the core. The gases are trapped in the middle sample and they cannot escape from the sample because of the fast solidification of the surface. Therefore, the porosity of the samples is concentrated in the centre of the samples.

Powder metallurgy (MA + SPS) results in a very fine microstructure of the investigated alloys (Figure 12). Authors described that mechanical alloying is a very effective process in obtaining of nanosized grains [37,38]. Refining of Ti_5Si_3 crystallites is probably due to a strong deformation and consequent recrystallization. The distribution of titanium silicides is much more homogeneous because the brittle powder is crushed to very fine particles during the first step of mechanical alloying. The elements in the obtained compounds are substituted mutually, exceeding the equilibrium solubility limits [39]. Figure 8 compares the hardness of silicide and aluminide phase in as-cast Ti-Al-Si alloys. Titanium aluminide $TiAl_3$ is harder than TiAl. This confirms the fact than $TiAl_3$ has covalent bond and the TiAl phase has a metallic bond, which causes the higher plasticity [40,41]. The hardness of titanium aluminide $TiAl_3$ is referred to as 450 HV [42], TiAl phase around 300 and 350 HV 5 [43]. The hardness of Ti_5Si_3 is referred to as 970 HV [44]. Calculated and measured results in Ti-Al-Si alloys are a little bit higher than the values reported in the literature. Hardness values of titanium silicides have a very small scattering of values, so it is possible to say that they have isotropic mechanical properties [36].

Powder metallurgy (MA + SPS) resulted in an enormous porosity reduction, which decreased from 26 vol. % in the case of as-cast TiAl10Si20 to under 1 vol. % by the same alloy (Figure 3, Figure 13). The other alloys recorded a similar decline. Also, the average equivalent diameter of pores decreases from the tens of micrometres to micrometre units. The applied the load, the sintering temperature and duration are the most important parameters, which influence the microstructure, porosity and pore size of the resulting compact sample. The better compaction of the powder and the smaller porosity and pore size are obtained from the higher temperature and applied load, but it may cause the grain coarsening [45]. In previous works, the optimal conditions of spark plasma sintering were studied so that the porosity and pore size were as low as possible [46].

The hardness, microhardness, and ultimate tensile strength (UTS) of the Ti-Al-Si alloys prepared by powder metallurgy (MA + SPS) are higher than the same alloys prepared by arc melting. The higher hardness and UTS corresponds to a more homogeneous and crackless microstructure, which is described above. The force during the mechanical alloying deforms the powder particles plastically, which leads to work hardening and fracture. Hardening, in combination with a fine-grained structure, increases the material hardness [45], but lowers the plasticity.

5. Conclusions

In this work, Ti-Al-Si were prepared by arc melting and powder metallurgy. Both technologies were compared. Ti-Al-Si alloys processed by arc melting are characterized by a porous structure with very coarse titanium silicide particles in a titanium aluminide matrix. Titanium silicides are full of cracks due to the relatively fast cooling. Reference samples processed by mechanical alloying and spark plasma sintering are characterized by a more homogeneous and finer structure with low porosity. These properties guarantee excellent mechanical properties. Vickers hardness of MA+SPS samples reaches double values, but Ti-Al-Si alloys processed by powder metallurgy are very brittle. The fracture toughness is worse than for technical ceramics. On the other hand, arc-melted samples have immeasurable fracture toughness using Palmqvist's method, the Vickers indentions do not cause the cracks, and crack propagation is stopped by the aluminide phase. For this reason, it is reasonable to say that TiAl15Si15 alloy prepared by melting metallurgy is the most promising alloy, as the formation of a coarse-grained structure improves the fracture toughness, the alloy is more ductile, and comparable ultimate tensile strength (as MA+SPS) predetermines this alloy for use in applications where high hardness is not the most important. However, it is necessary to improve the melting process (to optimize the conditions of melting and casting), or to use the hot isostatic pressing of the castings, in order to decrease the porosity of the material.

Author Contributions: A.K. was responsible for the theoretical background, writing the article, preparing samples and data analysis. P.N. was responsible for final correction. J.K. prepared samples by melting metallurgy. F.P. provided sintering process.

Funding: This research was financially supported by specific university research MSMT No 21-SVV/2019.

Conflicts of Interest: The authors declare no conflict of interest.

References

1. Clemens, H.; Mayer, S. Intermetallic titanium aluminides in aerospace applications–processing, microstructure and properties. *Mater. High Temp.* **2016**, *33*, 560–570. [CrossRef]
2. Brotzu, A.; Felli, F.; Marra, F.; Pilone, D.; Pulci, G. Mechanical properties of a TiAl-based alloy at room and high temperatures. *Mater. Sci. Technol.* **2018**, *34*, 1847–1853. [CrossRef]
3. Erdely, P.; Staron, P.; Maawad, E.; Schell, N.; Klose, J.; Clemens, H.; Mayer, S. Design and control of microstructure and texture by thermomechanical processing of a multi-phase TiAl alloy. *Mater. Des.* **2017**, *131*, 286–296. [CrossRef]
4. Schwaighofer, E.; Clemens, H.; Lindemann, J.; Stark, A.; Mayer, S. Hot-working behavior of an advanced intermetallic multi-phase γ-TiAl based alloy. *Mater. Sci. Eng. A* **2014**, *614*, 297–310. [CrossRef]
5. Liss, K.D.; Funakoshi, K.I.; Dippenaar, R.J.; Higo, Y.; Shiro, A.; Reid, M.; Suzuki, H.; Shobu, T.; Akita, K. Hydrostatic Compression Behavior and High-Pressure Stabilized β-Phase in γ-Based Titanium Aluminide Intermetallics. *Metals* **2016**, *6*, 165. [CrossRef]
6. Terner, M.; Biamino, S.; Baudana, G.; Penna, A.; Fino, P.; Pavese, M.; Ugues, D.; Badini, C. Initial Oxidation Behavior in Air of TiAl-2Nb and TiAl-8Nb Alloys Produced by Electron Beam Melting. *J. Mater. Eng. Perform.* **2015**, *24*, 3982–3988. [CrossRef]
7. Leyens, C.; Peters, M. *Titanium and Titanium Alloys: Fundamentals and Applications*; Wiley-VCH (John Wiley): Weinheim, Germany; Chichester, UK, 2003.
8. Novák, P.; Průša, F.; Šerák, J.; Vojtěch, D.; Michalcová, A. Oxidation resistance and thermal stability of Ti-Al-Si alloys produced by reactive sintering. In Proceedings of the Metal 2009, Hradec nad Moravicí, Czech Republic, 19-21 May 2009.
9. Lasalmonie, A. Intermetallics: Why is it so difficult to introduce them in gas turbine engines? *Intermetallics* **2006**, *14*, 1123–1129. [CrossRef]
10. Guan, Z.Q.; Pfullmann, T.; Oehring, M.; Bormann, R. Phase formation during ball milling and subsequent thermal decomposition of Ti–Al–Si powder blends. *J. Alloys Compd.* **1997**, *252*, 245–251. [CrossRef]
11. Suryanarayana, C. Synthesis of nanocomposites by mechanical alloying. *J. Alloys Compd.* **2011**, *509* (Suppl. 1), S229–S234. [CrossRef]
12. Novák, P.; Vojtěch, D.; Šerák, J.; Kubásek, J.; Průša, F.; Knotek, V.; Michalcová, A.; Novák, M. Synthesis of Intermediary Phases in Ti-Al-Si System by Reactive Sintering. *Chem. Listy* **2009**, *103*, 1022–1026.
13. Chen, G.; Peng, Y.; Zheng, G.; Qi, Z.; Wang, M.; Yu, H.; Dong, C.; Liu, C.T. Polysynthetic twinned TiAl single crystals for high-temperature applications. *Nat. Mater.* **2016**, *15*, 876. [CrossRef]
14. Tian, Y.; Ding, J.; Huang, X.; Zheng, H.R.; Song, K.; Lu, S.Q.; Zeng, X.G. Plastic deformation mechanisms of tension-compression asymmetry of nano-polycrystalline tial: Twin boundary spacing and temperature effect. *Comput. Mater. Sci.* **2020**, *171*, 109218. [CrossRef]
15. Appel, F.; Clemens, H.; Fischer, F.D. Modeling concepts for intermetallic titanium aluminides. *Prog. Mater. Sci.* **2016**, *81*, 55–124. [CrossRef]
16. Rao, K.P.; Du, Y.J.; Chung, J.C.Y.; Lau, K.C. In situ composite formation in Ti Al Si ternary system. *J. Mater. Process. Technol.* **1999**, *89–90*, 361–366. [CrossRef]
17. Kvanin, V.L.; Balikhina, N.T.; Vadchenko, S.G.; Borovinskaya, I.P.; Sychev, A.E. Preparation of γ-TiAl intermetallic compounds through self-propagating high-temperature synthesis and compaction. *Inorg. Mater.* **2008**, *44*, 1194–1198. [CrossRef]

18. Wu, X. Review of alloy and process development of TiAl alloys. *Intermetallics* **2006**, *14*, 1114–1122. [CrossRef]
19. Lapin, J. TiAl-based alloys: Present status and future perspectives. In Proceedings of the Metal 2009, Hradec nad Moravicí, Czech Republic, 19–21 May 2009.
20. Zemčík, L.; Dlouhý, A.; Król, S.; Pražmowskic, M. Vacuum Metallurgy of TiAl Intermetallics. In Proceedings of the Metal 2005; ISBN 80-86840-13-1.
21. Sugilal, G.; Jha, J.; Shashikumar; Rao, M.H.; Banerjee, K.; Dey, G.K. Indigenous development of induction skull melting technology for electromagnetic processing of refractory and reactive metals and alloys. *Mater. Today Proc.* **2016**, *3 Pt B*, 2942–2950. [CrossRef]
22. Guo, J.; Su, Y.; Jia, J.; Ding, H.; Liu, Y.; Ren, Z. Mechanism of skull formation during induction skull melting of intermetallic compounds. *Int. J. Cast Met. Res.* **1999**, *12*, 35–40.
23. Novák, P.; Michalcová, A.; Šerák, J.; Vojtěch, D.; Fabián, T.; Randáková, S.; Průša, F.; Knotek, V.; Novák, M. Preparation of Ti–Al–Si alloys by reactive sintering. *J. Alloys Compd.* **2009**, *470*, 123–126. [CrossRef]
24. Noda, T. Application of cast gamma TiAl for automobiles. *Intermetallics* **1998**, *6*, 709–713. [CrossRef]
25. Chesnutt, J.C. Titanium Aluminides for Aerospace Applications. Available online: https://www.tms.org/Superalloys/10.7449/1992/Superalloys_1992_381_389.pdf (accessed on 9 September 2019).
26. Knaislová, A.; Novák, P.; Cabibbo, M.; Průša, F.; Paoletti, C.; Jaworska, L.; Vojtěch, D. Combination of reaction synthesis and Spark Plasma Sintering in production of Ti-Al-Si alloys. *J. Alloys Compd.* **2018**, *752*, 317–326. [CrossRef]
27. Knaislová, A.; Novák, P.; Cygan, S.; Jaworska, L.; Cabibbo, M. High-Pressure Spark Plasma Sintering (HP SPS): A Promising and Reliable Method for Preparing Ti–Al–Si Alloys. *Materials* **2017**, *10*, 465. [CrossRef]
28. Novák, P.; Kříž, J.; Průša, F.; Kubásek, J.; Marek, I.; Michalcová, A.; Voděrová, M.; Vojtěch, D. Structure and properties of Ti–Al–Si-X alloys produced by SHS method. *Intermetallics* **2013**, *39*, 11–19. [CrossRef]
29. Novák, P. Příprava, vlastnosti a použití intermetalických sloučenin. *Chem. Listy* **2012**, *106*, 884–889.
30. Vojtěch, D.; Lejček, P.; Kopeček, J.; Bialasová, K. Směrová krystalizace eutektik systému Ti-Al-Si. In Proceedings of the Metal 2009, Hradec nad Moravicí, Czech Republic, 19-21 May 2009.
31. Knaislová, A.; Linhart, J.; Novák, P.; Průša, F.; Kopeček, J.; Laufek, F.; Vojtěch, D. Preparation of TiAl15Si15 intermetallic alloy by mechanical alloying and the spark plasma sintering method. *Powder Metall.* **2019**, *62*, 56–60. [CrossRef]
32. Wagner, W.C.; Chu, T.M. Biaxial flexural strength and indentation fracture toughness of three new dental core ceramics. *J. Prosthet. Dent.* **1996**, *76*, 140–144. [CrossRef]
33. Luo, Q.; Li, Q.; Zhang, J.Y.; Chen, S.L.; Chou, K.C. Experimental investigation and thermodynamic calculation of the Al–Si–Ti system in Al-rich corner. *J. Alloys Compd.* **2014**, *602*, 58–65. [CrossRef]
34. Rao, K.P.; Zhou, J.B. Characterization of mechanically alloyed Ti–Al–Si powder blends and their subsequent thermal stability. *Mater. Sci. Eng. A* **2002**, *338*, 282–298. [CrossRef]
35. Vyas, A.; Rao, K.P.; Prasad, Y.V.R.K. Mechanical alloying characteristics and thermal stability of Ti–Al–Si and Ti–Al–Si–C powders. *J. Alloys Compd.* **2009**, *475*, 252–260. [CrossRef]
36. Stoloff, N.S.; Sikka, V.K. *Physical Metallurgy and Processing of Intermetallic Compounds*; Chapman & Hall: London, UK, 1996.
37. Liang, G.; Meng, Q.; Li, Z.; Wang, E. Consolidation of nanocrystalline Al-Ti alloy powders synthesized by mechanical alloying. *Nanostruct. Mater.* **1995**, *5*, 673–678. [CrossRef]
38. Calderon, H.A.; Garibay-Febles, V.; Umemoto, M.; Yamaguchi, M. Mechanical properties of nanocrystalline Ti–Al–X alloys. *Mater. Sci. Eng. A* **2002**, *329–331*, 196–205. [CrossRef]
39. Haušild, P.; Karlík, M.; Čech, J.; Průša, F.; Nová, K.; Novák, P.; Minárik, P.; Kopeček, J. Preparation of Fe-Al-Si Intermetallic Compound by Mechanical Alloying and Spark Plasma Sintering. *Acta Phys. Pol. A* **2018**, *134*. [CrossRef]
40. Bester, G.; Fähnle, M. Interpretation of ab initio total energy results in a chemical language: II. Stability of TiAl3 and ScAl3. *J. Phys. Condens. Matter* **2001**, *13*, 11551. [CrossRef]
41. Music, D.; Schneider, J.M. Effect of transition metal additives on electronic structure and elastic properties of TiAl and Ti_3Al. *Phys. Rev. B* **2006**, *74*, 174110. [CrossRef]
42. Huy, T.D.; Fujiwara, H.; Yoshida, R.; Binh, D.T.; Miyamoto, H. Microstructure and Mechanical Properties of TiAl3/Al2O3 in situ Composite by Combustion Process. *Mater. Trans.* **2014**, *55*, 1091–1093. [CrossRef]

43. Tlotleng, M.; Lengopeng, T.; Pityana, S. Evaluation of the microstructure and microhardness of laser-fabricated titanium aluminate coatings. In Proceedings of the AMI Ferrous and Base Metals Development Network Conference 2016, Durban, South Africa, 19–21 October 2016.
44. Frommeyer, G.; Rosenkranz, R. In Structures and Properties of the Refractory Silicides Ti_5Si_3 and $TiSi_2$ and Ti-Si-(Al) Eutectic Alloys. In *Nato Science Series II. Mathematics, Physics and Chemistry*; Springer: Dordrecht, The Netherlands, 2004; pp. 287–308.
45. Molnárová, O.; Málek, P.; Veselý, J.; Minárik, P.; Lukáč, F.; Chráska, T.; Novák, P.; Průša, F. The Influence of Milling and Spark Plasma Sintering on the Microstructure and Properties of the Al7075 Alloy. *Materials* **2018**, *11*, 547. [CrossRef]
46. Knaislová, A.; Šimůnková, V.; Novák, P.; Průša, F.; Cygan, S.; Jaworska, L. The optimization of sintering conditions for the preparation of Ti-Al-Si alloys. *Manuf. Technol.* **2017**, *17*, 483–488.

Article

Influence of Heat Treatment on Microstructure and Properties of NiTi46 Alloy Consolidated by Spark Plasma Sintering

Pavel Salvetr [1,*], Jaromír Dlouhý [1], Andrea Školáková [2], Filip Průša [2], Pavel Novák [2], Miroslav Karlík [3,4] and Petr Haušild [3]

1. COMTES FHT, Prumyslova 995, 334 41 Dobrany, Czech Republic; jaromir.dlouhy@comtesfht.cz
2. Department of Metals and Corrosion Engineering, University of Chemistry and Technology, Technická 5, 166 28 Prague 6, Czech Republic; skolakoa@vscht.cz (A.Š.); prusaf@vscht.cz (F.P.); panovak@vscht.cz (P.N.)
3. Faculty of Nuclear Sciences and Physical Engineering, Department of Materials, Czech Technical University in Prague, Trojanova 13, 120 00 Prague 2, Czech Republic; miroslav.karlik@fjfi.cvut.cz (M.K.); petr.hausild@fjfi.cvut.cz (P.H.)
4. Faculty of Mathematics and Physics, Department of Physics of Materials, Charles University, Ke Karlovu 5, 121 16 Prague 2, Czech Republic

* Correspondence: pavel.salvetr@comtesfht.cz; Tel.: +420-220-44-4441

Received: 12 November 2019; Accepted: 3 December 2019; Published: 6 December 2019

Abstract: Ni-Ti alloys are considered to be very important shape memory alloys with a wide application area including, e.g., biomaterials, actuators, couplings, and components in automotive, aerospace, and robotics industries. In this study, the NiTi46 (wt.%) alloy was prepared by a combination of self-propagating high-temperature synthesis, milling, and spark plasma sintering consolidation at three various temperatures. The compacted samples were subsequently heat-treated at temperatures between 400 °C and 900 °C with the following quenching in water or slow cooling in a closed furnace. The influence of the consolidation temperature and regime of heat treatment on the microstructure, mechanical properties, and temperatures of phase transformation was evaluated. The results demonstrate the brittle behaviour of the samples directly after spark plasma sintering at all temperatures by the compressive test and no transformation temperatures at differential scanning calorimetry curves. The biggest improvement of mechanical properties, which was mainly a ductility enhancement, was achieved by heat treatment at 700 °C. Slow cooling has to be recommended in order to obtain the shape memory properties.

Keywords: Ni-Ti alloy; self-propagating high-temperature synthesis; spark plasma sintering; aging; compressive test; hardness; shape memory

1. Introduction

The Ni-Ti alloys named NiTinol (derived from nickel, titanium, and laboratory of discovery-Naval Ordnance Laboratory) are well-known shape memory alloys with good mechanical properties and high corrosion resistance, which enables usage as implants, medical devices, and other applications as biomaterials [1]. The shape memory effects occur due to the reversible solid-state transformation between the high-temperature and low-temperature phases. The high-temperature phase is called austenite with a high-symmetry structure-ordered body-centered cubic phase B2 (CsCl). The low-temperature and low-symmetry phase is called martensite with monoclinic B19′ lattice. Reversible strains at about 8% of the initial length are enabled due to the reversible phase transformation as well [1]. To achieve the desired shape memory effects-transformation temperatures, it is necessary to be careful about a chemical composition. The transformation temperatures are very sensitive for the nickel-titanium ratio and an increase in the content of nickel by 0.1 at a percentage causes a change of the transformation temperature A_F (austenite finish) up to 10 °C [2,3]. This fact makes the production of the Ni-Ti alloys more difficult because titanium and, subsequently, NiTi alloys have high affinity to oxygen from atmosphere and carbon from the melting crucible during vacuum induction melting (VIM) [4,5].

Spark plasma sintering (SPS) is a modern consolidation process, which is suitable for various materials such as ceramics and metals including many intermetallic systems (e.g., Ni-Ti and Fe-Al alloys [6–9]). The high heating rate and shortness of the whole process enable the use of SPS for the consolidation of nanocrystalline materials as well [10–12]. The process is based on the sintering of powder under the simultaneous influence of high electric current (direct or pulsed) and uniaxial pressure. The Joule heat is generated by passing the current through the graphite punch and die and between the powder's particles. The high heating rate was described as the route to reduce the amount of undesirable secondary phase such as the Ti_2Ni by the self-propagating high-temperature synthesis (SHS) [13]. Thus, the SPS process was examined as a heating source for initiating the SHS reaction between nickel and titanium elemental powders in the previous paper. However, the SPS process seems to be inapplicable for the initiation of the SHS reaction because the strongest increase of the temperature occurs on the surface of the particles. The formed intermetallic layers act further as diffusion barriers and separate unreacted nickel from titanium [14]. Therefore, the pre-alloyed Ni-Ti powder after mechanical alloying is usually sintered by SPS. This process can produce the highly dense NiTi materials [7] or the porous structure depending on the addition of the space holder (e.g., NH_4HCO_3) [15,16].

Heat treatment of the Ni-Ti alloys has a crucial effect on the properties of the samples. The parameters of heat treatment influence the microstructure, internal stresses, precipitation, shape memory, and mechanical properties [17]. The Ni-Ti alloys undergo homogenizing treatment at the temperature of about 800–1050 °C up to several hours of duration, which is followed by water quenching to get a homogeneous microstructure without the Ni-rich precipitates [17–22]. The second step of heat treatment represents the aging treatment, which usually occurs between 300 °C and 800 °C [23,24]. The grade of aging (density and size of the Ni_4Ti_3 precipitates) depends on the temperature and time. The precipitation process starts with the metastable Ni_4Ti_3 phase, which is transformed into the metastable Ni_3Ti_2 phase and the stable Ni_3Ti [25]. The precipitation process is accompanied by the hardness changes during aging due to formation, coarsening, and decomposition of the Ni_4Ti_3 and Ni_3Ti_2 phases. The addition of Al enhances the microstructural and hardness stability of Ni-rich Ni-Ti alloys until 500–600 °C [17,26]. The results of phase formation, stability, and transformation $Ni_4Ti_3 \rightarrow Ni_3Ti_2 \rightarrow Ni_3Ti$ during aging are presented by a time-temperature-transformation (T-T-T) diagram [17,25,27].

In this work, the NiTi46 (wt.%) alloy was processed by a combination of SHS, milling in a vibratory mill, and SPS consolidation at three temperatures to get fully dense materials. The prepared samples underwent the heat treatment in temperatures ranging from 400 to 900 °C. The characterization of samples was focused on phase composition, observing the transformation temperatures, and changing the mechanical properties depending on the heat treatment regime.

2. Materials and Methods

The metallic powders with the following particle sizes and purities were used as starting material for the NiTi46 (wt.%) alloy: nickel (particle size < 150 μm, 99.9 wt.% purity, Sigma-Aldrich, St. Louis, MO, USA) and titanium (particle size < 44 μm, 99.5 wt.% purity, STREM CHEMICALS, Newburyport, MA, USA). The powders were mixed manually corresponding to the chemical composition of the NiTi46 powder mixture, which was uniaxially compressed at room temperature to cylindrical green bodies of 12 mm in diameter at a pressure of 450 MPa for 5 min using LabTest 5.250SP1-VM universal loading machine (Labortech, Opava, Czech Republic). The SHS reaction of the pressed powder mixture was carried out in the fused silica ampoules evacuated to 10^{-2} Pa and sealed, which were placed in the preheated region to 1100 °C electric resistance furnace. The duration of the reaction was 20 min with the following cooling in air. The properties of the samples prepared this way were described in a previous paper [28]. The microstructure is composed of the two phases (NiTi austenite – cubic, Ti_2Ni – cubic), hardness 276 HV10, area fraction of the Ti_2Ni phase 11.7%, and transformation temperatures: A_S = 56 °C, A_F = 86 °C, and M_S = 21 °C. The SHS product was milled in a vibratory cylinder mill VM4 (OPS Přerov, Přerov, Czech Republic) in atmosphere with a duration of 7 min and the powder fraction with a particle size <355 µm was selected by sieving using Fritsch Analysette 3 device (FRITSCH GmbH, Germany). This pre-alloyed NiTi46 powder was consolidated by using the SPS method (FCT Systeme HP D 10, Frankenblicke, Germany) at three various temperatures (900, 1000, and 1100 °C) under the pressure of 50 MPa with a holding time of 10 min. The high heating rate was chosen as 300 °C/min at the beginning and the last 100 °C with the heating rate of 100 °C/min (for example, sintering temperature of 1000 °C: applied heating rate of 300 °C/min up to 900 °C, between the temperatures 900 and 1000 °C and the applied heating rate of 100 °C/min). The conditions of SPS consolidation as the temperature regime, compaction force, height reduction, and current flow are shown in Figure 1.

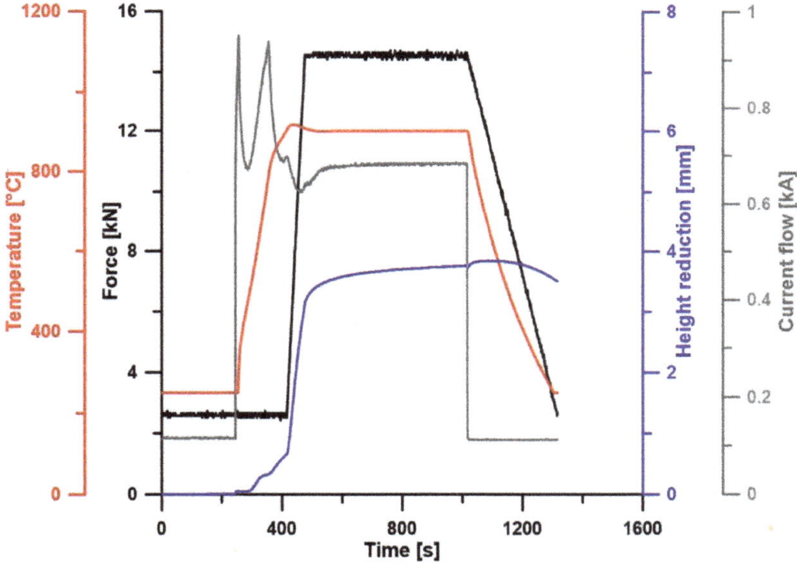

Figure 1. SPS parameters during consolidation at the temperature of 900 °C.

The heat treatment in the temperature range from 400 °C to 900 °C was applied to the SPS-ed samples. The duration of heat treatment was 60 min. Two variants of cooling were used including a high cooling rate with quenching in water and a slow cooling rate, which was provided by cooling in the closed furnace (average cooling rate approximately 2.5 °C/min between 700 °C and 300 °C).

The metallographic samples were prepared by grinding and polishing and the microstructure was revealed by etching in Kroll's reagent (5 mL HNO_3, 10 mL HF, and 85 mL H_2O). The microstructure was observed using scanning electron microscopes (SEM) equipped with the EDS (Energy Dispersive Spectroscopy) analyzers for identification of the chemical composition of the individual phases: VEGA 3 LMU (TESCAN, Brno, Czech Republic) equipped with the OXFORD Instruments X-max 20 mm^2 SDD EDS analyzer (Oxford Instruments, HighWycombe, UK) and JEOL IT 500 HR 500 (JEOL, Tokyo, Japan). The phase composition was analyzed by the X-ray diffraction analysis (XRD) using a X'Pert Pro (PANalytical, Almelo, The Netherlands) X-ray diffractometer with CuKα radiation and a LynxEye XE detector (PANalytical, Almelo, The Netherlands). The mechanical properties of the samples were evaluated by measuring Vickers hardness with a load of 10 kg and compression tests (LabTest 5.250SP1-VM universal loading machine Labortech, Opava, Czech Republic) with a strain rate of

$0.3\ s^{-1}$ on samples measuring 3.3 mm × 3.3 mm × 5 mm. Compression tests were conducted in both the direction (longitudinal and perpendicular) to the direction of SPS. A longitudinal direction is parallel to compressive force during the SPS process. Differential scanning calorimetry (DSC) analysis of the prepared alloys was performed using Setaram DSC 131 (Setaram, Caluire, France) to determine the transformation temperatures in products. Measurements for determining temperatures austenite start (A_S) and austenite finish (A_F) were carried out in the temperature range of −20 °C to 200 °C at a heating rate of 10 °C/min and cooling from 200 °C to −5 °C for detecting the martensite start (M_S) and martensite finish (M_F) temperatures.

The samples compacted by SPS at 900 °C and with following heat treatments at 600 and 700 °C for 1 h and slow cooling in the closed furnace were also investigated using transmission electron microscopy (JEOL JEM 2200FS, JEOL, Tokyo, Japan, accelerating voltage of 200 kV). Standard 3 mm samples prepared by a slow-speed diamond blade cutting were mechanically dimpled and ion polished in a Gatan PIPS 691 device (Pleasanton, CA, USA).

3. Results and Discussion

3.1. Microstructure, Phase Composition, and Phase Transformation

First, the influence of used sintering temperature on the quality of sintering individual particles was investigated by a porosity measurement. Since it is visible in Figure 2 and acquired by a light microscope, there are differences between samples compacted at various temperatures. The direction of observation plays an important role. The non-deformed grains were observed on the perpendicular cut (perpendicular to the direction of compression) whereas the elongated shape of grains after loading during SPS remained in the microstructure on the longitudinal cut (parallel to SPS compression). The highest values of porosity were determined at the samples sintered at 900 °C. The porosity was high at the perpendicular level and also at the longitudinal cut. The SPS temperature of 1000 °C was sufficient to the reduction of porosity in comparison to the temperature of 900 °C. Mainly in the longitudinal direction, the value of porosity decreased rapidly to a similar value, which was measured after sintering at 1100 °C. The values of porosity are compared in Table 1 and, based on the porosity measurement, it is clear that the higher temperature of the SPS process leads to superior sintering of individual particles.

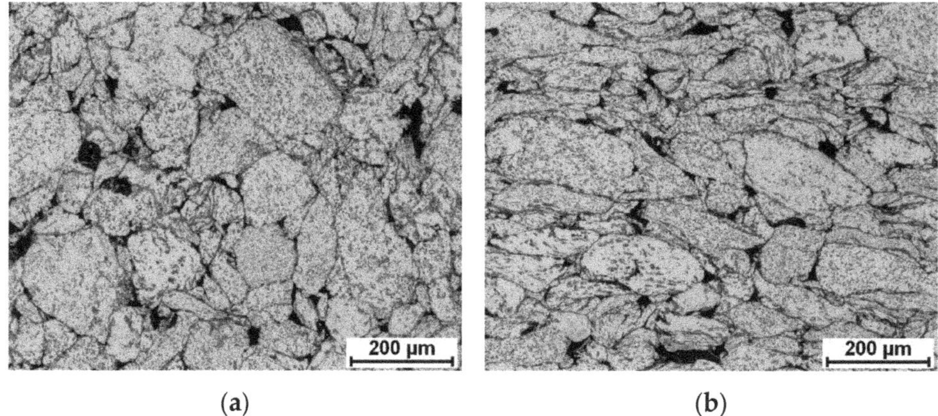

Figure 2. *Cont.*

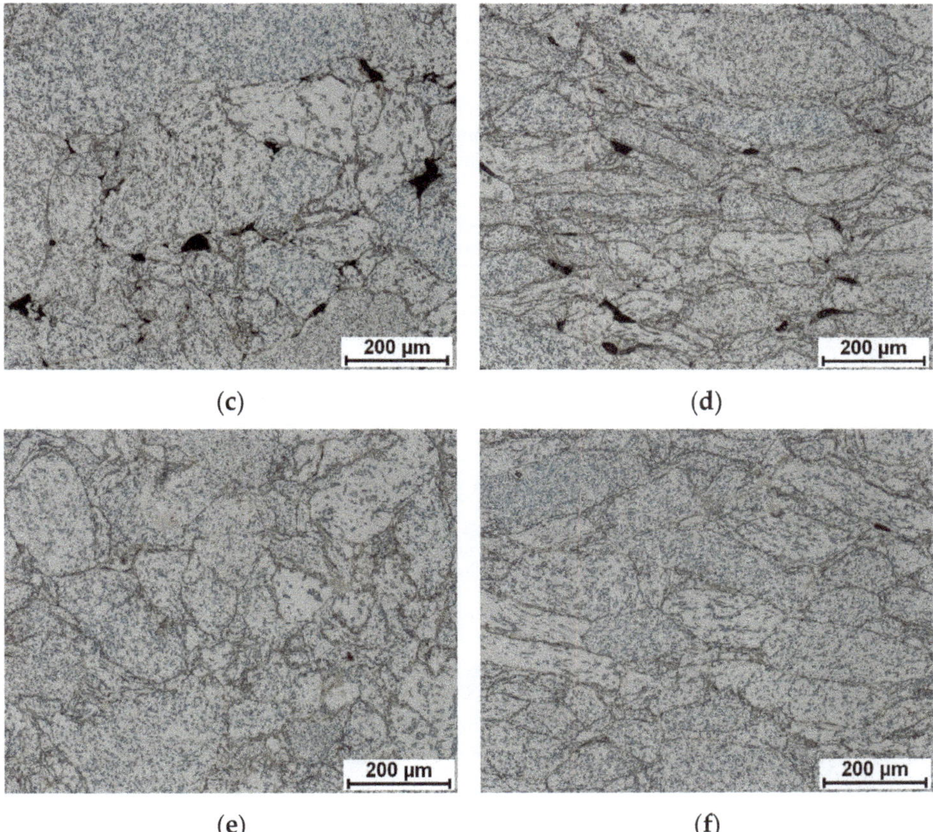

Figure 2. Microstructure and porosity of the NiTi46 alloy SPS consolidated at various temperatures: (**a**) 900 °C-perpendicular, (**b**) 900 °C-longitudinal, (**c**) 1000 °C-perpendicular, (**d**) 1000 °C-longitudinal, (**e**) 1100 °C-perpendicular, and (**f**) 1100 °C-longitudinal direction.

Table 1. Influence of SPS temperature on porosity of the sample and area fraction of the Ti_2Ni phase in the microstructure.

Parameter	Direction	Spark Plasma Sintering Temperature		
		900 °C	1000 °C	1100 °C
Porosity (%)	Perpendicular	1.6 ± 0.09	0.7 ± 0.16	< 0.1
	Longitudinal	1.4 ± 0.09	0.1 ± 0.06	< 0.1
Area fraction of the Ti_2Ni phase (%)		13.8 ± 2.41	17.0 ± 2.22	11.9 ± 1.22

The SPS temperature influences the phase composition and also mechanical properties. The effect of sintering temperature was investigated in the previous paper [29]. The area fraction of the undesirable Ti_2Ni and Ni_3Ti phases increased with an increasing sintering temperature. The high amount of the Ti_2Ni phase was formed along the boundaries of the sintered particles. It is necessary to point out that, in the previous paper, pulse current flow through the sample (another SPS device) was used while, in this paper, the regime of direct current flow through the sample was applied and it is the reason for different results. In this case, the lower amounts of the Ti_2Ni phase were measured generally and a growing trend with SPS temperature was not observed. The area fraction of the Ti_2Ni phase increased slightly by SPS consolidation at 900 °C and 1000 °C, but, after SPS sintering at 1100 °C, the amount of the Ti_2Ni phase was reduced to approximately 12%, which means a comparable value to result in samples prepared by the SHS method [29] (see Table 1).

At all SPS temperatures, the SEM observation was performed. Improving the fusion of grain boundaries was confirmed with increasing sintering temperature. The NiTi phase matrix with the fine Ni-rich precipitates in all samples was commonly found within the Ti_2Ni and Ni_3Ti phases. The microstructures after SPS are shown in Figure 3. In Table 2, there are summarized chemical compositions of individual areas observed by SEM. A good agreement in chemical compositions of the NiTi and Ti_2Ni phases to the binary Ni-Ti phase diagram was found out. The chemical composition of the area labelled 3 is close to the Ni_3Ti phase. Chemical composition of areas 7 and 10 is approaching the chemical composition of the Ni_4Ti_3 phase.

A more detailed observation of the microstructure was performed using a transmission electron microscope (TEM). The main goal of this experiment is based on investigating the fine needle-like particles in the NiTi matrix. Figure 4 shows TEM micrographs of sample SPS-ed at the temperature of 900 °C. The NiTi and Ti_2Ni phase were observed commonly with the Ni_4Ti_3 phase (determined by electron diffraction) in the NiTi matrix.

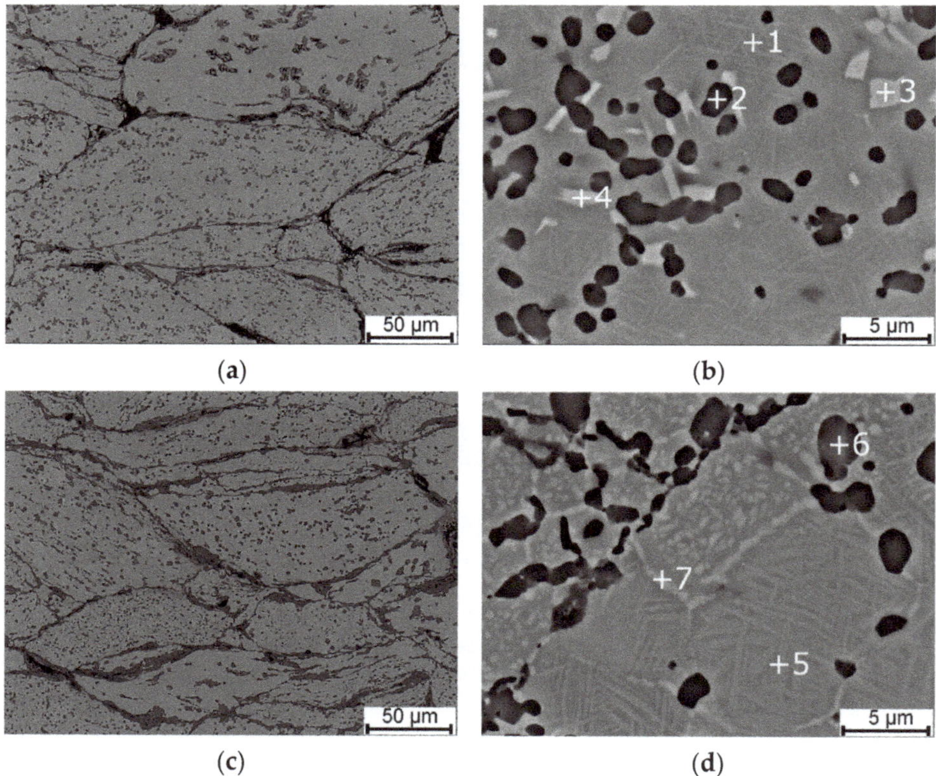

Figure 3. *Cont.*

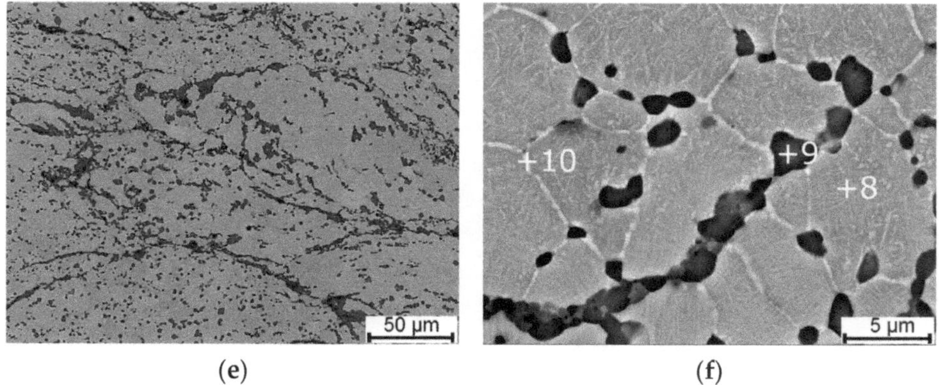

(e) (f)

Figure 3. SEM micrographs of SPS-ed samples at various temperatures: (**a**) 900 °C, (**b**) 900 °C-detail of Ni-rich precipitates, (**c**) 1000 °C, (**d**) 1000 °C-detail of Ni-rich precipitates, (**e**) 1100 °C, and (**f**) 1100 °C-detail of Ni-rich precipitates and formed the Ni_3Ti phase.

Table 2. Chemical composition of individual areas measured by EDS analysis.

Area	Ni (wt.%)	Ti (wt.%)
1	54.4	45.6
2	38.0	62.0
3	72.4	27.6
4	68.6	31.4
5	55.5	44.5
6	38.0	62.0
7	58.7	41.3
8	55.2	44.8
9	37.6	62.4
10	59.1	40.9

(a) (b)

Figure 4. *Cont.*

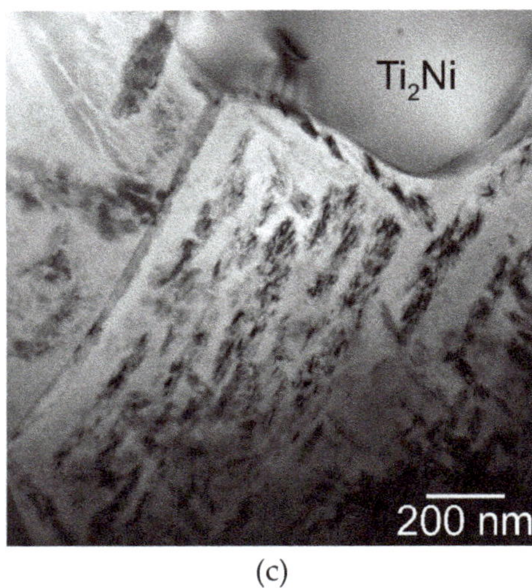

(c)

Figure 4. Ni_4Ti_3 phase in the matrix: (**a**) bright-field micrograph, g = 10-1, close to the [4,1,4] zone axis, (**b**) corresponding dark-field micrograph using Ni_4Ti_3 spot marked by a circle in the diffraction pattern in the inset, and (**c**) bright filed micrograph showing dark Ni_4Ti_3 particles in the matrix adjacent to a Ti_2Ni particle.

The effect of heat treatment on the microstructure and the phase composition was investigated at samples SPS-ed at the temperature of 900 °C. The phase compositions of SPS-ed samples were verified by XRD analysis. The diffraction lines were very similar through all SPS consolidation temperatures. The phase composition of SPS-ed samples consists of the NiTi phases (Cubic, Pm-3m), Ti_2Ni phase (Cubic, Fd-3 m), Ni_3Ti phase (Hexagonal, P63/mmc), and the Ni_4Ti_3 phase (Rhomboedral, R-3). XRD patterns are displayed. After heat treatment in the temperature range of 600–900 °C with a 1-h duration, there were no observed changes in the phase compositions of the samples (see Figure 5). This fact can seem to be strange in comparison with other studies. However, it is necessary to consider the initial state in individual studies (mostly after homogenization annealing, e.g., [17,27]). The initial state of these samples is after SPS consolidation at 900, 1000, and 1100 °C. The very high heating rate (approximately 300 °C/min) and a very high cooling rate (as shown in Figure 1) was applied during the SPS process, whereas the initial state of the Ni-Ti alloys is after homogenization or solution annealing for tens of minutes or several hours at temperatures at around 1000 °C [17,27,30]. Moreover, the transformations in the microstructure and phase composition occur in a bigger extent after heat treatment with longer duration since it is clear in this study [17]. The crystallite sizes of the NiTi phase after spark plasma sintering, which was determined by the means of the Sherrer's formula, range from 20 to 47 nm. The crystallite sizes increased after heat treatment and the highest values of the crystallite sizes were determined at samples heat-treated at a temperature of 700 °C. The significant increase of the crystallite sizes of the NiTi phase is related likely to the recrystallization process, grain growth, or the order-disorder transformation around the temperature of 600–700 °C reported in References [31,32]. The values of the crystallite sizes of the NiTi phase depending on the regime of the heat treatment are stated in Table 3.

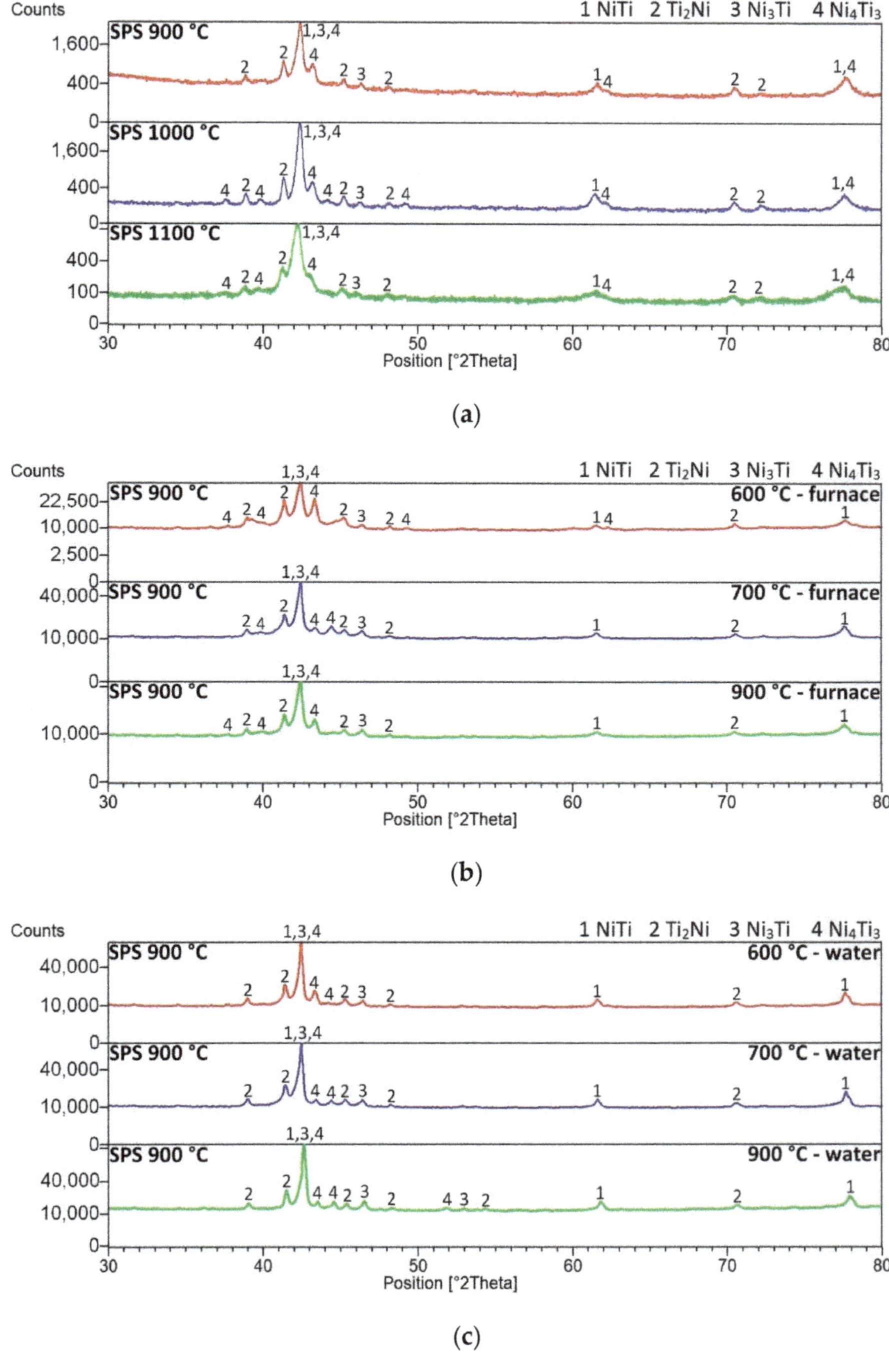

Figure 5. XRD patterns of the NiTi46 alloys: (**a**) spark plasma sintered at 900, 1000, and 1100 °C, (**b**) spark plasma sintered at 900 °C, heat-treated at 600–900 °C and slowly cooled, (**c**) spark plasma sintered at 900 °C, heat-treated at 600–900 °C, and cooled in water.

Table 3. Crystallite sizes of the NiTi phase depending on the regime of heat treatment.

Sample	Crystallite Size (nm)
SPS 900 °C	47
SPS 1000 °C	34
SPS 1100 °C	20
SPS 900 °C-HT 600 °C-furnace	25
SPS 900 °C-HT 700 °C-furnace	122
SPS 900 °C-HT 900 °C-furnace	28
SPS 900 °C-HT 600 °C-water	84
SPS 900 °C-HT 700 °C-water	129
SPS 900 °C-HT 900 °C-water	57

From the point of view of the microstructure, the particles of the Ti_2Ni, Ni_3Ti phases and Ni_4Ti_3 needles in the NiTi matrix were observed in Figure 6. A higher amount of the Ni_3Ti phase was formed in the sample heat-treated at 700 °C and cooled in water than in the samples with slow cooling in the closed furnace after heat treatment. The area fraction of the Ti_2Ni phase after heat treatment was almost invariable with values between 13% to 16%. Detailed observation of the Ni_4Ti_3 and other phases after heat treatment at 600 °C and 700 °C was carried out by TEM again (see Figure 7).

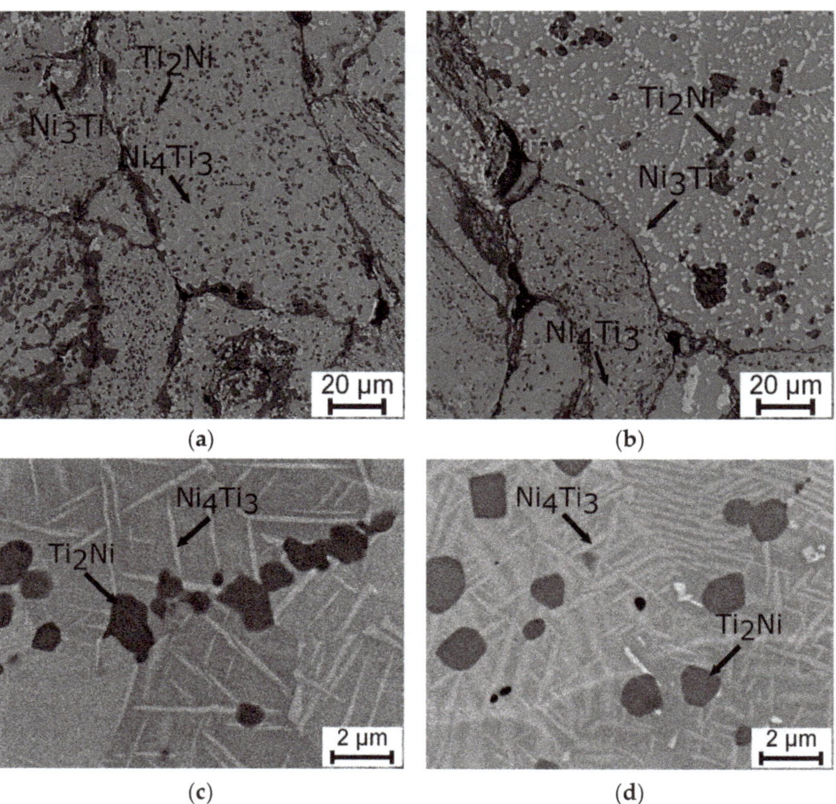

Figure 6. SEM microstructures of samples SPS-ed at 900 °C and heat-treated: (**a**,**c**) at 700 °C followed with slow cooling in the closed furnace, (**b**) at 700 °C and cooled in water, and (**d**) at 600 °C followed with slow cooling in the closed furnace.

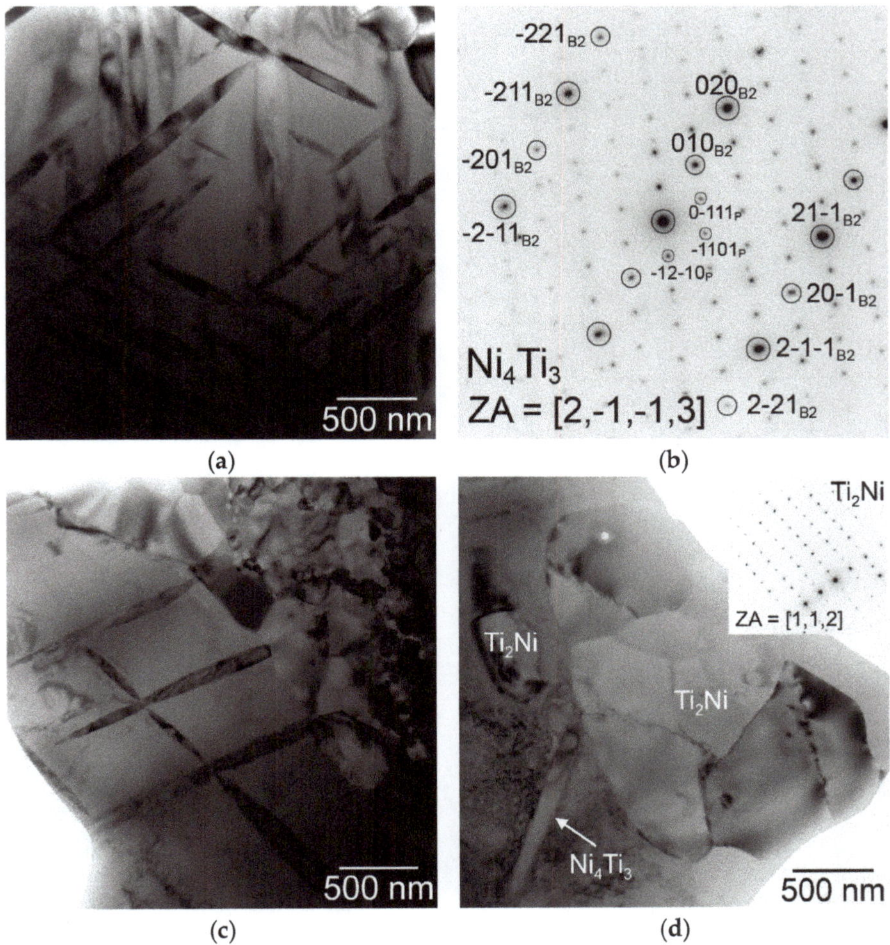

Figure 7. TEM observation of samples heat-treated at 600 °C (**a**,**b**) and at 700 °C (**c**,**d**) and slow cooled: (**a**) Ni_4Ti_3 precipitates with a typical lenticular cross-section, (**b**) corresponding diffraction pattern in the zone axis [2,-1-1,3] of the precipitate, which is coincident with the zone axis [1,0,2] of the B2 NiTi matrix, (**c**) Ni_4Ti_3 particles in the NiTi matrix adjacent to very fine grains of the Ti_2Ni phase, (**d**) Ni_4Ti_3 particle adjacent to a coarser grain of the Ti_2Ni phase. The inset shows a diffraction pattern of the Ti_2Ni phase in the [1,1,2] zone axis.

The phase transformation in the NiTi phase between austenite and the martensite structure was studied using differential scanning calorimetry. The straight lines were obtained for the samples as-SPS sintered at all temperatures and any phase transformation occurs in these samples. The change in the phase transformation behaviour was brought by heat treatment of the samples. When the samples were heat-treated, the temperature of heat treatment and way of cooling are important. Fast cooling in water did not cause the recovery of the phase transformation. Therefore, the microstructure observation and mechanical property investigation are focused on the samples cooled slowly in the closed furnace. The peaks on DSC curves were formed only at samples after heat treatment at temperatures of 600–700 °C, which were slow cooled in the closed furnace. The heating and cooling DSC curves of samples SPS-ed at 900 °C are shown in Figure 8.

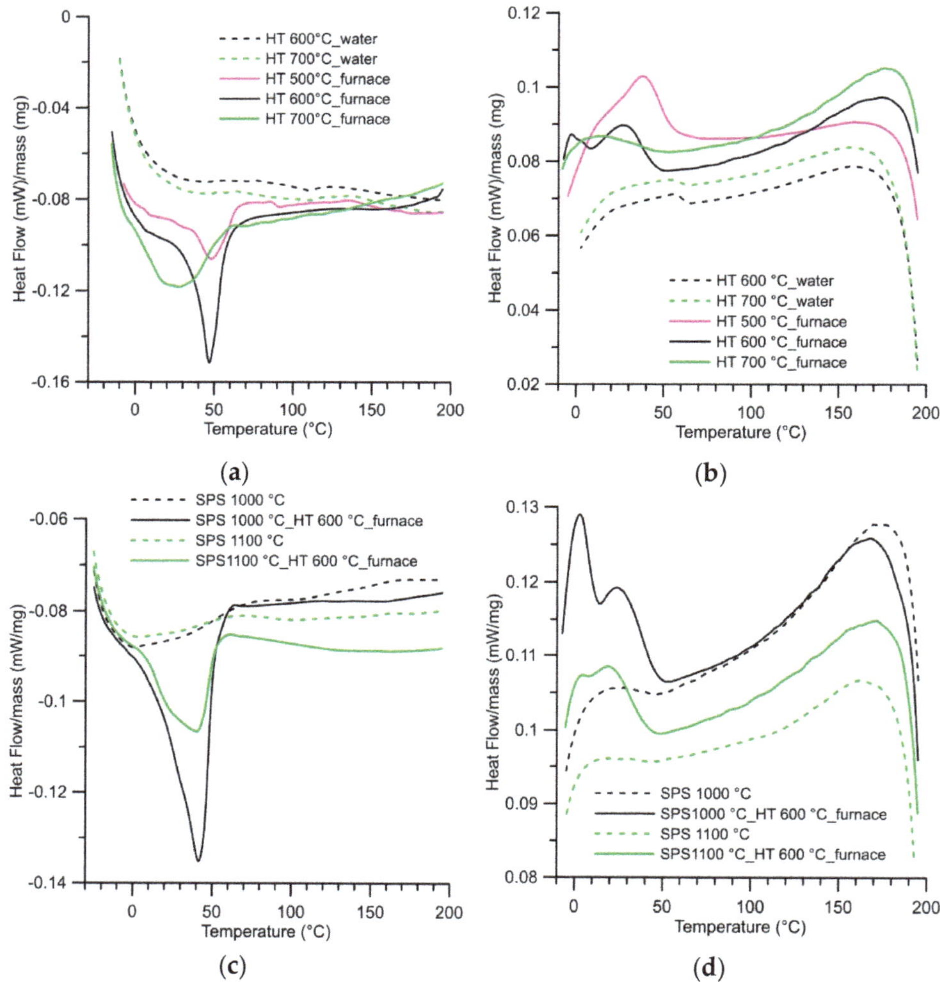

Figure 8. DSC heating and cooling curves: (**a**) Heating curves of samples SPS-ed at 900 °C and heat-treated between 500 and 700 °C. (**b**) Cooling curves of samples SPS-ed at 900 °C and heat-treated between 500 and 700 °C. (**c**) Heating curves of samples SPS-ed at 1000 and 1100 °C and SPS-ed at 1000 and 1100 °C with heat treatment at 600 °C. (**d**) Cooling curves of samples SPS-ed at 1000 and 1100 °C SPS-ed at 1000 and 1100 °C with heat treatment at 600 °C.

3.2. Mechanical Properties

The increase of hardness with the increasing temperature of SPS is similar to the previous results [29]. However, an increase of hardness was assigned to the rising value of the Ti_2Ni phase. It is contrary to the current study and to the decrease of the Ti_2Ni phase amount after SPS at 1100 °C. There are two possible reasons. Firstly, the hardness increases with the quality of powder sintering at higher SPS temperature, which is visible in Figure 3. Secondly, the precipitation process of Ni-rich phases occurs during the SPS process while heat treatment at the temperatures of 900–1000 °C lead to improved hardness [17]. Values of hardness as SPS-ed samples are stated in Table 4.

Table 4. Summary of hardness and compressive stress-strain test of samples after spark plasma sintering at various temperatures. Longitudinal direction is parallel to the direction of compressive force by the SPS process.

SPS Temperature		900 °C	1000 °C	1100 °C
Hardness (HV 10)/Std. dev. (±)		562/25	596/20	624/23
Longitudinal	UCS (MPa)	1903	2116	2315
	Agt (%)	8.7	8.7	8.7
Perpendicular	UCS (MPa)	1953	2212	2243
	Agt (%)	7.4	9.4	8.6

The stress-strain behavior was analyzed in tandem with hardness and the same influence (increasing values of UCS) with an increasing temperature of the SPS process was observed. As visible in Figure 9, the samples after SPS reach high values of ultimate compressive strength (UCS) 1900–2300 MPa, but there are not the areas of plastic deformation on stress-strain curves and the samples fail in a brittle manner at a maximum load perpendicularly as well as in the longitudinal direction. The values of elongation at maximum force (Agt) were between 7.4–9.4% for all SPS temperatures and both directions. In the case of the compression test, the lower porosity of samples SPS-ed at higher temperatures is likely connected with increasing values of UCS.

After heat treatment with cooling in the closed furnace, the decrease of hardness was found and the lowest value was measured after heat treatment at the temperature of 700 °C. Currently, with the decrease of hardness, increased ductility and plasticity of samples was measured by a compressive test (Table 5). The evolution of hardness depending on the temperature of heat treatment is shown in Figure 9 for samples SPS sintered at 900 °C. In case of samples SPS sintered, a 1000 °C and 1100 °C, hardness after heat treatment at 600 °C dropped to values of 504 HV 10 and 509 HV 10. By the compressive test after heat treatment, the difference between annealing temperatures of 600 °C and 700 °C was observed. In case of annealing temperature of 600 °C, the similar values of Agt were measured at longitudinal and perpendicular directions between 9.5% and 10.5%, whereas, at an annealing temperature of 700 °C, the increase of Agt was found and the difference between longitudinal (16.3–19.1%) and perpendicular (9.3–13.6%) direction grew up. The UCS of the samples SPS-ed at 1000 °C and 1100 °C increased after heat treatment about 100 MPa in comparison with the state after SPS. Generally, the values of UCS and Agt increased with an increasing temperature of the SPS process.

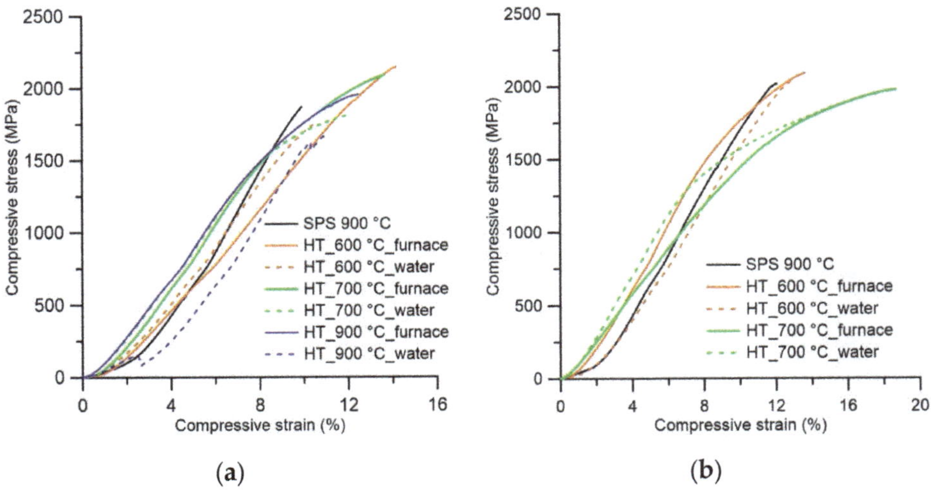

Figure 9. *Cont.*

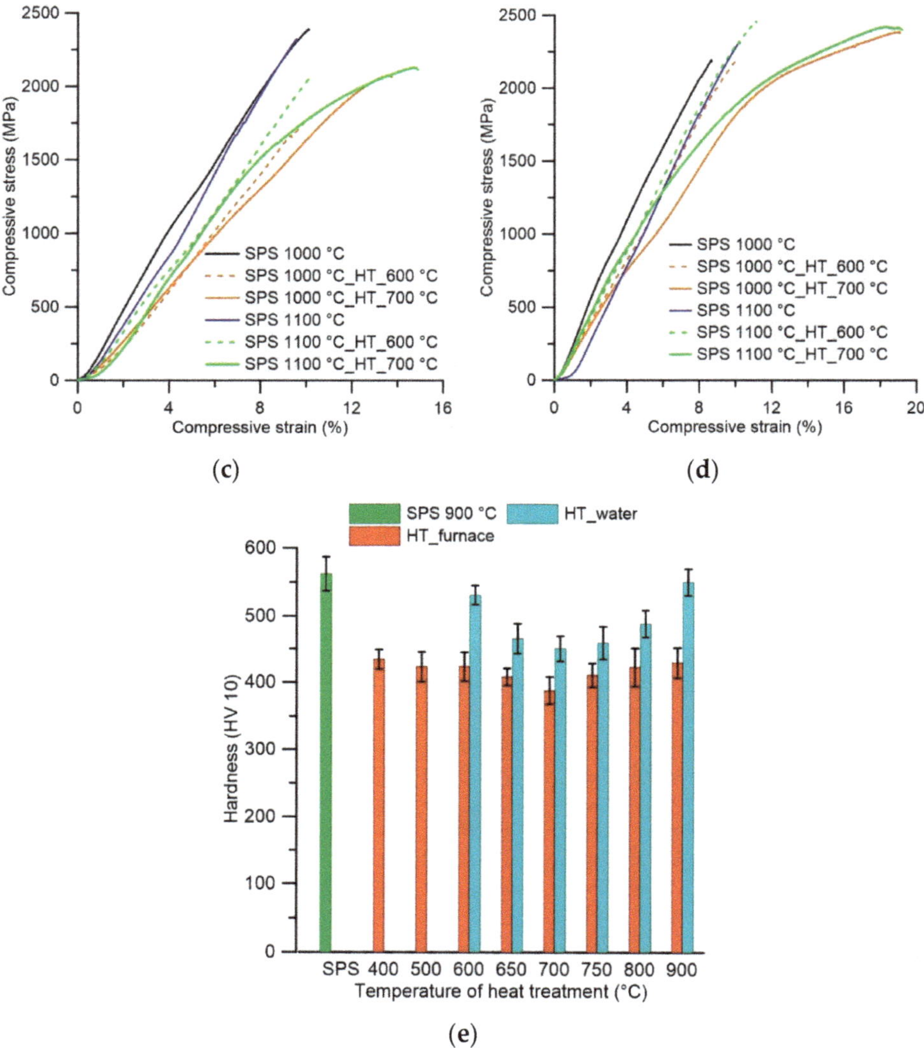

Figure 9. Compressive stress-strain curves and evolution of hardness: (**a**) Stress-strain curves of samples SPS-ed at 900 °C and heat-treated-perpendicular direction. (**b**) Stress-strain curves of samples SPS-ed at 900 °C and heat-treated-longitudinal direction. Stress-strain curves of samples SPS-ed at 1000 °C, 1100 °C, and heat-treated-perpendicular (**c**) and longitudinal (**d**) direction. (**e**) Evolution of hardness during heat treatment of sample SPS-ed at 900 °C.

Table 5. Summary of mechanical properties of heat-treated samples consolidated by the SPS process.

Sample	Heat Treatment Regime	Hardness (HV 10)/Std. dev. (±)	Direction	UCS (MPa)	Agt (%)
SPS 900 °C	600 °C-furnace	423/21	Perpendicular	2163	10.5
			Longitudinal	2089	11.6
SPS 900 °C	600 °C-water	531/14	Perpendicular	1508	7.6
			Longitudinal	1994	9.9
SPS 900 °C	700 °C-furnace	388/20	Perpendicular	1430	9.3
			Longitudinal	1907	16.3
SPS 900 °C	700 °C-water	450/19	Perpendicular	1566	9.3
			Longitudinal	1878	13.7
SPS 900 °C	900 °C-furnace	429/22	Perpendicular	1957	11.1
SPS 900 °C	900 °C-water	550/20	Perpendicular	1670	-
SPS 1000 °C	600 °C-furnace	504/21	Perpendicular	2089	9.5
			Longitudinal	2235	9.2
SPS 1000 °C	700 °C-furnace	444/14	Perpendicular	2099	12.5
			Longitudinal	2355	18.5
SPS 1100 °C	600 °C-furnace	509/22	Perpendicular	2295	10.0
			Longitudinal	2488	10.6
SPS 1100 °C	700 °C-furnace	448/10	Perpendicular	2163	13.6
			Longitudinal	2465	19.1

The explanation of the rapid decrease of hardness could be caused due to removing the deformation strengthening from the milling of an SHS product. This type of decrease has to have the same scale for all samples. The dependence of hardness on the temperature of heat treatment is undeniable. Thus, the different processes and changes in microstructure must occur during heat treatment at various temperatures. However, these changes were not observed in phase composition and microstructure due to the short time of annealing during heat treatment. In this study [17], the longer time heat treatment was applied and the following changes in the microstructure occurred. The high UCS, hardness, and low ductility of samples were attributed to the precipitates of Ni_3Ti_4 and Ni_3Ti_2 phases, while, when the microstructure contains the Ni_3Ti phase or combination of the Ni_3Ti and Ni_3Ti_2 phases (corresponding to heat treatment between 600 and 800 °C while a longer time must be applied at a temperature of 600 °C), high mechanical properties and good ductility were obtained. Both studies [17,24] show similar evolution of the hardness values during heat treatment when compared to our results.

4. Conclusions

The fabrication process composed of Self-propagating High-temperature Synthesis (SHS) and Spark Plasma Sintering (SPS) was chosen to obtain a completely dense material. The highest temperature of SPS (1100 °C) led to the highest values of ultimate compressive strength without the formation of an excessive number of undesirable phases. The following heat treatment is necessary to obtain a material with a good combination of strength and ductility. The heat treatment leads to the disappearance of the deformation strengthening coming from milling in the vibration mill and to recover the phase transformation between the austenite and martensite structure of the NiTi phase, which was detected by differential scanning calorimetry (DSC). The heat treatment near the temperatures of 600–700 °C with slow cooling is recommended to obtain good ductility, strength, and probable shape memory properties.

Author Contributions: P.S. and P.N. designed the experiment and evaluated the phase composition. P.S. and A.Š. provided sample preparation by the SHS reaction, measurement of mechanical properties, and DSC analysis. F.P. consolidated samples by the SPS method. P.S. and J.D. analyzed the microstructure by LM and SEM. M.K. and P.H. observed the microstructure by SEM and TEM. P.S. wrote the paper. P.N. and M.K. reviewed and edited the paper.

Funding: European Regional Development Fund (projects Pre-Application Research of Functionally Graduated Materials by Additive Technologies, No. CZ.02.1.01/0.0/0.0/17_048/0007350 and Nanomaterials Centre for Advanced

Applications, No. CZ.02.1.01/0.0/0.0/15_003/0000485) and specific university research (MSMT No. 21-SVV/2019) supported the research.

Conflicts of Interest: The authors declare no conflict of interest.

References

1. Van Humbeeck, J. Shape Memory Alloys: A Material and a Technology. *Adv. Eng. Mater.* **2001**, *3*, 837–850. [CrossRef]
2. Khalil-Allafi, J.; Dlouhy, A.; Eggeler, G. Ni_4Ti_3-precipitation during aging of NiTi shape memory alloys and its influence on martensitic phase transformations. *Acta Mater.* **2002**, *50*, 4255–4274. [CrossRef]
3. Tang, W.; Sundman, B.; Sandström, R.; Qiu, C. New modelling of the B2 phase and its associated martensitic transformation in the Ti–Ni system. *Acta Mater.* **1999**, *47*, 3457–3468. [CrossRef]
4. Nayan, N.; Saikrishna, C.N.; Ramaiah, K.V.; Bhaumik, S.K.; Nair, K.S.; Mittal, M.C. Vacuum induction melting of NiTi shape memory alloys in graphite crucible. *Mater. Sci. Eng. A* **2007**, *465*, 44–48. [CrossRef]
5. Zhang, Z.; Frenzel, J.; Neuking, K.; Eggeler, G. On the reaction between NiTi melts and crucible graphite during vacuum induction melting of NiTi shape memory alloys. *Acta Mater.* **2005**, *53*, 3971–3985. [CrossRef]
6. Chuvildeev, V.N.; Panov, D.V.; Boldin, M.S.; Nokhrin, A.V.; Blagoveshchensky, Y.V.; Sakharov, N.V.; Shotin, S.V.; Kotkov, D.N. Structure and properties of advanced materials obtained by Spark Plasma Sintering. *Acta Astronaut.* **2015**, *109*, 172–176. [CrossRef]
7. Velmurugan, C.; Senthilkumar, V.; Biswas, K.; Yadav, S. Densification and microstructural evolution of spark plasma sintered NiTi shape memory alloy. *Adv. Powder Technol.* **2018**, *29*, 2456–2462. [CrossRef]
8. Průša, F.; Šesták, J.; Školáková, A.; Novák, P.; Haušild, P.; Karlík, M.; Minárik, P.; Kopeček, J.; Laufek, F. Application of SPS consolidation and its influence on the properties of the FeAl20Si20 alloys prepared by mechanical alloying. *Mater. Sci. Eng. A* **2019**, *761*, 138020. [CrossRef]
9. Novák, P.; Vanka, T.; Nová, K.; Stoulil, J.; Pruša, F.; Kopeček, J.; Haušild, P.; Laufek, F. Structure and properties of Fe-Al-Si alloy prepared by mechanical alloying. *Materials* **2019**, *12*, 2463. [CrossRef] [PubMed]
10. Prusa, F.; Vojtech, D.; Kucera, V.; Bernatikova, A. Preparation of Ultrafine-Grained and Nano-crystalline Materials by Mechanical Alloying and Spark Plasma Sintering. *Chem. Listy* **2017**, *111*, 314–321.
11. Novák, P.; Kubatík, T.; Vystrčil, J.; Hendrych, R.; Kříž, J.; Mlynár, J.; Vojtěch, D. Powder metallurgy preparation of Al–Cu–Fe quasicrystals using mechanical alloying and Spark Plasma Sintering. *Intermetallics* **2014**, *52*, 131–137. [CrossRef]
12. Marek, I.; Vojtěch, D.; Michalcová, A.; Kubatík, T.F. High-strength bulk nano-crystalline silver prepared by selective leaching combined with spark plasma sintering. *Mater. Sci. Eng. A* **2015**, *627*, 326–332. [CrossRef]
13. Novák, P.; Mejzlíková, L.; Michalcová, A.; Čapek, J.; Beran, P.; Vojtěch, D. Effect of SHS conditions on microstructure of NiTi shape memory alloy. *Intermetallics* **2013**, *42*, 85–91. [CrossRef]
14. Novák, P.; Školáková, A.; Pignol, D.; Průša, F.; Salvetr, P.; Kubatík, T.F.; Perriere, L.; Karlík, M. Finding the energy source for self-propagating high-temperature synthesis production of NiTi shape memory alloy. *Mater. Chem. Phys.* **2016**, *181*, 295–300. [CrossRef]
15. Zhang, L.; Zhang, Y.Q.; Jiang, Y.H.; Zhou, R. Superelastic behaviors of biomedical porous NiTi alloy with high porosity and large pore size prepared by spark plasma sintering. *J. Alloys Compd.* **2015**, *644*, 513–522. [CrossRef]
16. Zhao, Y.; Taya, M.; Kang, Y.; Kawasaki, A. Compression behavior of porous NiTi shape memory alloy. *Acta Mater.* **2005**, *53*, 337–343. [CrossRef]
17. Adharapurapu, R.R.; Jiang, F.; Vecchio, K.S. Aging effects on hardness and dynamic compressive behavior of Ti–55Ni (at.%) alloy. *Mater. Sci. Eng. A* **2010**, *527*, 1665–1676. [CrossRef]
18. Khamei, A.A.; Dehghani, K. A study on the mechanical behavior and microstructural evolution of Ni60wt.%–Ti40wt.% (60Nitinol) intermetallic compound during hot deformation. *Mater. Chem. Phys.* **2010**, *123*, 269–277. [CrossRef]
19. Safdel, A.; Zarei-Hanzaki, A.; Shamsolhodaei, A.; Krooß, P.; Niendorf, T. Room temperature superelastic responses of NiTi alloy treated by two distinct thermomechanical processing schemes. *Mater. Sci. Eng. A* **2017**, *684*, 303–311. [CrossRef]

20. Sun, B.; Fu, M.W.; Lin, J.; Ning, Y.Q. Effect of low-temperature aging treatment on thermally- and stress-induced phase transformations of nanocrystalline and coarse-grained NiTi wires. *Mater. Des.* **2017**, *131*, 49–59. [CrossRef]
21. Kocich, R.; Szurman, I.; Kursa, M.; Fiala, J. Investigation of influence of preparation and heat treatment on deformation behaviour of the alloy NiTi after ECAE. *Mater. Sci. Eng. A* **2009**, *512*, 100–104. [CrossRef]
22. Karaca, H.E.; Kaya, I.; Tobe, H.; Basaran, B.; Nagasako, M.; Kainuma, R.; Chumlyakov, Y. Shape memory behavior of high strength Ni54Ti46 alloys. *Mater. Sci. Eng. A* **2013**, *580*, 66–70. [CrossRef]
23. Khoo, Z.X.; An, J.; Chua, C.K.; Shen, Y.F.; Kuo, C.N.; Liu, Y. Effect of heat treatment on repetitively scanned SLM NiTi shape memory alloy. *Materials* **2019**, *12*, 77. [CrossRef] [PubMed]
24. Saedi, S.; Turabi, A.S.; Taheri Andani, M.; Haberland, C.; Karaca, H.; Elahinia, M. The influence of heat treatment on the thermomechanical response of Ni-rich NiTi alloys manufactured by selective laser melting. *J. Alloys Compd.* **2016**, *677*, 204–210. [CrossRef]
25. Otsuka, K.; Ren, X. Physical metallurgy of Ti–Ni-based shape memory alloys. *Prog. Mater. Sci.* **2005**, *50*, 511–678. [CrossRef]
26. Chen, H.; Zheng, L.J.; Zhang, F.X.; Zhang, H. Thermal stability and hardening behavior in superelastic Ni-rich Nitinol alloys with Al addition. *Mater. Sci. Eng. A* **2017**, *708*, 514–522. [CrossRef]
27. Nishida, M.; Wayman, C.M.; Honma, T. Precipitation processes in near-equiatomic TiNi shape memory alloys. *Metall. Trans. A* **1986**, *17*, 1505–1515. [CrossRef]
28. Salvetr, P.; Novák, P.; Moravec, H. Ni-Ti alloys produced by powder metallurgy. *Manuf. Technol.* **2015**, *15*, 689–694.
29. Salvetr, P.; Kubatík, T.F.; Pignol, D.; Novák, P. Fabrication of Ni-Ti Alloy by Self-Propagating High-Temperature Synthesis and Spark Plasma Sintering Technique. *Metall. Mater. Trans. B* **2017**, *48*, 772–778. [CrossRef]
30. Stroz, D.; Kwarciak, J.; Morawiec, H. Effect of ageing on martensitic transformation in NiTi shape memory alloy. *J. Mater. Sci.* **1988**, *23*, 4127–4131. [CrossRef]
31. Sadrnezhaad, S.K.; Mirabolghasemi, S.H. Optimum temperature for recovery and recrystallization of 52Ni48Ti shape memory alloy. *Mater. Des.* **2007**, *28*, 1945–1948. [CrossRef]
32. Xu, Y.; Shimizu, S.; Suzuki, Y.; Otsuka, K.; Ueki, T.; Mitose, K. Recovery and recrystallization processes in Ti-Pd-Ni high-temperature shape memory alloys. *Acta Mater.* **1997**, *45*, 1503–1511. [CrossRef]

Article

Application of a $Dy_3Co_{0.6}Cu_{0.4}H_x$ Addition for Controlling the Microstructure and Magnetic Properties of Sintered Nd-Fe-B Magnets

Katerina Skotnicova [1,*], Pavel A. Prokofev [2,3], Natalia B. Kolchugina [3], Gennady S. Burkhanov [3], Alexander A. Lukin [2], Yurii S. Koshkid'ko [3,4], Tomas Cegan [1], Henryk Drulis [4], Tatyana Romanova [4] and Nikolay A. Dormidontov [3]

1 VSB–Technical University of Ostrava, Faculty of Materials Science and Technology, 70800 Ostrava, Czech Republic; tomas.cegan@vsb.cz

2 Joint Stock Company «Spetsmagnit», Moscow 127238, Russia; pav3387@yandex.ru (P.A.P.); lukinaalukin@rambler.ru (A.A.L.)

3 Baikov Institute of Metallurgy and Materials Science, Russian Academy of Sciences, Moscow 119334, Russia; natalik014@yandex.ru (N.B.K.); genburkh@imet.ac.ru (G.S.B.); yurec@mail.ru (Y.S.K.); ontip@mail.ru (N.A.D.)

4 Institute of Low Temperature and Structure Research, Polish Academy of Sciences, 50–422 Wroclaw, Poland; HDrulis@int.pan.pl (H.D.); T.Romanova@int.pan.wroc.pl (T.R.)

* Correspondence: katerina.skotnicova@vsb.cz; Tel.: +420-596-993-404

Received: 14 November 2019; Accepted: 15 December 2019; Published: 17 December 2019

Abstract: The focus of new technologies on the formation of inhomogeneous distributions of heavy rare-earth metals (REMs) in hard magnetic Nd–Fe–B materials is of scientific importance to increase their functional properties, along with preserving existing sources of heavy REMs. This paper focused on the coercivity enhancement of $Nd_2Fe_{14}B$-based magnets by optimizing the microstructure, which includes the processes of grain boundary structuring via the application of a $Dy_3Co_{0.6}Cu_{0.4}H_x$ alloy added to the initial Nd–Fe–B-based powder mixtures in the course of their mechanical activation. We have studied the role of alloying elements in the formation of phase composition, microstructure, the fine structure of grains, and the hysteretic properties of hard magnetic $Nd(R)_2Fe_{14}B$-based materials. It was shown that the Dy introduction via the two-component blending process (the hydrogenated $Dy_3Co_{0.6}Cu_{0.4}$ compound is added to a powder mixture) resulted in the formation of the core-shell structure of 2–14–1 phase grains. The efficient improvement of the coercivity of Nd(RE)–Fe–B magnets, with a slight sacrifice of remanence, was demonstrated.

Keywords: grain boundary diffusion; Nd–Fe–B magnets; hydrogenation; microstructure; magnetic properties

1. Introduction

Researchers have made many attempts to reduce the heavy rare-earth (RE) consumption Nd–Fe–B sintered magnets with high-coercivity. Some progress has been achieved using Dy and/or Tb in various forms to realize approaches named grain boundary diffusion (GBD) [1–3] and grain boundary structuring (GBS) [4–8]. The application of binary mixtures allows one to improve the structure of the boundary phases and grain boundaries of the main magnetic phase and to realize the diffusion of a required component of the alloy directly through the boundaries. It has been demonstrated that by controlling the process time and temperature of GBD processes, the coercivity of the magnet can be greatly enhanced, without sacrificing the remanence.

It was shown in our previous studies that hydrogenated Tb and Dy additions allowed us to enhance the coercivity with a slight decrease in the remanence [9] and increase the stability of the magnet properties during annealing at the low-temperature [10], respectively.

The grain boundary restructuring, with rare-earth-rich low-melting compounds added to low-alloyed Nd–Fe–B-based compositions in the course of technological processing, was realized when using $(Pr,Nd)_6Fe_{13}Cu$ [4], $Dy_{32.5}Fe_{62}Cu_{5.5}$ [5], $Dy_{69}Ni_{31}$ [6], $Dy_{88}Mn_{12}$ (wt.%) [11], $Pr_{34.4}Co_{65.6}$ (wt.%) [12], and $Dy_{82.3}Co_{17.7}$ (wt.%) [13], which is a low-melting eutectic composition. It was shown that the intrinsic coercivity evidently increased when using $Dy_{82.3}Co_{17.7}$ and the maximum intrinsic coercivity was achieved when its content was 2 wt.%. At the same time, the remanence and maximum-energy product decreased slightly as the $Dy_{82.3}Co_{17.7}$ content increased. By adding a small amount of $Dy_{82.3}Co_{17.7}$, the coercivity improved greatly, and the irreversible loss decreased sharply. The increase in the Curie temperature of the magnets suggests that Co atoms have been incorporated into the 2:14:1 main phase. A well-developed a core–shell structure is formed in these magnets.

The experiments with REM-M-H compounds (rare earth metal-transition metal(s)-hydrogen), which are added at the stage of mechanical milling and alloying, were performed to realize the optimum microstructure, nano-heterogeneous distribution of heavy REMs (Dy or Tb) within a grain, and economically alloyed composition of magnets, which assumes, in particular, the distribution of heavy REMs within the near-grain boundary areas. Such a heavy-REM distribution allows us to (1) locally increase the coercive force and decrease the probability of the formation of reverse domains at grain boundaries; (2) limit the substitution of heavy REM for neodymium in the matrix phase and, thus, decrease the probability of decreasing magnetization and remanence; and (3) decrease the amount of heavy REMs, which is required to reach the given increase in the coercive force. The latter circumstance determines the possibility of the development of physico-chemical and technological foundations of resource-saving technology, the possibility of decreasing the material costs and prices of products manufactured from the new alloys, and the possibility of substantially widening the functionality of the materials.

Thus, by applying compositions with a heavy rare-earth metal, the outer region of the $Nd_2Fe_{14}B$ matrix grains was enriched during the sintering process and substitutes for Nd were used in the matrix grains to form the $(Nd,Dy)_2Fe_{14}B$ core–shell phase.

This paper focused on optimizing the microstructure of the near-stoichiometric $Nd_2Fe_{14}B$-based magnet, which included the grain boundary diffusion and grain boundary structuring processes via the application of a hydrogenated $Dy_3Co_{0.6}Cu_{0.4}H_x$ composition added to a powder mixture.

2. Experimental

The strip casting technique was used for the preparation of the base Nd-24.0, Pr-6.5, Dy-0.5, B-1.0, Al-0.2, Fe-balance alloy (wt.%). The strip-cast alloy was subsequently subjected to hydrogen decrepitation process, which was realized during heating to 270 °C in a hydrogen flow at a pressure of 0.1 MPa and holding at this temperature for 1 h.

The $Dy_3(Co_{1-x}Cu_x)$ alloy with $x = 0.4$ was produced by the arc melting of the starting components (distilled Dy of 99.9% purity, Co of ≥ 99.25% purity, and oxygen-free Cu of 99.95% purity) in an argon atmosphere using a water-cooled copper bottom and a non-consumable tungsten electrode. The ingot was homogenized at 600 °C for 90 h and subjected to hydrogenation under conditions used for the strip-casting alloy, namely, upon heating to 270 °C in a hydrogen flow at a pressure of 0.1 MPa and subsequent 1 h heating at this temperature (Regime 1 was used to manufacture the magnet), and upon heating to 700 °C in a high-purity hydrogen atmosphere and holding at this temperature for 1 h in a glass Sieverts-type apparatus (Regime 2 was used for investigations). In the case of heating at 700 °C, the hydrogenation up to the $Dy_3Co_{0.6}Cu_{0.4}H_x$ composition with $x = 8.26$ was realized. It is expected that such a hydrogen content accords with the complete hydrogenation of dysprosium to a dysprosium hydride.

The mixture of hydrogen-decrepitated strip-cast Nd(RE)–Fe–B alloy and the $Dy_3Co_{0.6}Cu_{0.4}H_x$ alloy (Regime 1) was milled for 40 min to an average particle size of 3 μm using a vibratory mill and isopropyl alcohol medium. After wet pressing of the pulp in a transverse magnetic field of 1500 kA/m, compacts were sintered at 1080 °C for 2 h and optimally heat treated (HT) at 500 °C for 2 h. Then, samples of the magnet were subjected to low-temperature heat treatment in the temperature range 400–900 °C, with subsequent quenching in N_2.

The phase composition of the $Dy_3Co_{0.6}Cu_{0.4}$ and $Dy_3Co_{0.6}Cu_{0.4}H_x$ ($x = 8.26$) alloys was investigated by X-ray diffraction (XRD) analysis using an Ultima IV (Rigaku») diffractometer (equipped with a "D/teX" detector, CuKα radiation) and a Philips X'Pert 1 diffractometer, respectively; the scanning step was 0.001°. X-ray diffraction patterns were processed, and the phase composition of the alloy was determined using PowderCell software. Data on the crystal structure type, lattice parameters, and the crystallographic positions of atoms in the Dy–Co, Dy–Cu, and H–Dy systems [14–16] were used to simulate theoretical XRD patterns.

An Quanta 450 FEG high-resolution field emission gun scanning electron microscope (FEI Company, Fremont, USA) equipped with an energy-dispersive spectroscopy (EDS, EDAX Inc., Mahwah, USA) microprobe was used to investigate the structure, chemical composition, and distribution of magnet components (X-ray mapping) of the addition and magnet sample. The mean particle size was evaluated by means of a MasterSizer 3000 laser diffraction particle size analyzer (Malvern Panalytical Ltd, Malvern, United Kingdom). The hysteretic properties of the magnet sample were measured at room temperature (RT) using an automatic hysteresis graph system MH-50 (Walker Scientific Inc., Worcester, USA). The differential thermal analysis (DTA) and thermogravimetric analysis were performed under an argon atmosphere with a heating/cooling rate of 30 °C/min using a STA 449 F3 Jupiter installation (Netzsch Holding, Selb, Germany).

3. Results and Discussion

3.1. X-Ray Diffraction Analysis

Figure 1 shows the X-ray diffraction pattern of the $Dy_3Co_{0.6}Cu_{0.4}$ alloy subjected to prolonged annealing in an argon atmosphere. The reflections belong to the main $Dy_3(Co,Cu)$ phase and the Dy(Cu,Co) phase based on DyCu [14,15]. The analysis of the crystal structures of the found compounds and theoretical XRD patterns constructed for the simulated structures allowed us to determine variations in the lattice parameters of the $Dy(Cu_{1-y}Co_y)$ and $Dy_3(Co_{1-x}Cu_x)$ phases alloyed with Co and Cu, respectively (see Table 1). As seen, the alloying of the binary compounds with Co and Cu did not change the crystal structure type of the compounds. In accordance with the binary phase diagrams [14,15], the phases present in the alloy are alloyed compositions of the binary compounds.

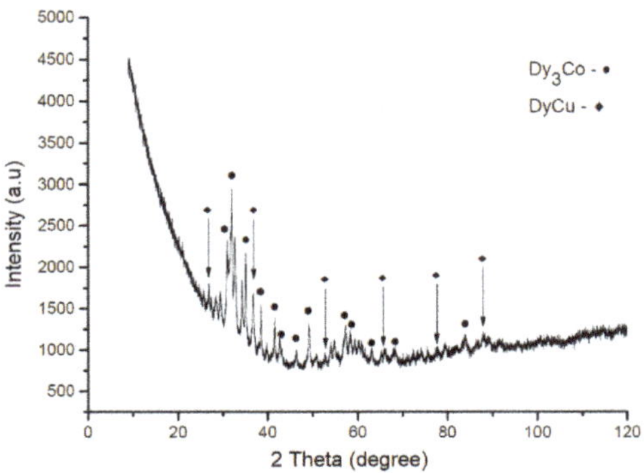

Figure 1. X-ray diffraction pattern of the $Dy_3Co_{0.6}Cu_{0.4}$ alloy.

Table 1. The crystal structure type and lattice parameters of the phases in the $Dy_3(Co_{0.6}Cu_{0.4})$ alloy.

Compound	Space Group	C	*a* (nm)	*b* (nm)	*c* (nm)	References
Dy_3Co	Pnma	Fe_3C	0.69650	0.93410	0.62330	[14]
$Dy_3(Co_{1-x}Cu_x)$	Pnma	Fe_3C	0.69331	0.93847	0.62564	This work
DyCu	$Pm\bar{3}\bar{m}$	CsCl	0.34610	0.34610	0.34610	[15]
$Dy(Cu_{1-y}Co_y)$	$Pm\bar{3}\bar{m}$	CsCl	0.34522	0.34522	0.34522	This work

The phase composition of the alloy was also confirmed by the EDS microanalysis, see Figure 2 and Table 2. The microstructure consisted of $Dy_3(Co_{1-x}Cu_x)$ ($x \sim 0.4$) dendrites (point 1 in Figure 2) and $Dy(Cu_{1-y}Co_y)$ + $Dy_3(Co_{0.6}Cu_{0.4})$ mixture (point 2 in Figure 2) found in the interdendritic regions. The composition of the $Dy(Cu_{1-y}Co_y)$ phase cannot be accurately determined by the EDS analysis because of its small size, since the surrounding matrix is analyzed along with this very small inclusion. However, the increased content of copper is evident in this mix area.

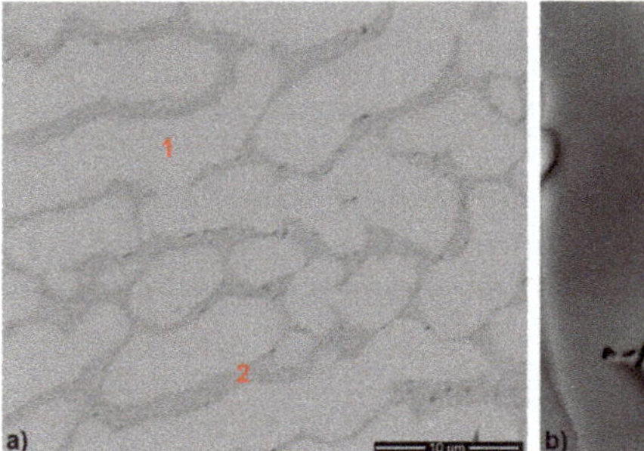

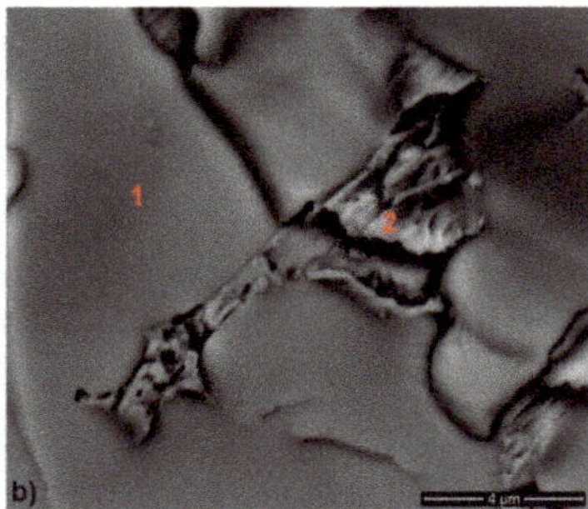

Figure 2. Scanning electron microscopy images of the microstructure of $Dy_3Co_{0.6}Cu_{0.4}$ alloy subjected to prolonged annealing in an argon atmosphere: (**a**) metallographic section, and (**b**) fracture surface.

Table 2. The chemical composition (at.%) of phases found in the microstructure of the $Dy_3Co_{0.6}Cu_{0.4}$ alloy (the average value from three analysis).

Element/phase	Dy	Co	Cu
Point 1—$Dy_3(Co_{0.6}Cu_{0.4})$	75.5	15.1	9.5
Point 2—$Dy(Co_{1-y}Cu_y)$+$Dy_3(Co_{0.6}Cu_{0.4})$	68.1	9.0	22.9

As is shown in Table 1, the substitution of Cu for Co in $Dy_3(Co_{1-x}Cu_x)$ (with regard to the solubility of Cu and Co in Dy_3Co and DyCu, respectively) changed the lattice parameters: the lattice parameters *b* and *c* increased as the radius of Cu atoms (0.128 nm) was higher than that of the Co atoms (0.125 nm), whereas the lattice parameter *a* decreased. This is likely to be due to the fact that copper atoms substitute for cobalt atoms only at certain sites.

We assumed that the solidification of the alloy occurs via the primary formation of the Dy_3Co-based phase by peritectic reaction; the DyCu-based compound is the secondary phase. According to the Co–Dy phase diagram, the solidification path may include the formation of the $Dy_{12}Co_7$-based phase by peritectic reaction.

3.2. Interaction of $Dy_3(Co,Cu)$ Alloy with Hydrogen

The saturation of the $Dy_3Co_{0.6}Cu_{0.4}$ alloy with hydrogen led to the embrittlement of the alloy (i.e., the powder material suitable for further introduction of the composition into the Nd–Fe–B magnetic alloy powder during cooperative milling was obtained). Figure 3a shows the X-ray diffraction analysis data for the $Dy_3Co_{0.6}Cu_{0.4}$ alloy subjected to hydrogenation (Regime 2). The hydrogenated composition contained DyH_2 [17] and DyH_3 [18] hydrides. Other reflections corresponded to the $Dy_3(Co,Cu)$ phase; it is likely that small quantities of the $Dy_3(Co,Cu)$ and $Dy(Cu,Co)$ phases did not react with hydrogen. After hydrogenation, copper and cobalt may be present in the form of a fine mixture.

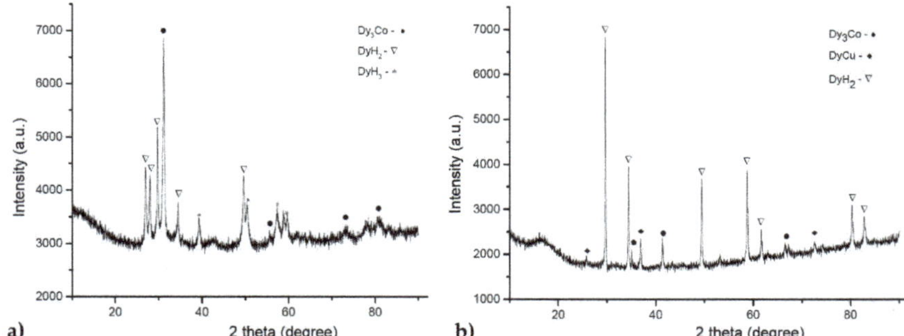

Figure 3. X-ray diffraction pattern of the $Dy_3(Co,Cu)$ alloy after saturation with hydrogen (Regime 2) (**a**) and after the thermal dehydrogenation process (DTA) (**b**).

Figure 3b shows the X-ray diffraction analysis data of the alloy $Dy_3Co_{0.6}Cu_{0.4}H_x$ subjected to thermal dehydrogenation (upon heating during DTA). The sample was heated up to 700 °C (Figure 3). After heating, the presence of DyH_2 and small quantities of the $Dy_3(Co,Cu)$ and $Dy(Cu,Co)$ phases were detected; DyH_3 was absent. The presence of a thin mechanical mixture of Cu and Co is also possible.

According to the DTA data (Figure 4), the decomposition of DyH_3 started at a temperature of ~314 °C, which agreed with the literature data [16]. Between ~314 °C and ~690 °C, no thermal effects were identified. Above ~690 °C, in accordance with the Dy–H [16] diagram, the solid solution of hydrogen in dysprosium decomposed to form dysprosium. However, the thermal effects at temperatures above 600 °C can correspond to the melting of one of the metallic phases of the alloy; nevertheless, the thermal effect corresponding to ~690 °C is accompanied by a significant weight loss. The observed formation of Dy hydrides indicates the possibility of the hydrogenated $Dy_3Co_{0.6}Cu_{0.4}$ alloy to be used as additions in manufacturing sintered Nd–Fe–B magnets.

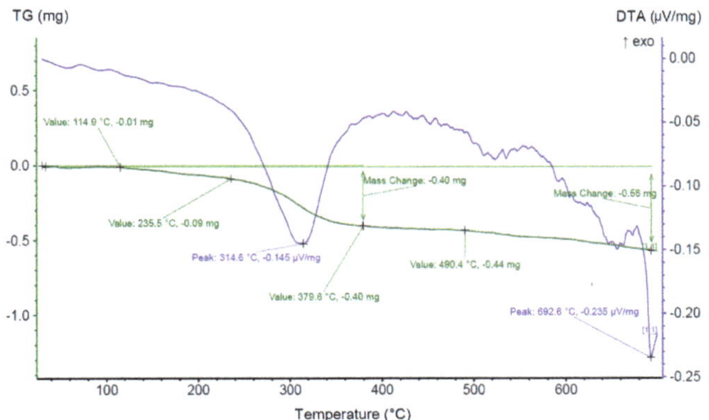

Figure 4. Thermogravimetric analysis (TG) and differential thermal analysis (DTA) curves of the $Dy_3(Co,Cu)H_{8.26}$ sample.

3.3. Microstructure and Electron Microprobe Analysis of Sintered NdFeB-Based Magnet

In accordance with the microprobe analysis data shown in Table 3, the microstructure of a magnet prepared from the powder mixture with 2 wt.% $Dy_3Co_{0.6}Cu_{0.4}H_x$ (Regime 1) was characterized by the presence of four structural components differing in the chemical composition, see Figure 5 (the phases are indicated by red numbers).

Table 3. Chemical composition of phases observed in the structure of the Nd–Fe–B sintered magnet prepared from a powder mixture with 2 wt.% of $Dy_3Co_{0.6}Cu_{0.4}H_x$ (the values averaged for three measurements are presented).

Element/Phase	O	Dy	Al	Nb	Pr	Nd	Fe	Co	Cu
					(at.%)				
Phase_1		**1.0**	**0.6**	**0.2**	**3.1**	**10.3**	**83.0**	1.2	0.5
Phase_2		2.8	0.7	1.4	16.9	44.2	28.1	2.3	3.5
Phase_3_1	47.7	0.9	0.2	0.2	7.4	22.3	20.2	0.5	0.6
Phase_3_2	67.0	1.3	0.0	0.1	7.2	20.9	2.9	0.3	0.4
Phase_3_3	64.9	1.4	0.0	0.2	7.7	21.8	3.7	0.3	0.2
Phase_4		0.4	0.2	47.7	0.6	1.8	48.9	0.3	0.3

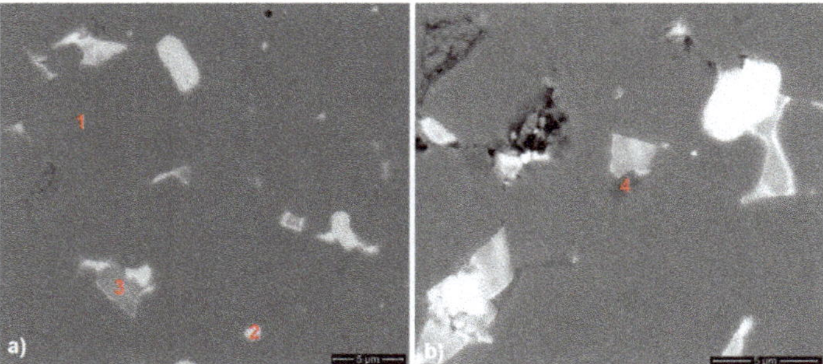

Figure 5. The microstructure of the Nd–Fe–B sintered magnet prepared from the powder mixture with 2 wt.% $Dy_3Co_{0.6}Cu_{0.4}H_x$; (**a**) phase 1—2:14:1 phase grains, phase 2—Nd-rich phase, phase 3—oxide phases; (**b**) phase 4—a phase based on Fe–Nb; (scanning electron microscopy; marked phases correspond to those in Table 2).

The chemical composition of matrix grains (Phase 1 in Figure 5a) was close to the stoichiometric $(Nd,R)_2Fe_{14}B$ composition. The presence of Dy in the matrix alloy did not allow us to unambiguously conclude the formation of the core–shell structure, but the presence of cobalt in 2:14:1 phase grains demonstrates the possibility of micro-alloying through the use of hydrogenated low-melting Co-containing compounds (the melting temperature was lower than the sintering temperature of Nd–Fe–B magnets). The Nd-rich phase (Phase 2 in Figure 5a) was characterized by a variable composition. Phase 3 (Figure 5a) corresponded to the oxide phases. In accordance with the literature data [19,20], they may be based on NdO, Nd_2O_3, or NdO_2. The presence of a phase based on Fe–Nb in triple junctions (TJ) was observed (Phase 4, Figure 5b). This fact may be related to impurities in the industrially prepared alloy matrix.

The distribution of rare earth elements, Co and Cu in the matrix grains, and in the intergranular Nd-rich phases (phase 2 in Figure 5a) in the sintered magnets prepared from the powder mixture with 2 wt.% of $Dy_3Co_{0.6}Cu_{0.4}H_x$ addition was also investigated by X-ray mapping (see Figure 6). The nonuniform Dy distribution within the 2:14:1 phase grains could be observed. The depletion of triple junctions of Co and their enrichment in Cu should be noted in the case of the addition of $Dy_3Co_{0.6}Cu_{0.4}H_x$. The presence of reactive Dy powder (originating from DyH_2 that was decomposed during sintering) ensures the diffusion of Dy atoms to the 2:14:1 phase lattice, since the atomic radius of Dy atoms is lower than that of Nd atoms. This led to ousting Nd atoms to peripheral areas.

The diffusion coefficient of Nd atoms is lower than that of Dy atoms [21]; thus, the diffusion of Dy is more significant. Such an inequality of diffusion flows of atoms caused lattice stresses and resulted in the inhomogeneous Dy and Nd(Pr) distribution over the 2:14:1 phase grains. The core–shell structure (Dy-enriched shell and Dy-depleted core) is evident in Figure 6.

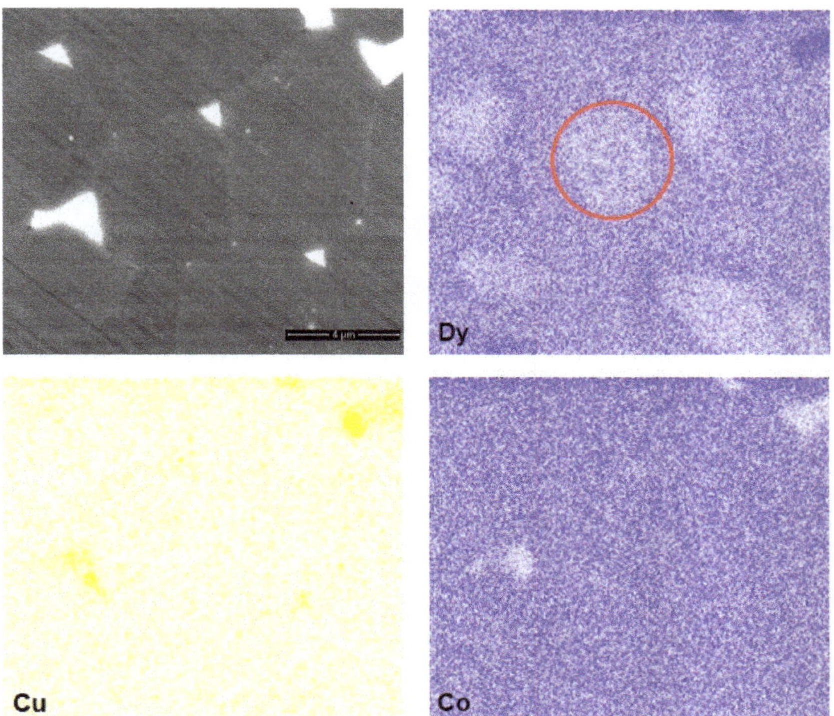

Figure 6. Co, Cu, and Dy mapping in 2:14:1 phase grains and triple junction phases of the Nd–Fe–B sintered magnet prepared from the powder mixture with 2 wt.% Dy_3(Co,Cu). The red circle indicates the depletion of 2:14:1 phase grain in Dy (i.e. the formation of core-shell structure).

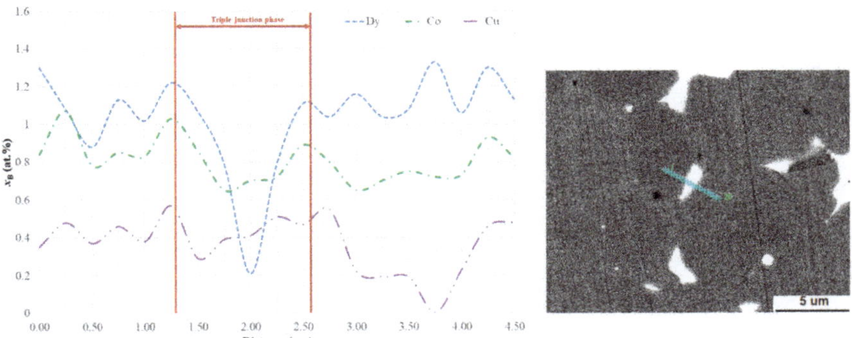

Figure 7. **(left)** Line chemical analysis over the triple junction phase and (**right**) corresponding SEM image with the analysis direction marked.

The other components of the $Dy_3Co_{0.6}Cu_{0.4}H_x$ composition (i.e., Cu and Co) are also useful additions for Nd–Fe–B-based magnets. It is evident from Figures 6 and 7 that Co evinced the tendency to incorporate the 2:14:1 phase grains, while the Cu enriched triple junction phases. The role of Cu in the grain-boundary restructuring and positive effects of Co on the coercivities of Nd–Fe–B magnets were reported in our previous work [22] and were also considered in [23–31].

3.4. Dependence of the Coercive Force ($_jH_c$) on the Heat Treatment Temperature

The magnetic properties ($_jH_c$) of the magnets (see Table 4 and Figure 8) prepared with the hydrogenated $Dy_3Co_{0.6}Cu_{0.4}$ alloy were lower than those in the case of the application of the addition of the DyH_2 [31]. One of the causes is the incomplete hydrogenation of the alloy (see Figure 3, XRD data) and, therefore, the incomplete occurrence of the grain boundary diffusion of the available Dy. The small quantity of the Dy_3(Co,Cu) phase present in the $Dy_3Co_{0.6}Cu_{0.4}$ alloy was subjected to hydrogenation. However, the value of B_r in the case of $Dy_3Co_{0.6}Cu_{0.4}H_x$ was higher than that in the case of DyH_2, which may be due to a difference in the Dy content in the chemical composition of the 2:14:1 phase. The difference in the rare-earth metal and Cu contents in the Nd-rich phases provided a lower value of H_k in the case of magnets with 2 wt.% $Dy_3Co_{0.6}Cu_{0.4}H_x$. The hysteretic properties of the Nd–Fe–B magnet, without the addition of hydride after optimal HT, are also shown in Table 4 for comparison.

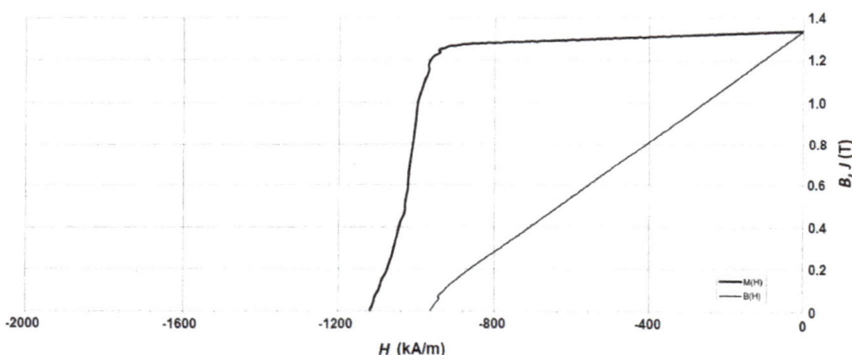

Figure 8. Magnetization reversal portions of hysteresis loop for the Nd–Fe–B sintered magnets prepared from the powder mixture with 2 wt.% $Dy_3Co_{0.6}Cu_{0.4}H_x$.

We assumed that the optimal HT for magnets of this type was in the range of 475 to 500 °C, as in the case of the magnets considered in [32–34]. Subsequent HT in this temperature range, which is performed after the optimal heat treatment (500 °C), will lead to an increase in the coercive force of magnets with 2 wt.% $Dy_3Co_{0.6}Cu_{0.4}H_x$.

Table 4. Hysteretic properties of sintered magnets prepared from the powder mixtures with 2 wt.% $Dy_3Co_{0.6}Cu_{0.4}H_x$ and DyH_2 and optimally heat treated at 500 °C for 2 h; B_r = remanence of magnetic flux density; $_jH_c$ = coercivity of magnetic polarization; H_k = parameter adopted as a criterion of coercivity (i.e., the magnetic field determined at $0.9 \times B_r$); $(BH)_{max}$ = maximum energy product; HT = heat treatment.

Addition/Annealing Conditions	B_r (T)	$_jH_c$ (kA/m)	H_k (kA/m)	$(BH)_{max}$ (kJ/m^3)
$Dy_3Co_{0.6}Cu_{0.4}H_x$/optimal HT	1.34	1120	968	336
DyH_2/optimal HT	1.29	1309	1262	322
0 wt.% of addition/optimal HT * [9]	1.36	1000	850	358

* The initial Nd–Fe–B alloy contains 0.5 wt.% Dy.

Figure 9 shows the variations of the coercive force ($_{j}H_{c}$) with changing heat treatment (HT) temperature. As can be seen from the data, after low-temperature HT in a range of 475–500 °C, $_{j}H_{c}$ demonstrated an abrupt increase.

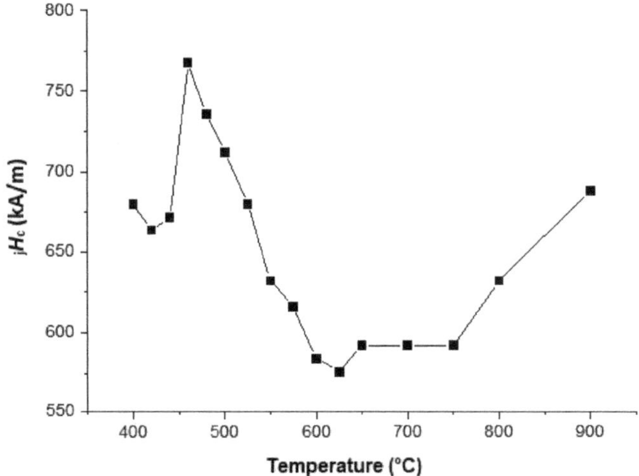

Figure 9. The dependence of $_{j}H_{c}$ on the heat treatment temperature of Nd–Fe–B-based magnet prepared from the powder mixture with 2 wt.% of $Dy_3Co_{0.6}Cu_{0.4}H_x$.

4. Conclusions

The phase composition of the $Dy_3Co_{0.6}Cu_{0.4}$ alloy in the initial homogenized and hydrogenated states was studied. The alloy in the homogenized state was multiphase and contained the Dy_3(Co,Cu) and Dy(Cu,Co) phases. During the hydrogenation of the alloy, the disproportionation or hydrogenolysis process took place, which, regardless of the multiphase composition of the initial alloy, resulted in the formation of DyH_{2-3} hydride and a fine (Co + Cu) mixture with small trace quantities of Dy_3(Co,Cu) and Dy(Cu,Co).

The study of the sintered Nd(RE)–Fe–B magnet prepared from the strip-cast alloy showed that Dy introduction via the two-component blending method (the hydogenated $Dy_3Co_{0.6}Cu_{0.4}$ compound was added to the powder mixture) resulted in the formation of the core–shell structure of 2–14–1 phase grains. The efficient enhancement of the coercivity of Nd(RE)–Fe–B magnets, with a slight sacrifice of remanence, was demonstrated.

The positive effect of REM-alloy hydrogenated additions to the Nd–Fe–B powder mixture allows the possibility of introducing various components to the permanent magnets (heavy REMs, elements structuring grain boundaries, and restricting the magnet grain growth) at the preparation stage, rather than at the alloy-melting stage. This gives the possibility of using a unified initial alloy for the manufacture of magnets with improved (high-coercive or high-performance) magnetic characteristics.

Author Contributions: Writing—original draft, K.S., P.A.P., N.B.K.; writing—review and editing, K.S., P.A.P., N.B.K.; investigation (structure, phase composition, thermal dehydrogenation, magnetic properties), K.S., P.A.P., N.B.K., T.C., T.R.; conceptualization and methodology, A.A.L.; software, Y.S.K.; validation of results of hydrogenation, H.D.; resources, N.A.D.; supervision and project administration, G.S.B.

Funding: This study was carried out within the project "Development of physico-chemical and engineering foundations for the initiation of innovative resources-economy technology of high-power and high-coercivity (Nd,r)–Fe–B (R = Pr, Tb, Dy, Ho) low-REM permanent magnets", projects No. LTARF18031 funded by the Ministry of Education, Youth and Sports of the Czech Republic and No. 14.616.21.0093 (unique identification number: RFMEFI61618x0093) funded by the Ministry of Science and Higher Education of the Russian Federation. The SEM/EDS investigation and particle size analysis were performed using the research infrastructure of the Regional Materials Science and Technology Centre, VSB–Technical university of Ostrava (Czech Republic), and

XRPD analysis, DTA/TG analysis, and the study of magnetic characteristics were carried out at the Center of Collaborative Access for Functional Nanomaterials and High-Purity Substances, Baikov Institute of Metallurgy and Materials Science, Russian Academy of Sciences. The MasterSizer 3000 particle size analyzer (Malvern) was acquired within the "Development of research and development basis of RMSTC" project (No. CZ.1.05/2.1.00/19.0387) within the frame of the operational program "Research and Development for Innovations" financed by structural funds and the state budget of the Czech Republic.

Conflicts of Interest: The authors declare no conflict of interest.

References

1. Liu, W.Q.; Sun, H.; Yi, X.F.; Liu, X.C.; Zhang, D.T.; Yue, M.; Zhang, J.X. Coercivity enhancement in Nd–Fe–B sintered permanent magnet by Dy nanoparticles doping. *J. Alloys Compd.* **2010**, *501*, 67–69. [CrossRef]
2. Sepehri-Amin, H.; Liu, L.; Ohkubo, T.; Yano, M.; Shoji, T.; Kato, A.; Schrefl, T.; Hono, K. Microstructure and temperature dependent of coercivity of hot-deformed Nd–Fe–B magnets diffusion processed with Pr–Cu alloy. *Acta Mater.* **2015**, *99*, 297–306. [CrossRef]
3. Komuro, M.; Satsu, Y.; Suzuki, H. Increase of coercivity and composition distribution in fluoride-diffused NdFeB sintered magnets treated by fluoride solutions. *IEEE Trans. Magn.* **2010**, *46*, 3831–3833. [CrossRef]
4. Ni, J.; Ma, T.; Yan, M. Improvement of corrosion resistance in Nd–Fe–B magnets through grain boundaries restructuring. *Mater. Lett.* **2012**, *75*, 1–3. [CrossRef]
5. Liang, L.; Man, T.; Zhang, P.; Jin, J.; Yan, M. Coercivity enhancement of NdFeB sintered magnets by low melting point $Dy_{32.5}Fe_{62}Cu_{5.5}$ alloy modification. *J. Magn. Magn. Mater.* **2014**, *355*, 131–135. [CrossRef]
6. Liu, X.; Wang, X.; Liang, L.; Zhang, P.; Jin, J.; Zhang, Y.; Ma, T.; Yan, M. Rapid coercivity increment of Nd–Fe–B sintered magnets by $Dy_{69}Ni_{31}$ grain boundary restructuring. *J. Magn. Magn. Mater.* **2014**, *370*, 76–80. [CrossRef]
7. Guo, S.; Chen, R.J.; Ding, Y.; Yan, G.H.; Lee, D.; Yan, A.R. Effect of DyH_x addition on the magnetic properties and microstructure of $Nd_{14.1}Co_{1.34}Cu_{0.04}Fe_{bal.}B_{5.84}$ magnets. *J. Phys. Conf. Ser.* **2011**, *266*, 16–19. [CrossRef]
8. Kim, T.-H.; Lee, S.-R.; Kim, H.-J.; Lee, M.-W.; Jang, T.-S. Magnetic and microstructural modification of the Nd–Fe–B sintered magnet by mixed DyF_3/DyH_x powder doping. *J. Appl. Phys.* **2014**, *115*, 17A763. [CrossRef]
9. Lukin, A.; Kolchugina, N.B.; Burkhanov, G.S.; Klyueva, N.E.; Skotnicova, K. Role of terbium hydride additions in the formation of microstructure and magnetic properties of sintered Nd-Pr-Dy-Fe-B magnets. *Inorg. Mater. Appl. Res.* **2013**, *4*, 256–259. [CrossRef]
10. Burkhanov, G.S.; Lukin, A.A.; Kolchugina, N.B.; Koshkid'ko, Y.S.; Dormidontov, A.G.; Skotnicova, K.; Zivotsky, O.; Cegan, T.; Sitnov, V.V. Effect of low-temperature annealing on the structure and hysteretic properties of Nd–Fe–B magnets prepared with hydride-containing mixtures. In *23rd International Workshop on Rare-Earth and Future Permanent Magnets and Their Applications*; REPM: Annapolis, MD, USA, 2014; pp. 367–369.
11. Li, X.; Liu, S.; Cao, X.; Zhou, B.; Chen, L.; Yan, A.; Yan, G. Coercivity and thermal stability improvement in sintered Nd–Fe–B permanent magnets by intergranular addition of Dy–Mn alloy. *J. Magn. Magn. Mater.* **2016**, *407*, 247–251. [CrossRef]
12. Jin, C.; Chen, R.; Yin, W.; Tang, X.; Wang, Z.; Ju, J.; Lee, D.; Yan, A. Magnetic properties and phase evolution of Nd–Fe–B magnets with intergranular addition of Pr-Co alloy. *J. Alloys Compd.* **2016**, *670*, 72–77. [CrossRef]
13. Zhang, X.; Guo, S.; Yan, C.-J.; Cai, L.; Chen, R.; Lee, D.; Yan, R. Improvement of the thermal stability of sintered Nd–Fe–B magnets by intergranular addition of $Dy_{82.3}Co_{17.7}$. *J. Appl. Phys.* **2014**, *115*, 17A757. [CrossRef]
14. Massalski, T.B.; Okamoto, H.; Subramanian, P.R.; Kacprzak, L. (Eds.) *Binary Alloy Phase Diagrams*; ASM International: Cleveland, OH, USA, 1990; Volume 2, pp. 971–2104.
15. Baker Hughes. *ASM Handbook Volume 3: Alloy Phase Diagrams*; Hiroaki, O., Mark, E.S., Erik, M.M., Eds.; ASM International: Cleveland, OH, USA, 1992; Volume 3, p. 1741.
16. Predel, B. *Phase Equilibria, Crystallographic and Thermodynamic Data of Binary Alloys' of Landolt-Börnstein—Group IV. Physical Chemistry*, 1st ed.; Springer: Berlin, Germany, 1995; p. 337.
17. The Materials Project. 2019. Available online: https://materialsproject.org/materials/mp-1191571/ (accessed on 20 April 2019).
18. The Materials Project. 2019. Available online: https://materialsproject.org/materials/mp-24151 (accessed on 20 April 2019).

19. Kim, T.-H.; Lee, S.-R.; Namkumg, S.; Jang, T.-S. A study on the Nd-rich phase evolution in the Nd–Fe–B sintered magnet and its mechanism during post-sintering annealing. *J. Alloys Compd.* **2012**, *537*, 261–268. [CrossRef]
20. Wang, S.C.; Li, Y. In situ TEM study of Nd-rich phase in NdFeB magnet. *J. Magn. Magn. Mater.* **2005**, *285*, 177–182. [CrossRef]
21. Cook, B.A.; Harringa, J.L.; Laabs, F.C.; Dennis, K.W.; Russell, A.M.; McCallum, R.W. Diffusion of Fe, Co, Nd, and Dy in $R_2(Fe_{1-x}Co_x)_{14}B$ where R=Nd or Dy. *J. Magn. Magn. Mater.* **2001**, *23*, 136–141. [CrossRef]
22. Skotnicova, K.; Burkhanov, G.S.; Kolchugina, N.B.; Kursa, M.; Cegan, T.; Lukin, A.A.; Zivotsky, O.; Prokofev, P.A.; Jurica, J.; Li, Y. Structural and magnetic engineering of (Nd, Pr, Dy, Tb)–Fe–B sintered magnets with $Tb_3Co_{0.6}Cu_{0.4}H_x$ composition in the powder mixture. *J. Magn. Magn. Mater.* **2020**, *498*, 166220. [CrossRef]
23. Pan, M.; Ge, H.; Zhang, P.; Zhu, J.; Jiao, Z.; Zhao, Y. Effects of cobalt addition on the coercivity of sintered NdFeb magnets prepared by hd method. *Zhongguo Xitu Xuebao* **2010**, *28*, 247–251.
24. Nishio, S.; Goto, R.; Matsuura, M.; Tezuka, N.; Sugimoto, S. Wettability between $Nd_2Fe_{14}B$ and Nd-Rich Phase in Nd-Fe-B Alloy System. *J. Jpn. Inst. Met.* **2008**, *72*, 1010–1014. [CrossRef]
25. Goto, R.; Nishio, S.; Matsuura, M.; Sugimoto, S.; Tezuka, N. Wettability and interfacial microstructure between $Nd_2Fe_{14}B$ and Nd-rich phases in Nd-Fe-B alloys. *IEEE Trans. Magn.* **2008**, *44*, 4232–4234. [CrossRef]
26. Cui, X.G.; Yan, M.; Ma, T.Y.; Yu, L.Q. Effects of Cu nanopowders addition on magnetic properties and corrosion resistance of sintered Nd–Fe–B magnets. *Physica B* **2008**, *403*, 4182–4185. [CrossRef]
27. Liu, Y.L.; Liang, J.; He, Y.C.; Li, Y.F.; Wang, G.F.; Ma, Q.; Liu, F.; Zhang, Y.; Zhang, X. FThe effect of CuAl addition on the magnetic property, thermal stability and corrosion resistance of the sintered NdFeB magnets. *AIP Adv.* **2018**, *8*, 056227. [CrossRef]
28. Kim, T.-H.; Lee, S.-R.; Kim, J.W.; Kim, Y.D.; Kim, H.-J.; Lee, M.-W.; Jang, T.-S. Optimization of the post-sintering annealing condition for the high Cu content Nd-Fe-B sintered magnet. *J. Appl. Phys.* **2014**, *115*, 17A770. [CrossRef]
29. Lee, S.; Kwon, J.; Cha, H.-R.; Kim, K.M.; Kwon, H.-W.; Lee, J.; Lee, D. Enhancement of coercivity in sintered Nd-Fe-B magnets by grain-boundary diffusion of electrodeposited Cu-Nd alloys. *Met. Mater. Int.* **2016**, *22*, 340–344. [CrossRef]
30. Zhong, H.; Fu, Y.; Li, G.; Liu, T.; Cui, W.; Liu, W.; Zhang, Z.; Wang, Q. Enhanced coercivity thermal stability realized in Nd–Fe–B thin films diffusion-processed by Nd–Co Alloys. *J. Mag. Mag. Mater.* **2017**, *426*, 550–553. [CrossRef]
31. Lee, M.-W.; Bae, K.-H.; Lee, S.-R.; Kim, H.-J.; Jang, T.-S. Microstructure and magnetic properties of NdFeB sintered magnets diffusion-treated with Cu/Al mixed DyCo alloy-powder. *Arch. Metall. Mater.* **2017**, *62*, 1263–1266. [CrossRef]
32. Burkhanov, G.S.; Kolchugina, N.B.; Lukin, A.A.; Koshkid'ko, Y.S.; Cwik, J.; Skotnicova, K.; Sitnov, V.V. Structure and magnetic properties of Nd–Fe–B magnets prepared from DyH_2-containing powder mixtures. *Inorg. Mater. Appl. Res.* **2018**, *9*, 509–516. [CrossRef]
33. Burkhanov, G.S.; Kolchugina, N.B.; Koshkid'ko, Y.S.; Cwik, J.; Skotnicova, K.; Cegan, T.; Prokofev, P.A.; Drulis, H.; Hackemer, A. Structure and phase composition of $Tb_3Co_{0.6}Cu_{0.4}$ alloys for efficient additions to Nd–Fe–B sintered magnets. In Proceedings of the METAL 2017—26th International Conference on Metallurgy and Materials, EU: Conference Proceedings, Brno, Czech Republic, 24–26 May 2017; pp. 1775–1781.
34. Prokofev, P.; Kolchugina, N.B.; Burkhanov, G.S.; Lukin, A.A.; Koshkid'ko, Y.S.; Skotnicova, K.; Cegan, T.; Zivotsky, O.; Kursa, M. Multiphase characterization of phase equilibria in the Tb-rich corner of the Co-Cu-Tb system. *J. Phase Equilibria Diffus.* **2019**, *40*, 403–412. [CrossRef]

Article

Densification of Magnesium Aluminate Spinel Using Manganese and Cobalt Fluoride as Sintering Aids

Ali Talimian [1], Vaclav Pouchly [2,3], Karel Maca [2,3] and Dusan Galusek [1,4,*]

1 Centre for Functional and Surface Functionalised Glass, Alexander Dubcek University of Trencin, 91150 Trencin, Slovakia; ali.talimian@tnuni.sk
2 CEITEC BUT, Brno University of Technology, Purkynova 123, 62100 Brno, Czech Republic; vaclav.pouchly@ceitec.vutbr.cz (V.P.); maca@fme.vutbr.cz (K.M.)
3 Faculty of Mechanical Engineering, Brno University of Technology, Technicka 2, 62100 Brno, Czech Republic
4 Joint Glass Centre of the IIC SAS, TnUAD and FChPT STU, 91150 Trencin, Slovakia
* Correspondence: dusan.galusek@tnuni.sk; Tel.: +421-32-7400-590

Received: 3 December 2019; Accepted: 20 December 2019; Published: 24 December 2019

Abstract: Highly dense magnesium aluminate spinel bodies are usually fabricated using pressure-assisted methods, such as spark plasma sintering (SPS), in the presence of lithium fluoride as a sintering aid. The present work investigates whether the addition of transition metal fluorides promotes the sintering of $MgAl_2O_4$ bodies during SPS. At the same time, such fluorides can act as a source of optically active dopants. A commercial $MgAl_2O_4$ was mixed with 0.5 wt% of LiF, MnF_2, and CoF_2 and, afterwards, consolidated using SPS at 1400 °C. Although MnF_2 and CoF_2 promote the densification as effectively as LiF, they cause significant grain growth.

Keywords: $MgAl_2O_4$; lithium fluoride; cobalt fluoride; manganese fluoride; spark plasma sintering; grain growth

1. Introduction

Magnesium aluminate spinel is a material of interest for optical applications due to its excellent mechanical and optical properties. $MgAl_2O_4$ has low density (3.58 g cm^{-3}), typical fracture toughness of 1.9 MPa·$m^{0.5}$, and high optical transmissivity in the visible to mid-infrared ranges [1–7]. Moreover, the spinel structure can host optically active elements, e.g., transition metal ions [8–10]. Having a symmetrical-cubic structure, transparent $MgAl_2O_4$ ceramics with high optical homogeneity can be fabricated by removing the scattering centers, such as pores and impurities [7,11–13]. Fabricating highly dense $MgAl_2O_4$ is, however, difficult because of the slow diffusion of oxygen. Therefore, spinel is usually densified by two-stage sintering, i.e., pressure-less sintering followed by hot isostatic pressing (post HIPing). Alternatively, spinel can be produced via single-stage pressure-assisted sintering, such as hot pressing (HP) or spark plasma sintering (SPS) [2,14–17]. Using the SPS method makes it possible to fabricate highly dense spinel bodies at a significantly lower temperature and a shorter time as compared with the other methods; this enables suppressing grain growth and producing high-quality samples.

Lithium fluoride (LiF) is a conventional sintering aid in processing $MgAl_2O_4$; it promotes the densification by producing transient liquid at low temperatures and introducing cation defects into the spinel structure. Moreover, LiF removes carbon contamination by forming volatile CF_x species. [18–22]. However, lithium incorporation into the $MgAl_2O_4$ structure can have a detrimental effect on optical properties, especially when spinel is doped with optically active elements [22,23].

Transition metal fluorides that melt at low temperatures are other suitable candidates to be used as additives for sintering of magnesium aluminate spinel. Such dopant provides double benefits. They assist densification through the formation of a transient liquid and, at the same time, introduce an optically active element into the $MgAl_2O_4$ structure. In this study, magnesium aluminate spinel bodies were fabricated by spark plasma sintering of a commercial aluminate spinel powder using LiF, MnF_2, and CoF_2 as sintering additives. The effect of sintering aids on densification behavior and final microstructure was investigated.

2. Materials and Methods

A commercial magnesium aluminate spinel powder, Baikalox S30CR (Baikowski, Paris, France) was used as the starting material in this study. The powder is characterized by a BET specific surface area of 26 m^2g^{-1} and a median particle size (d_{50}) of 0.2 μm according to the data provided by the supplier. The spinel powder contains minute amounts of impurities, mainly, S(600), Na(41), and Ca(15) in wt. ppm. Lithium fluoride (LiF), manganese fluoride (MnF_2), and cobalt fluoride (CoF_2), ACS grade > 99.0, purchased from Sigma-Aldrich (St. Louis, MO, USA) were used as sintering aids.

$MgAl_2O_4$ ceramics doped with the sintering aids (0.5 wt%) were prepared by dispersing and mixing of powders in isopropanol, ACS grade >99.0%, using an ultrasonic homogenizer (UW2200, BANDELIN, Berlin, Germany). Then, the mixtures were transferred to a rotary evaporator and dried. Ready-to-press (RTP) powder was prepared by passing the dried mixture through a sieve with a screen mesh of 500 μm.

Samples were consolidated using a spark plasma sintering machine (Dr. SINTER SPS-625, FUJI, Tokyo, Japan). The RTP powder was filled in a graphite die with an inner diameter of ca. 12 mm. The powder was separated from the die by graphite paper placed between powder, punches, and the die wall. The die was then wrapped in a carbon felt and placed between the moving rams of the SPS. The sintering schedules consisted of fast increases of the temperature to 600 °C in 3 min followed by heating of the sample at a constant heating rate of 100 °C min^{-1} to 1400 °C at which the shrinkage stops; therefore, sintering processes were carried out with no dwelling time to avoid unnecessary grain growth. The sintering was carried out under vacuum (5 to 9 Pa). A constant uniaxial pressure of 75 MPa was applied above 800 °C.

The displacement of punches and the temperature were recorded during the whole heating/cooling step. The pellets' temperature was measured constantly by using an optical pyrometer focused on the hole drilled into the die wall. The coefficient of thermal expansion (CTE) of the system (e.g., graphite die, paper, and punches) was determined separately (i.e., in a run without the specimen) throughout the temperature range of this study (600 to 1400 °C) in order to account for the instrumental error. The sintered pellets were subsequently subjected to a heat treatment at 800 °C (heating rate 2.5 °C min^{-1}) for 60 min in air in a muffle furnace to remove the residual carbon from the surfaces.

The bulk density and apparent porosity of sintered bodies were measured using Archimedes' method in deionized water according to the ASTM standard (C329-88(2016)) [24]. All provided values are the means of at least 10 independent measurements.

The melting temperature of sintering aids and their reactions with the spinel powder were studied using thermal analysis. The measurements were performed by a simultaneous thermal analyzer (STA 449 F1 Jupiter, Netzsch, Selb, Germany) in Differential Thermal Analysis, DTA, configuration, using alumina crucibles in flowing N_2 (20 mL min^{-1}). Thermal Gravimetric analysis, TG, was performed simultaneously. Data were collected on ca. 100 mg of mixtures containing 10 wt% of sintering aids upon heating at a constant rate of 20 °C min^{-1} to 1350 °C.

The samples' microstructure was examined using a scanning electron microscope, SEM, (JEOL 7600F, JEOL, Tokyo, Japan) equipped with an energy dispersive X-ray spectrometer (EDXS, Oxford Instruments, Abingdon, UK). Small fragments were collected from the fractured surface of samples and fixed on aluminium sample holders using conductive adhesive tape and coated with carbon to prevent charging.

3. Results

Figure 1 shows the SEM micrograph of the magnesium aluminate spinel powder; the powder consists of submicron agglomerates comprised of smaller nanoparticles with a median diameter of 90 ± 15 nm. However, the specific surface area indicates a somewhat smaller primary particle size of approximately 64 nm. Similarly, Maca et al. examined the primary particle size of the same commercial $MgAl_2O_4$ powder, and reported an average particle size of 58 nm, by assuming that the primary particles have a spherical shape. The median particle size provided by the producer is, therefore, related to the size of agglomerates [25].

Figure 1. Scanning electron microscope (SEM) image of magnesium aluminate spinel powder.

Table 1 summarizes the measured density and porosity of samples produced from the powder mixture containing 0.5 wt% of additives and the additive-free sample; the theoretical density of samples was calculated using the density of magnesium aluminate spinel (3.58 g cm^{-1}) and the density of a respective additive following the rule of mixtures. The density of LiF, MnF_2, and CoF_2 are 2.64, 3.98, and 2.70 g cm^{-3}, respectively. The measured density of an additive containing samples is within the range of experimental error comparable to the density of additive-free samples. While the residual porosity of additive-free samples is almost zero, the doped samples are characterized by limited amounts of closed porosity. Such behavior can be related to the evaporation of additives at high temperatures.

Table 1. Relative density and apparent porosity of additive-free and transition metal fluoride-doped samples produced by spark plasma sintering (SPS) at 1400 °C (no isothermal dwell). The numbers in parenthesis represent standard errors.

Sample	Relative Density (%)	Apparent Porosity (%)
Additive-free	99.90 (0.02)	0.07 (0.03)
0.5 wt% LiF	99.6 (0.1)	0.43 (0.00)
0.5 wt% MnF_2	99.4 (0.7)	0.57 (0.06)
0.5 wt% CoF_2	99.7 (0.2)	0.33(0.07)

Figure 2 shows the temperature dependence of shrinkage and shrinkage rates of the powder mixtures containing 0.5 wt% of the additives and of the additive-free spinel powder during SPS; the shrinkage was determined by measuring the punch displacement upon heating at the constant heating rate of 100 °C min^{-1}, between 600 °C and 1400 °C. The shrinkage rate was calculated point-by-point, using Equation (1):

$$\frac{d(\frac{\Delta l}{l_0})}{dt} = \frac{1}{l_0}\frac{\Delta l_{T+\delta T} - \Delta l_{T-\delta T}}{2\delta T}\dot{T} \quad (1)$$

where Δl represents linear shrinkage measured at the temperature T, l_o is the original length of the sample, t represents time, and the variable $\dot{T}$ stands for heating rate. The shrinkage curves are characterized by two main regions, a rapid shrinkage of ~5%, occurring around 800 °C, followed by continuous shrinkage up to ~23%, after which the curve reaches a plateau. While the latter is related to the densification by sintering, the former is attributed to the powder particles' rearrangement when pressure was applied [26]. The densification of all samples starts at around 850 °C. The sintering aids clearly decrease the temperature at which the densification is completed. The shrinkage curve of additive-free spinel reaches a plateau indicating the end of shrinkage, at 1350 °C, whereas the shrinkage of samples doped with LiF, CoF, and MnF_2 stops at 1170, 1195, and 1250 °C, respectively. Moreover, the shrinkage rate of doped samples is significantly higher than that of the pure spinel, particularly at temperatures higher than 1000 °C (Figure 2b).

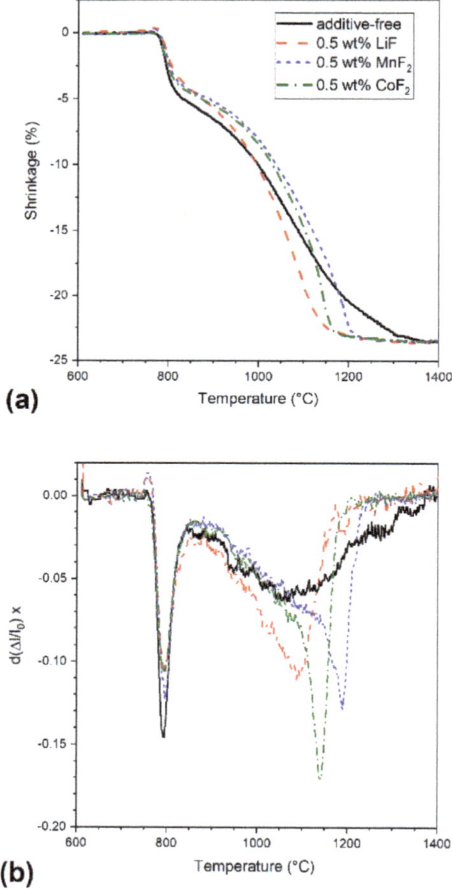

Figure 2. (**a**) Relative shrinkage of additive-free spinel and spinels doped with LiF, MnF_2, and CoF_2 (0.5 wt%) with temperature and (**b**) first derivative of the shrinkage calculated using Equation (1).

Figure 3 summarizes the results of DTA and TG analyses of powder mixtures containing 10 wt% of fluorides in the temperature interval between 600 to 1350 °C. These were carried out as reference measurements elucidating thermal processes on fluoride doped powders. The DTA curve of LiF-doped samples is characterized by a sharp endothermic peak at ~830° attributed to chemical reactions and melting of lithium fluoride, as discussed below (Equations (6) and (7)). In contrast, the sample containing MnF_2 exhibits no clear endothermic effect at the melting temperature of MnF_2 (856 °C). The behavior of the CoF_2 containing sample is similar, showing no thermal effect, which could be attributed to melting of CoF_2 (i.e., at ca 930 °C). However, all samples exhibit an endothermic peak at 1240 °C, attributed to the eutectic melting of magnesium fluoride, indicating chemical reactions between sintering aids and $MgAl_2O_4$, yielding MgF_2, as pointed out in the following text. The TG curves show that the weight of LiF samples decreases rapidly above 1050 °C, while samples containing MnF_2 and CoF_2 exhibit slower weight loss in the following two steps: a slow decline above 850 °C followed by a rapid decrease over 1050 °C. The onset of weight loss can be correlated with the small endothermic effect on DTS curves associated with melting of MgF_2 (1263 °C). The observed weight loss was then associated with evaporation of MgF_2 from the melt. On the basis of the literature data, vapor pressure of molten MgF_2 reaches ~13 Pa at 1270 °C and ~130 Pa at 1434 °C, so its loss is expected to be significant.

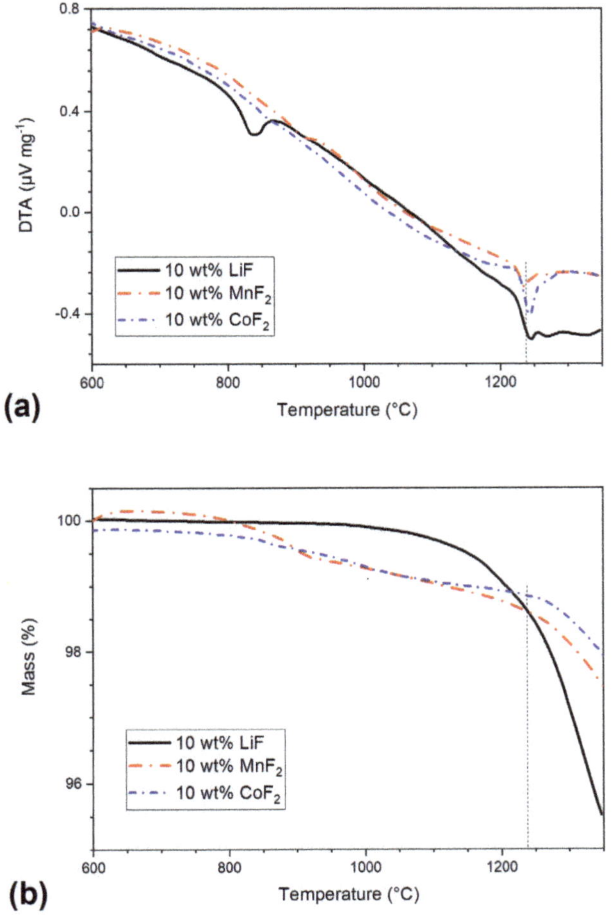

Figure 3. DTA (**a**) and TG (**b**) records of samples comprising 10 wt% of LiF, MnF_2, and CoF_2, heating rate = 20 °C min^{-1}. Vertical line at 1230 °C represents onset of melting for TM fluorides doped spinel ceramics.

Figure 4 shows the X-ray diffraction pattern of pellets produced by SPS at 1400 °C; the XRD pattern of as-received spinel powder is also shown for comparison. According to the XRD experiments, magnesium aluminate spinel is the only crystalline phase present in the samples; the sintered samples are characterized by sharp and narrow diffraction maxima that imply the sintering procedure (heating up to 1400 °C with the heating rate of 100 °C min^{-1}, with no dwell time) increases the size of coherently diffracting domains (crystallites). The XRD patterns were analyzed further by using Rietvel refinement [27,28]. The lattice parameter of additive-free spinel is estimated to be 8.0798 ± 0.0002 Å while the lattice parameter for the samples doped with LiF, MnF_2, and CoF_2 are 8.0814 ± 0.0001 Å, 8.0833 ± 0.0003 Å, and 8.0833 ± 0.0001 Å, respectively. The incorporation of dopants into the spinel structure results in a slight increase of the lattice parameter due to size mismatch of doping cations and Mg^{2+} and Al^{3+} in the spinel crystal lattice.

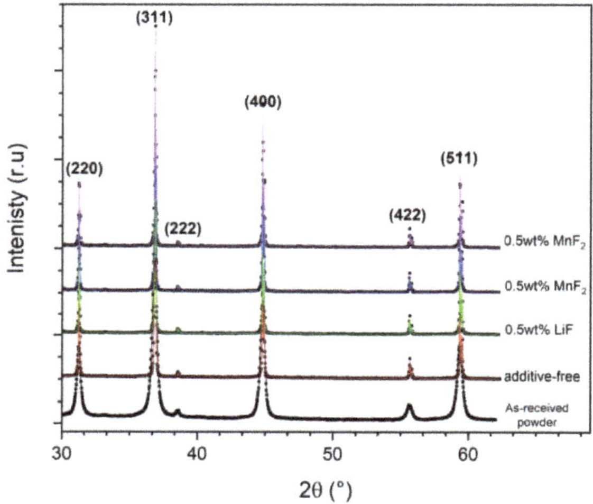

Figure 4. XRD patterns of additive-free and doped samples with 0.5 wt% (LiF, MnF_2, and CoF_2), spark plasma sintered at 1400 °C. The diffraction pattern of additive-free powder is also shown for comparison. The experimental data are fitted by the model patterns obtained by Rietveld refinement of experimental data.

Figure 5 shows the fracture surface of additive-free and doped samples with the 0.5 wt% of LiF, MnF_2, and CoF_2. All doped samples exhibit significant grain coarsening. The LiF-doped spinel is, interestingly, characterized by a smaller grain size as compared with the spinel doped with MnF_2 and CoF_2.

Bright spots observed on fracture surfaces were studied by EDX. Figure 6 shows a typical EDX spectrum collected from a bright spot in a CoF_2 doped sample. The spots also contain, along with doping ions, a significant concentration of Sulphur, implying that the sulphate impurities in the spinel powder reacted with the dopant yielding sulfate phases during sintering. However, the content of sulphates was below the detection limit of X-ray diffraction, and the size of sulphate inclusions too small to be identified by EBSD.

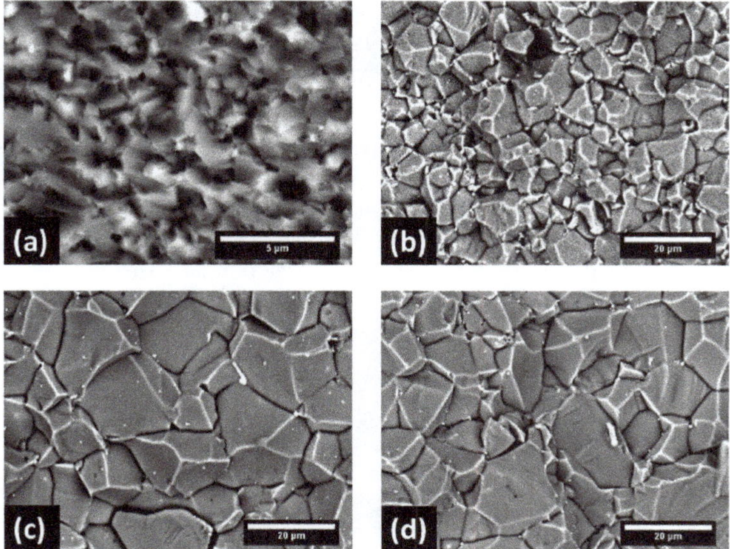

Figure 5. Scanning electron micrographs of the fracture surfaces after spark plasma sintering at 1400 °C. The images were recorded in a BSE mode from (**a**) additive-free sample, and the samples doped with 0.5 wt% of (**b**) LiF, (**c**) MnF, and (**d**) CoF_2.

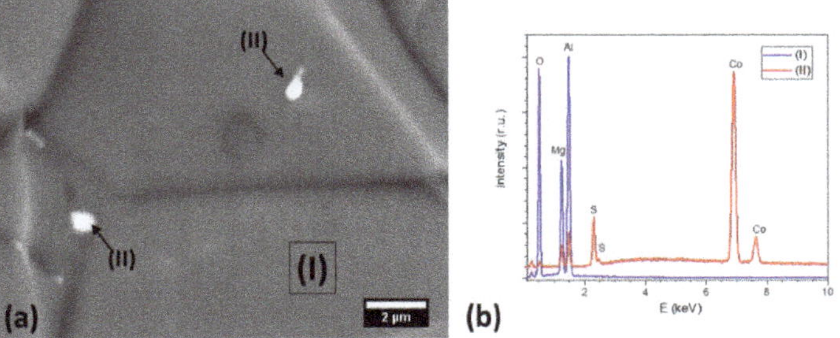

Figure 6. (**a**) Scanning electron micrograph showing in detail bright spots in CoF_2 doped spinel after spark plasma sintering and (**b**) the EDX spectrum of the selected areas/spots: (I), spinel grain and (II), bright spot.

4. Discussion

Figure 2 shows that the onset of densification of all studied samples occurs at a lower temperature than in conventional sintering. The application of pressure during spark plasma influences densification in two ways. First, powder particles rearrange under pressure. Secondly, the densification mechanism is also affected, due to grain boundary sliding [24,27]. Consequently, the maximum in densification rate of SPS samples is achieved at a lower temperature as compared with that reported for conventional sintering of the same powder (1350 °C, S30CR, Baikowski) [29,30].

The LiF-doping results in a higher densification rate, and larger grains than in the additive-free spinel are formed. The interaction between LiF and $MgAl_2O_4$ above the melting LiF point, 840 °C, and the formation of liquid are described as follows [31,32]:

$$LiF(l) + MgAl_2O_4(s) \rightarrow LiF : MgF_2(l) + LiAlO_2(s) \tag{2}$$

The transient liquid enhances the densification through two main mechanisms, liquid redistribution facilitates particle rearrangement and the fluorine rich liquid enhances the mass transport. Moreover, lithium aluminate spinel can produce solid solution with magnesium aluminate spinel and introduces structural defects according to:

$$LiAlO_2 \overset{MgAl_2O_4}{\longrightarrow} Li_{Mg}^{/} + V_{Al}^{///} + Al_{Al}^{\times} + 2O_O^{\times} + 2V_O^{\cdot\cdot} \tag{3}$$

Introduction of structural defects, such as oxygen vacancies, facilitates the movement of ions, particularly oxygen ions and, in turn, promotes the densification, as well as the grain growth. With the further increase of temperature, the evaporation rate of the transient liquid accelerates (above ca. 1100 °C, Figure 3b) and the liquid phase is effectively removed from the system. Consequently, the densification rate decreases.

$$LiF : MgF_2(l) \rightarrow LiF(g) + MgF_2(g) \tag{4}$$

Finally, gaseous MgF_2 reacts with $LiAlO_2$ forming spinel again:

$$2LiAlO_2(s) + MgF_2(g) \rightarrow 2LiF(g) + MgAl_2O_4(s) \tag{5}$$

Considering densification behavior of MnF_2 and CoF_2 similar to that of the LiF doped samples, a similar sintering mechanism can also be expected for the transition metal fluorides. Surprisingly, although CoF_2 and MnF_2 have higher melting temperatures than LiF, the CoF_2 and MnF_2 doped spinel exhibits more extensive grain coarsening than the LiF doped ones.

Interestingly, the DTA records show no endothermic effect, which would indicate melting of pure transition metal fluoride additives around the expected temperatures of their melting. Only a small endothermic effect corresponding to melting of MgF_2 implies chemical reactions between MnF_2 or CoF_2 and $MgAl_2O_4$ yielding transition phases. Due to the similar ionic radius of magnesium, manganese and cobalt ($r_{Mg^{2+}} = 72$ pm, $r_{Mn^{2+}} = 70$ pm, and $r_{Co^{2+}} = 75$ pm), it can be assumed that the transition metal ions replace magnesium ions in the spinel crystal lattice, producing MgF_2:

$$MnF_2(s) + MgAl_2O_4(s) \rightarrow MnAl_2O_4(s) + MgF_2(s) \tag{6}$$

$$CoF_2(s) + MgAl_2O_4(s) \rightarrow CoAl_2O_4(s) + MgF_2(s) \tag{7}$$

The formation of a solid-solution with different divalent cations within the spinel structure results in spinel structure strain, as well as introduction of point defects, such as oxygen vacancies as a result of hosting divalent ions in octahedral sites. [7,33]. The TG results, Figure 3b, confirm that the weight loss of MnF_2 and CoF_2 doped samples begins at higher temperatures than in the LiF doped material (1250 °C vs. 1100 °C). The transient liquid is, therefore, present at grain boundaries for a longer time, providing a faster diffusion path for the elements, and resulting not only in efficient densification but also in more pronounced grain growth. This resulted in the increase of the median grain size from 0.8 μm in undoped ceramic to 10.3 μm, 14.0 μm, and 11.6 μm in LiF, MnF_2, and CoF_2 doped spinel, respectively, as shown in Figure 5. Apart from the presence of transient liquid in spinels doped by transition metal fluorides, the finer grain size of LiF doped samples can be attributed also to Zener pinning effect of $LiAlO_2$ (Equation (1)) precipitated at grain boundaries [34].

Transition metal fluorides act as sintering aid during the densification of magnesium aluminate spinel and produce spinel structures containing optically active ions, e.g., Mn^{2+} and Co^{2+}, that can be used in applications such as white LEDs or Q-switches. [35,36] Further studies are required to evaluate whether and how the addition of transition metal fluorides affects the densification and final properties of magnesium aluminate spinel ceramics.

5. Conclusions

Highly dense magnesium aluminate spinel bodies doped with LiF, MnF_2, and CoF_2 were produced using spark plasma sintering. Although the contribution of CoF_2 and MnF_2 to the densification of $MgAl_2O_4$ is more complicated as compared with LiF, they promote the densification almost as efficiently as LiF, despite the higher melting points of transition metal fluorides. The MnF_2 and CoF_2 containing samples exhibit larger grains as compared with LiF-doped spinel spark plasma sintered under the same conditions.

Author Contributions: Conceptualization, D.G. and A.T.; methodology, A.T. and V.P.; formal analysis, A.T., V.P., and D.G.; investigation, A.T. and V.P.; resources, D.G. and K.M.; data curation, A.T. and V.P.; writing—original draft preparation, A.T. and V.P.; writing—review and editing, D.G. and K.M.; visualization, A.T. and V.P.; supervision, D.G. and K.M.; project administration, D.G.; funding acquisition, K.M and D.G. All authors have read and agreed to the published version of the manuscript.

Funding: This research was funded by the European Union's Horizon 2020 research and innovation programme, grant agreement no. 739566. Financial support of this work by the grants SAS-MOST JRP 2015/6 and VEGA 2/0026/17 is gratefully acknowledged. We appreciate the support provided by the Czech Ministry of Education under grants LTT18013 (Inter-Transfer) and CEITEC 2020 (LQ1601).

Acknowledgments: This paper is a part of the dissemination activities of the project FunGlass.

Conflicts of Interest: The authors declare no conflict of interest.

References

1. Ganesh, I. A review on magnesium aluminate ($MgAl_2O_4$) spinel: Synthesis, processing and applications. *Int. Mater. Rev.* **2013**, *58*, 63–112. [CrossRef]
2. Goldstein, A. Correlation between $MgAl_2O_4$-spinel structure, processing factors and functional properties of transparent parts (progress review). *J. Eur. Ceram. Soc.* **2012**, *32*, 2869–2886. [CrossRef]
3. Reimanis, I.; Kleebe, H.-J. A Review on the Sintering and Microstructure Development of Transparent Spinel ($MgAl_2O_4$). *J. Eur. Ceram. Soc.* **2009**, *92*, 1472–1480. [CrossRef]
4. Gilde, G.; Patel, P.; Patterson, P.; Blodgett, D.; Duncan, D.; Hahn, D. Evaluation of Hot Pressing and Hot Isostastic Pressing Parameters on the Optical Properties of Spinel. *J. Eur. Ceram. Soc.* **2005**, *88*, 2747–2751. [CrossRef]
5. Krell, A.; Waetzig, K.; Klimke, J. Influence of the structure of MgO·nAl_2O_3 spinel lattices on transparent ceramics processing and properties. *J. Eur. Ceram. Soc.* **2012**, *32*, 2887–2898. [CrossRef]
6. Hosseini, S.M. Structural, electronic and optical properties of spinel $MgAl_2O_4$ oxide. *Phys. Status Solidi* **2008**, *245*, 2800–2807. [CrossRef]
7. Sickafus, K.E.; Wills, J.M.; Grimes, N.W. Structure of Spinel. *J. Eur. Ceram. Soc.* **1999**, *82*, 3279–3292. [CrossRef]
8. Tomita, A.; Sato, T.; Tanaka, K.; Kawabe, Y.; Shirai, M.; Tanaka, K.; Hanamura, E. Luminescence channels of manganese-doped spinel. *J. Lumin.* **2004**, *109*, 19–24. [CrossRef]
9. Jouini, A.; Yoshikawa, A.; Brenier, A.; Fukuda, T.; Boulon, G. Optical properties of transition metal ion-doped $MgAl_2O_4$ spinel for laser application. *Phys. Status Solidi C* **2007**, *4*, 1380–1383. [CrossRef]
10. Hanamura, E.; Kawabe, Y.; Takashima, H.; Sato, T.; Tomita, A. Optical properties of transition-metal doped spinels. *J. Nonlinear Opt. Phys. Mater.* **2003**, *12*, 467–473. [CrossRef]
11. Sokol, M.; Ratzker, B.; Kalabukhov, S.; Dariel, M.P.; Galun, E.; Frage, N. Transparent Polycrystalline Magnesium Aluminate Spinel Fabricated by Spark Plasma Sintering. *Adv. Mater.* **2018**, *30*, 1706283. [CrossRef] [PubMed]
12. Wang, S.F.; Zhang, J.; Luo, D.W.; Gu, F.; Tang, D.Y.; Dong, Z.L.; Tan, G.E.B.; Que, W.X.; Zhang, T.S.; Li, S.; et al. Transparent ceramics: Processing, materials and applications. *Prog. Solid State Chem.* **2013**, *41*, 20–54. [CrossRef]
13. Krell, A.; Klimke, J.; Hutzler, T. Transparent compact ceramics: Inherent physical issues. *Opt. Mater.* **2009**, *31*, 1144–1150. [CrossRef]
14. Morita, K.; Kim, B.-N.; Yoshida, H.; Hiraga, K. Densification behavior of a fine-grained $MgAl_2O_4$ spinel during spark plasma sintering (SPS). *Scr. Mater.* **2010**, *63*, 565–568. [CrossRef]
15. Kim, J.-M.; Kim, H.-N.; Park, Y.-J.; Ko, J.-W.; Lee, J.-W.; Kim, H.-D. Fabrication of transparent $MgAl_2O_4$ spinel through homogenous green compaction by microfluidization and slip casting. *Ceram. Int.* **2015**, *41*, 13354–13360. [CrossRef]

16. Gajdowski, C.; Böhmler, J.; Lorgouilloux, Y.; Lemonnier, S.; d'Astorg, S.; Barraud, E.; Leriche, A. Influence of post-HIP temperature on microstructural and optical properties of pure $MgAl_2O_4$ spinel: From opaque to transparent ceramics. *J. Eur. Ceram. Soc.* **2017**, *37*, 5347–5351. [CrossRef]
17. Sokol, M.; Halabi, M.; Kalabukhov, S.; Frage, N. Nano-structured $MgAl_2O_4$ spinel consolidated by high pressure spark plasma sintering (HPSPS). *J. Eur. Ceram. Soc.* **2017**, *37*, 755–762. [CrossRef]
18. Rubat du Merac, M.; Reimanis, I.E.; Smith, C.; Kleebe, H.-J.; Müller, M.M. Effect of Impurities and LiF Additive in Hot-Pressed Transparent Magnesium Aluminate Spinel. *Int. J. Appl. Ceram. Technol.* **2013**, *10*, E33–E48. [CrossRef]
19. Bonnefont, G.; Fantozzi, G.; Trombert, S.; Bonneau, L. Fine-grained transparent $MgAl_2O_4$ spinel obtained by spark plasma sintering of commercially available nanopowders. *Ceram. Int.* **2012**, *38*, 131–140. [CrossRef]
20. Waetzig, K.; Hutzler, T. Highest UV-vis transparency of $MgAl_2O_4$ spinel ceramics prepared by hot pressing with LiF. *J. Eur. Ceram. Soc.* **2017**, *37*, 2259–2263. [CrossRef]
21. Rozenburg, K.; Reimanis, I.E.; Kleebe, H.-J.; Cook, R.L. Sintering Kinetics of a $MgAl_2O_4$ Spinel Doped with LiF. *J. Eur. Ceram. Soc.* **2008**, *91*, 444–450. [CrossRef]
22. Reimanis, I.E.; Kleebe, H.-J. Reactions in the sintering of $MgAl_2O_4$ spinel doped with LiF. *IJMR* **2007**, *98*, 1273–1278. [CrossRef]
23. Villalobos, G.R.; Sanghera, J.S.; Aggarwal, I.D. Degradation of Magnesium Aluminum Spinel by Lithium Fluoride Sintering Aid. *J. Eur. Ceram. Soc.* **2005**, *88*, 1321–1322. [CrossRef]
24. C21 Committee. *Test Method for Specific Gravity of Fired Ceramic Whiteware Materials*; ASTM International: West Conshohocken, PA, USA, 1988.
25. Maca, K.; Trunec, M.; Chmelik, R. Processing and Properties of Fine-Grained Transparent $MgAl_2O_4$ Ceramics. Available online: http://www.ceramics-silikaty.cz/index.php?page=cs_detail_doi&id=502 (accessed on 3 December 2019).
26. Chaim, R.; Marder, R.; Estournés, C.; Shen, Z. Densification and preservation of ceramic nanocrystalline character by spark plasma sintering. *Adv. Appl. Ceram.* **2012**, *111*, 280–285. [CrossRef]
27. Scardi, P.; Lutterotti, L.; Maggio, R.D. Size-Strain and Quantitative Phase Analysis by the Rietveld Method. *Adv. X-ray Anal.* **1991**, *35*, 69–76. [CrossRef]
28. Lutterotti, L.; Matthies, S.; Wenk, H.R. *MAUD: A Friendly Java Program for Material Analysis Using Diffraction*; (IUCr) Newsletter: Buffalo, NY, USA, 1999; pp. 14–15.
29. Benameur, N.; Bernard-Granger, G.; Addad, A.; Raffy, S.; Guizard, C. Sintering Analysis of a Fine-Grained Alumina–Magnesia Spinel Powder. *J. Eur. Ceram. Soc.* **2011**, *94*, 1388–1396. [CrossRef]
30. Talimian, A.; Pouchly, V.; El-Maghraby, H.F.; Maca, K.; Galusek, D. Impact of high energy ball milling on densification behaviour of magnesium aluminate spinel evaluated by master sintering curve and constant rate of heating approach. *Ceram. Int.* **2019**, *45*, 23467–23474. [CrossRef]
31. Meir, S.; Kalabukhov, S.; Froumin, N.; Dariel, M.P.; Frage, N. Synthesis and Densification of Transparent Magnesium Aluminate Spinel by SPS Processing. *J. Eur. Ceram. Soc.* **2009**, *92*, 358–364. [CrossRef]
32. Rozenburg, K.; Reimanis, I.E.; Kleebe, H.-J.; Cook, R.L. Chemical Interaction Between LiF and $MgAl_2O_4$ Spinel During Sintering. *J. Eur. Ceram. Soc.* **2007**, *90*, 2038–2042. [CrossRef]
33. Ting, C.-J.; Lu, H.-Y. Defect Reactions and the Controlling Mechanism in the Sintering of Magnesium Aluminate Spinel. *J. Eur. Ceram. Soc.* **1999**, *82*, 841–848. [CrossRef]
34. Mordekovitz, Y.; Shelly, L.; Halabi, M.; Kalabukhov, S.; Hayun, S. The Effect of Lithium Doping on the Sintering and Grain Growth of SPS-Processed, Non-Stoichiometric Magnesium Aluminate Spinel. *Materials* **2016**, *9*, 481. [CrossRef] [PubMed]
35. Goldstein, A.; Loiko, P.; Burshtein, Z.; Skoptsov, N.; Glazunov, I.; Galun, E.; Kuleshov, N.; Yumashev, K. Development of Saturable Absorbers for Laser Passive Q-Switching near 1.5 μm Based on Transparent Ceramic Co^{2+}:$MgAl_2O_4$. *J. Eur. Ceram. Soc.* **2016**, *99*, 1324–1331. [CrossRef]
36. Sai, Q.; Xia, C.; Rao, H.; Xu, X.; Zhou, G.; Xu, P. Mn, Cr-co-doped $MgAl_2O_4$ phosphors for white LEDs. *J. Lumin.* **2011**, *131*, 2359–2364. [CrossRef]

Article

Rapidly Solidified Aluminium Alloy Composite with Nickel Prepared by Powder Metallurgy: Microstructure and Self-Healing Behaviour

Alena Michalcová *, Anna Knaislová, Jiří Kubásek, Zdeněk Kačenka and Pavel Novák

Department of Metals an Corrosion Engineering, Faculty of Chemical Technology, University of Chemistry and Technology, Prague 16628, Czech Republic; knaisloa@vscht.cz (A.K.); kubasekj@vscht.cz (J.K.); kacenkaz@vscht.cz (Z.K.); panovak@vscht.cz (P.N.)

* Correspondence: michalca@vscht.cz

Received: 13 November 2019; Accepted: 11 December 2019; Published: 13 December 2019

Abstract: Composite material prepared by spark plasma sintering (SPS) from a powder mixture of AlCrFeSi rapidly solidified alloy and 5 wt. % of Ni particles was studied in this work. It was proven that during SPS compaction at 500 °C, no intermetallic phases formed on the surface of Ni particles. The material exhibited sufficient mechanical properties obtained by tensile testing (ultimate tensile stress of 203 ± 4 MPa, ductility of 0.8% and 0.2% offset yield strength of 156 ± 2 MPa). Tensile samples were pre-stressed to 180 MPa and annealed at 450 and 550 °C for 1 h. Annealing at 450 °C did not lead to any recovery of the material. Annealing at 550 °C caused the full recovery of 0.2% offset yield strength, while the ductility was decreased. The self-healing behaviour originates from the growth of intermetallic phases between the Ni particle and the Al matrix. The sequence of NiAl, Ni_2Al_3 and $NiAl_3$ intermetallic phases formation was observed. In particular, the morphology of the $NiAl_3$ phase, growing in thin dendrites into the Al matrix, is suitable for the closing of cracks, which pass through the material.

Keywords: self-healing; aluminium alloy; microstructure

1. Introduction

Self-healing materials are designed in order to enrich structural materials by the ability of closing crack that was formed during material utilization. Closing of the crack will lead to the restoration of the mechanical cohesion of the material. This may decrease the impacts of an accident caused by material failure.

Several mechanisms of the self-healing behaviours have been proposed. Precipitation mechanism at low temperature was observed in an Fe-Au alloy [1]. In several other alloys, precipitation self-healing of the crack occurred after an exposure to elevated temperature [2–5]. Phase transformation may also be a driving force for the self-healing behaviour. Composite materials with the shape memory alloy (SMA) reinforcement and the matrix with off-eutectic composition were described in [6,7]. In this case, the phase transformation of the SMA (smart material, in this case shape memory alloy) reinforcement leads to the crack closing and a partial matrix melting heals the already closed crack. Another type of the self-healing behaviour leading to crack closing was observed in Al-Ag alloy, where the phase transformation was between the Ag_2Al phase and the fcc-Al [8]. A different strategy for the crack healing is a usage of an encapsulated healing agent. A low melting metal or alloy is filled into a hollow diffusion barrier, typically composed of oxides [5]. The idea is that the spreading crack will break the diffusion barrier and enter the area with the low melting healing agent. After annealing, the low melting healing agent will fill the crack and the hollow diffusion barrier remains as a spherical pore in the microstructure.

The idea presented in this paper is to modify the encapsulated-healing-agent strategy by adding Ni powder into rapidly solidified Al alloy. The diffusion barrier will be formed by nickel aluminides and may provide the material with a better cohesion, compared to the oxides. During annealing, intermetallic phases will be formed by a reaction of Ni and Al. This process is accompanied by volume changes [9,10], so a decreased level of porosity may be expected compared to the self-healing by the encapsulated low-melting metal/alloy mechanism. By this innovative process, self-healing materials will be prepared by the two-step powder metallurgy processing (1) preparation of rapidly solidified alloy, (2) compaction. No further steps are necessary to supply self-healing agent to the material. Simple powder mixing before compaction is required.

2. Materials and Methods

Powder alloy with intended composition AlCr6Fe2Si1 (given in wt. %) was prepared by a gas atomisation and mixed with 5 wt. % of a gas atomized Ni. In both cases, the finest granulometric fraction of the size less than 63 μm was used. The powder mixture was consolidated using the spark plasma sintering (SPS) device (FCT Systeme HP D 10, Rauenstein, Germany) forming a bulk sample of 40 mm in diameter and approximately 13 mm in height. Chemical composition measured by X-Ray fluorescence spectroscopy (XRF, ARL 9400 XP, Thermo ARL, Switzerland) of the sintered sample denoted as "AlCrFeSi + Ni" is given in Table 1.

Table 1. Chemical composition of AlCrFeSi + Ni.

Element (wt. %)	Al	Si	Cr	Fe	Ni
AlCrFeSi + Ni	86.42	1.53	5.47	1.86	4.68
error	0.17	0.06	0.11	0.07	0.11

Bulk material was observed by scanning electron microscopy (SEM, TESCAN VEGA 3 equipped by energy dispersive spectroscopy (EDS) detector by Oxford Instruments, Abingdon, UK). SEM samples were prepared by mechanical grinding (P320-P4000), by polishing using diamond paste (D3, D0.7) and by etching in 0.5 % HF solution. TEM samples were prepared by ion polishing using a Gatan PIPs device (Gatan, Pleasanton, CA, USA). Observation by the transmission electron microscopy (TEM) was done by Jeol 2200 FS equipped by EDS detector by Oxford Instruments. Tensile testing was performed by the Instron 5882 universal loading machine with the strain rate of 1 mm / min. Dog-bone shaped samples with the length of 16 mm, width of 4 mm, and height of 2 mm with 20 mm radius were used. Phase composition was determined by XRD using diffractometer PANanalytical X´Pert PRO, Co lamp (Almelo, The Netherlands).

3. Results and Discussion

3.1. Microstructure of As-Sintered AlCrFeSi + Ni Material

Bulk sintered material was composed of Al matrix, $Al_{13}Cr_2$ and $Al_{13}Fe_4$ intermetallic phases and Ni as proven by XRD pattern, shown in Figure 1.

Figure 2 shows the SEM micrograph of the as-sintered AlCrFeSi + Ni composite material. In the Al matrix, fine particles of intermetallic phases $Al_{13}Cr_2$ (round) and $Al_{14}Fe_4$ (needles) are visible. The initial particles of Al alloy were deformed during the SPS process, while Ni particles remained spherical.

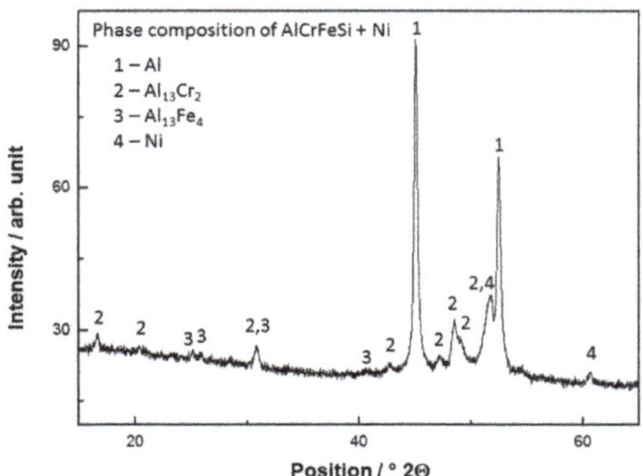

Figure 1. X-Ray diffraction pattern of as-sintered AlCrFeSi + Ni.

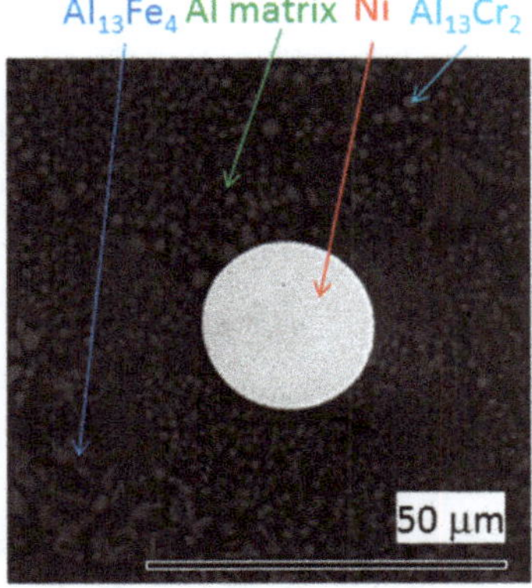

Figure 2. Micrograph of as-sintered AlCrFeSi + Ni (SEM/BSE).

To describe the boundary between the Al matrix and the Ni particle, it is necessary to use the TEM. The TEM micrograph is given in Figure 3. The light grey Al matrix contains fine intermetallic particles. The internal structure of the Ni particle is dendritic.

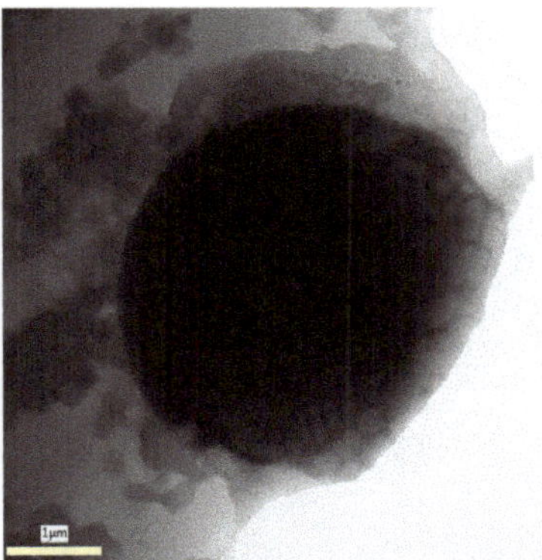

Figure 3. Detailed micrograph of as-sintered AlCrFeSi + Ni (TEM).

More detailed view was obtained in STEM mode (Figure 4). Sufficient contact of the Al matrix with the Ni particle was observed, but the connection was not uniform. EDS line profiles were measured in a location with the good connection (Line 1) and in a location where the gap is evident (Line 2).

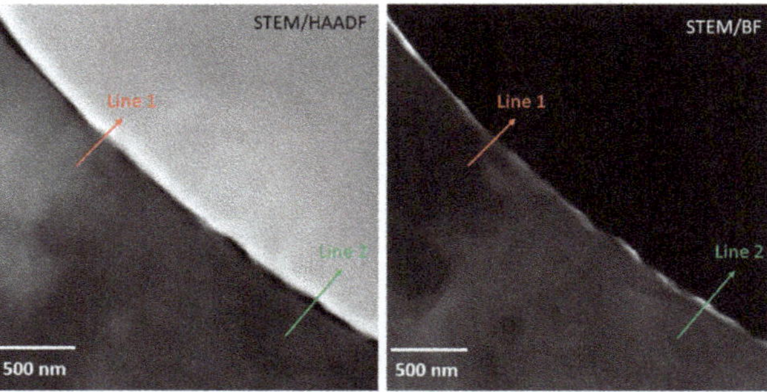

Figure 4. Detailed micrograph of boundary between Al matrix and Ni particle in the as-sintered AlCrFeSi + Ni (STEM) with indication of line profiles.

The EDS profile of the Line 1 is plotted in Figure 5. Although it is the location with a good connection, no plateau in any profile was observed. This means that no intermetallic phase with a stoichiometric composition was formed. Enrichment of the Ni particle surface by Cr was observed. This impurity might originate from the initial Ni powder.

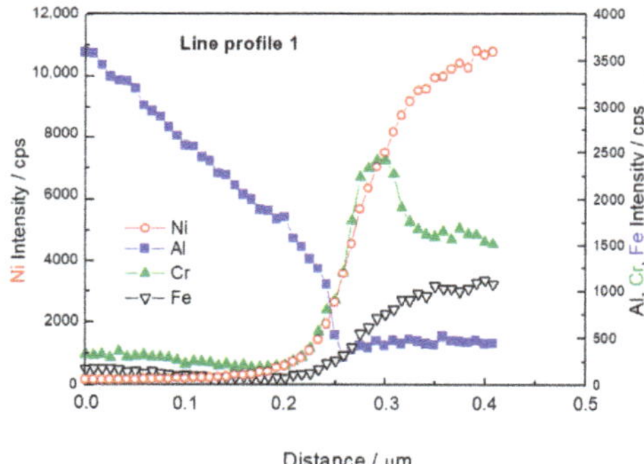

Figure 5. Line profile 1, location shown in Figure 4.

The EDS profile of the Line 2 is plotted in Figure 6. At the position of 0.3 µm from the beginning of the line, the minimum of all the elements' profile lines can be seen. This is the position of the gap. No plateau was observed again in the vicinity of the gap. This proves that the sintering process caused only partial connection of Al matrix and Ni particles. The other possible explanation of the presence of a gap would be a formation of the gap caused by different thermal expansion of intermetallic interlayer between Al matrix and Ni particle during cooling of sintered samples. As the composition changes gradually (Figure 5), there is no evidence of phase transformation during the SPS process.

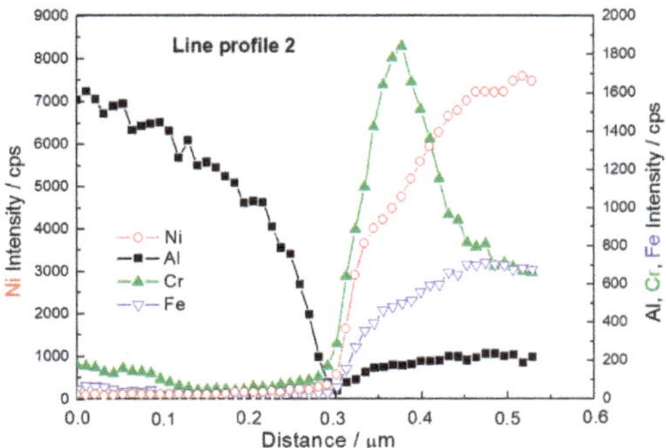

Figure 6. Line profile 2, location shown in Figure 3.

3.2. *Microstructure of AlCrFeSi + Ni Material Annealed at 550 °C for 1 h*

Figure 7 illustrates microstructure of AlCrFeSi + Ni composite after annealing at 550 °C for 1 h. The composite behaved like diffusion couple that was already well described in several works [10–16]. The Ni residual core is covered by a thin layer of NiAl phase. A large layer of Ni_2Al_3 is on the top of the NiAl layer. Dendrites of $NiAl_3$ grow into Al matrix. Porosity caused by diffusion (Kirkendall porosity) can be observed between $NiAl_3$ dendrites. The intermetallic particle has an irregular shape compared to the initial Ni sphere. This observation verifies the volume change during formation of intermetallic phases, which is a sign of the self-healing behaviour potential of the AlCrFeSi + Ni material.

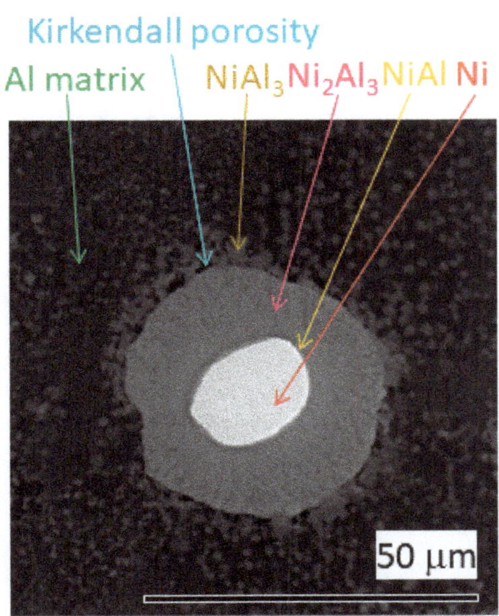

Figure 7. Micrograph of AlCrFeSi + Ni annealed at 550 °C for 1 h (SEM/BSE).

The phase composition was determined by point EDS analysis. Actual positions of each measured spectrum are shown in Figure 8. Theoretical content of Al in NiAl is 50 at. %, in Ni_2Al_3 60 at. % and in $NiAl_3$ 75 at. %. The values of Al content presented in Table 2 are slightly higher that the teoretical ones. It is due to interaction volume of the EDS point spectra measurement (several cubic micrometers). It is highly probable that the EDS spectra contain a partial signal from the Al matrix, which increases the measured values.

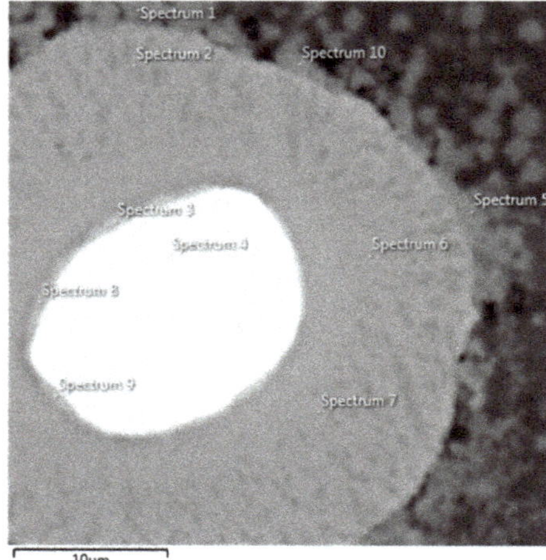

Figure 8. Micrograph of AlCrFeSi + Ni annealed at 550 °C for 1 h with labelled EDS point spectra.

Table 2. Chemical composition of points labelled in Figure 8 given in at. % (EDS).

Element (at. %)	Al	Cr	Fe	Ni	Phase
Spectrum 1	84.15	7.85	2.44	5.55	$NiAl_3$
Spectrum 2	75.94	2.66	1.74	19.65	Ni_2Al_3
Spectrum 3	51.6	5.75	3.85	38.8	$NiAl_3$
Spectrum 4	2.34	10.65	7.57	79.43	Ni
Spectrum 5	83.16	7.27	2.44	7.13	$NiAl_3$
Spectrum 6	74.99	2.68	1.77	20.56	Ni_2Al_3
Spectrum 7	74.54	2.78	1.54	21.14	Ni_2Al_3
Spectrum 8	64.76	4.53	3.00	27.71	NiAl
Spectrum 9	58.93	5.14	3.52	32.41	NiAl
Spectrum 10	82.36	5.61	2.94	9.09	$NiAl_3$

3.3. *Tensile Testing*

Figure 9 plots the results of tensile testing of AlCrFeSi + Ni composite material. The black curve corresponds to a testing of the material in the as-sintered state. The results give the essential description of behaviour of the material. The ultimate tensile stress was 203 ± 4 MPa, the ductility was 0.8 %. The yield stress cannot be estimated clearly, so the 0.2% offset yield strength was measured and it reaches the value of 156 ± 2 MPa. Subsequently, the samples were pre-stressed to 180 MPa, where the tensile test was stopped. The pre-stressed samples were annealed for 1 h at temperatures of 450 °C and 550 °C. The pre-stressed and annealed samples underwent the tensile testing under the same conditions as the as-sintered AlCrFeSi + Ni material. Annealing at 450 °C caused an increase in ductility, but the 0.2% offset yield strength was lower. This means that the most important property form the structural point of view was not preserved. The material annealed at 550 °C had a lower ductility, but the 0.2% offset yield strength was exactly the same as the one of the as-sintered material.

For the self-healing materials with SMA reinforcement, is was described that the material after healing reached 95% of UTS value and the ductility decrease from 6.4% to 2.2% (corresponding to 34% of the original ductility value) [6]. In our case, 63% of UTS value was achieved but what is more important by preserving the value of 0.2% offset yield strength. The value of ductility decreased in our case from 0.8% to 0.27%. This means that the self-healed material had ductility reaching 33% of the value of sintered material, which is in good agreement with other materials with self-healing properties [6].

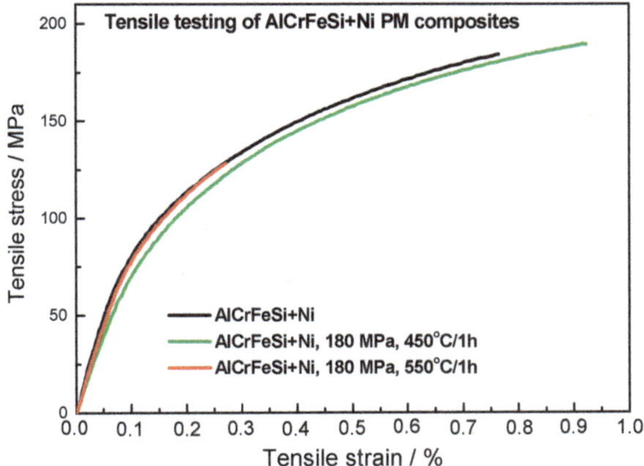

Figure 9. Tensile testing of AlCrFeSi + Ni, black—tensile tested in as-sintered state, green—prestressed up to 180 MPa, annealed at 450 °C / 1 h and tensile tested, red—prestressed up to 180 MPa, annealed at 550 °C/1 h and tensile tested.

Figure 10 shows the SEM micrographs of fracture surfaces of the tensile tested materials. The Ni particle visible in the as-sintered material (a) has a dendritic surface morphology. The SPS process was successful, as the Ni particle is an integral part of the material and does not serve as stress concentrator (the Al matrix is still surrounding the Ni particle). The morphology of the Ni particle after annealing at 450 °C was not changed (b) and the whole fracture surface looks similar to the as-sintered material. The fracture surface of the material annealed at 550 °C is significantly different. The fracture surface of Ni_2Al_3 particle is shown in the middle (c) with small round island of NiAl. Dendrites of $NiAl_3$ are interconnecting Ni_2Al_3 particle with Al matrix in the vicinity. In the areas where initial Ni particles were close (top left corner), the $NiAl_3$ are not observed. In these places, polyhedrons of Ni_2Al_3 are formed. This may be the reason for the decrease of the ductility of the material annealed at 550 °C.

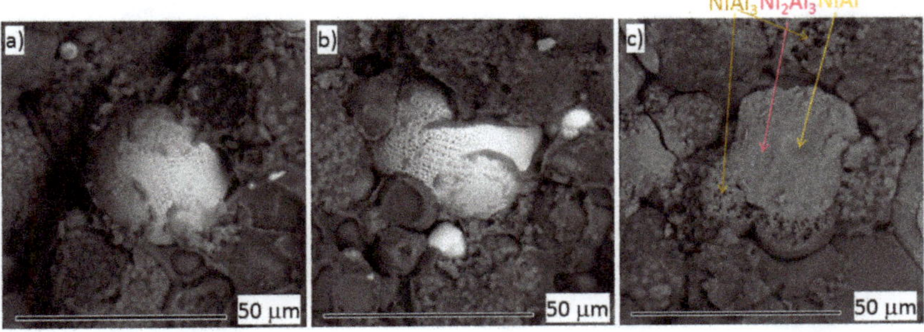

Figure 10. Fractography of AlCrFeSi + Ni, (**a**) tensile tested in as-sintered state, (**b**) prestressed up to 180 MPa, annealed at 450 °C / 1 h and tensile tested, (**c**) prestressed up to 180 MPa, annealed at 550 °C / 1 h and tensile tested (SEM).

4. Conclusions

AlCrFeSi + Ni composite material was successfully prepared by the SPS process. During sintering, no intermetallic phase was formed on the surface of Ni particles, as proved by TEM observation. The results of tensile testing showed sufficient mechanical properties of the AlCrFeSi + Ni material (ultimate tensile stress: 203 ± 4 MPa, ductility: 0.8%, 0.2% offset yield strength: 156 ± 2 MPa).

The tensile testing samples were pre-stressed to 180 MPa (the value above 0.2% offset yield strength) and annealed. On the annealed samples, the tensile testing was performed again. The sample annealed at 450 °C, which is under the temperature of intermetallic formation, exhibits a slightly higher ductility, but the 0.2% offset yield strength did not reach the value of the as-sintered material. The value of 0.2% offset yield strength for the sample annealed at 550 °C was similar to the value of the as-sintered material. The self-healing behaviour was enabled by the formation of a sequence of nickel aluminides from initial Ni particle into Al matrix. The formation of Ni_2Al_3 polyhedrons caused the decrease of materials ductility. This problem might be solved by an optimization of the Ni content in the material.

Author Contributions: A.M. wrote the paper and performed TEM measurements. A.K. performed SEM measurements. J.K. performed tensile testing. Z.K. prepared samples for TEM. P.N. analysed the data connected with intermetallic growth.

Funding: This research was financially supported by the Czech Science Foundation, project No. GJ17-25618Y. The authors acknowledge the assistance in TEM sample preparation provided by the Research Infrastructure NanoEnviCz, supported by the Ministry of Education, Youth, and Sports of the Czech Republic under Project No. LM2015073.

Conflicts of Interest: The authors declare no conflict of interest.

References

1. Zhang, S.; Kwakernaak, C.; Sloof, W.; Bruck, E.; van der Zwaag, S.; van Dijk, N. Self Healing of Creep Damage by Gold Precipitation in Iron Alloys. *Adv. Eng. Mater.* **2015**, *17*, 598–603. [CrossRef]
2. Hautakangas, S.; Schut, H.; Rivera, P.; van Dijk, N.H.; van der Zwaag, S. A first step towards self healing in aluminum alloys. In Proceedings of the International Conference on New Frontiers in Light Metals, Delft, The Netherlands, 9 November 2006.
3. Hautakangas, S.; Schut, H.; van der Zwaag, S.; Rivera Diaz del Castillo, P.E.J.; van Dijk, N.H. The Role of the Aging Temperature on the Self-Healing Kinetics in an Underaged AA2024 Aluminium alloy. In Proceedings of the First International Conference on Self Healing Materials, Noordwijk aan Zee, The Netherlands, 18–20 April 2007.
4. Hautakangas, S.; Schut, H.; van Dijk, N.H.; del Rivera Díaz Castillo, P.E.J.; van der Zwaag, S. Self healing of deformation damage in underaged Al-Cu-Mg alloys. *Scr. Mater.* **2008**, *58*, 719–722. [CrossRef]
5. Ferguson, J.B.; Schultz, B.F.; Rohatgi, P.K. Self-Healing Metals and Metal Matrix Composites. *JOM* **2014**, *66*, 866–871. [CrossRef]
6. Manuel, M.V.; Olson, G.B. Biomimetic self-healing metals. In Proceedings of the 1st International Conference on Self-Healing Materials, Noordwijik aan Zee, The Netherlands, 18–20 April 2007.
7. Rohatgi, P.K. Al-shape memory alloy self-healing metal matrix composite. *Mater. Sci. Eng. A* **2014**, *619*, 73–76. [CrossRef]
8. Michalcová, A.; Marek, I.; Knaislová, A.; Sofer, Z.; Vojtěch, D. Phase Transformation Induced Self-Healing Behavior of Al-Ag Alloy. *Materials* **2018**, *11*, 199–205. [CrossRef]
9. Jansen, M.M.P. Diffusion in the Nickel-Rich Part of the Ni-AI System at 1000 °C to 1300 °C Ni3AI Layer Growth, Diffusion Coefficients, and Interface Concentrations. *Metall. Trensactions* **1973**, *4*, 1623–1633.
10. Fujiwara, K.; Horita, Z. Measurement of intrinsic diffusion coefficients of Al and Ni in Ni3Al using Ni/NiAl diffusion couples. *Acta Mater.* **2002**, *50*, 1571–1579. [CrossRef]
11. Čelko, L.; Klakurková, L.; Švejcar, J. Diffusion in Al-Ni and Al-NiCr Interfaces at Moderate Temperatures. *Defect Diffus. Forum* **2010**, *297–301*, 771–777. [CrossRef]
12. Marino, K.A.; Carter, E.A. Ni and Al diffusion in Ni-rich NiAl and the effect of Pt additions. *Intermetallics* **2010**, *18*, 1470–1479. [CrossRef]
13. Zhao, J.C.; Zhengb, X.; Cahillb, D.G. Thermal conductivity mapping of the Ni–Al system and the beta-NiAl phase in the Ni–Al–Cr system. *Scr. Mater.* **2018**, *66*, 935–938. [CrossRef]
14. Kwiecien, I.; Bobrowski, P.; Wierzbicka-Miernik, A.; Litynska-Dobrzynska, L.; Wojewoda-Budka, J. Growth Kinetics of the Selected Intermetallic Phases in Ni/Al/Ni System with Various Nickel Substrate Microstructure. *Nanomaterials* **2019**, *9–27*, 134–152. [CrossRef] [PubMed]

15. Adadi, M.; Amadeh, A.A. Formation mechanisms of Ni–Al intermetallics during heat treatment of Ni coating on 6061 Al substrate. *Trans. Nonferrous Met. Soc. China* **2015**, *25*, 3959–3966. [CrossRef]
16. Jozwik, P.; Polkowski, W.; Bojar, Z. Applications of Ni3Al Based Intermetallic Alloys—Current Stage and Potential Perceptivities. *Materials* **2015**, *8*, 2537–2568. [CrossRef]

Article

Zn-Mg Biodegradable Composite: Novel Material with Tailored Mechanical and Corrosion Properties

Jiří Kubásek [1,*], Drahomír Dvorský [1], Jaroslav Čapek [2], Jan Pinc [2] and Dalibor Vojtěch [1]

[1] Department of Metals and Corrosion Engineering, Faculty of Chemical Technology, University of Chemistry and Technology, Prague Technická 5, Dejvice, 166 28 Prague 6, Czech Republic; dvorskyd@vscht.cz (D.D.); Vojtechd@vscht.cz (D.V.)

[2] Department of Functional Materials, Institute of Physics of the Czech Academy of Sciences, Na Slovance 1999/2, 182 21 Prague 8, Czech Republic; capekj@fzu.cz (J.Č.); Pincik789@gmail.com (J.P.)

* Correspondence: kubasekj@vscht.cz

Received: 7 November 2019; Accepted: 24 November 2019; Published: 27 November 2019

Abstract: Zinc-based alloys represent one of the most highly developed areas regarding biodegradable materials. Despite this, some general deficiencies such as cytotoxicity and poor mechanical properties (especially elongation), are not properly solved. In this work, a Zn-5Mg (5 wt.% Mg) composite material with tailored mechanical and superior corrosion properties is prepared by powder metallurgy techniques. Pure Zn and Mg are mixed and subsequently compacted by extrusion at 200 °C and an extrusion ratio of 10. The final product possesses appropriate mechanical properties (tensile yield strength = 148 MPa, ultimate tensile strength = 183 MPa, and elongation = 16%) and decreased by four times the release of Zn in the initial stage of degradation compared to pure Zn, which can highly decrease cytotoxicity effects and therefore positively affect the initial stage of the healing process.

Keywords: biomaterials; metallic composites; powder technology; zinc

1. Introduction

Binary zinc–magnesium alloys have been studied with regard to the improvement of corrosion protection of steel [1–3], and the development of arterial stents [4–10] and various biodegradable fixation devices [11,12]. As a relatively new biodegradable material, they can compete with magnesium products [7–10]. Although zinc (ρ = 7.14 g·cm^{-3}) is much heavier than Mg (ρ = 1.74 g·cm^{-3}), it is less susceptible to corrosion in the human environment. Various Zn-based materials with superior mechanical and corrosion properties have been developed and studied [9,10,13,14] but pure Zn (even in low doses) is relatively toxic to cells [15] compared to Mg. The daily zinc requirement for an adult is estimated to be 15 mg/day [7,16]. On the contrary, even up to 700 mg/day of Mg can be tolerated [7,16]. The degradation of Zn-based materials is connected to the production of Zn^{2+} ions, which are released to the surrounding extracellular space and spread into different parts of the organism. Zn^{2+} ions play an important role in various cellular processes (cell proliferation, differentiation, and signaling). Among other things, a high concentration of Zn^{2+} can induce apoptosis or necrosis or destroy ion-dependent intracellular signaling pathways [14]. On the contrary, zinc ions in adequate doses can enhance regulation of genes, cell survival/growth and differentiation, extracellular matrix (ECM) mineralization, and osteogenesis [17]. Magnesium is well tolerated by the organism in larger doses than zinc and affects activation of many enzymes, co-regulation of protein synthesis and muscle contraction, and stabilization of DNA and RNA [7]. Although the results of in vitro cytotoxicity tests are generally inconsistent [10,15,18] and long term in vivo tests do possess some good results [10], the lower initial release of Zn from the degradable implant is desirable to improve the osseointegration process during the initial stage of healing and suppress the possible formation of necrotic tissue. Unfortunately, alloying generally leads to a minor decrease in the corrosion rate of Zn-based alloys [10,14]. In this work, we selected a non-traditional way of producing a Zn-5Mg composite material by powder metallurgical methods. Although powder metallurgy is a powerful method for the production of materials with superior mechanical and corrosion properties, it has been very rarely used to prepare biodegradable Zn-based materials intended for medical devices [19,20]. However, our presented results demonstrate that this technique enables the preparation of materials with tailored mechanical and corrosion properties and improved biocompatibility.

2. Materials and Methods

2.1. Sample Preparation

Powders of pure Mg in the form of spherical particles 50–200 μm in diameter and pure Zn in the form of elongated irregular particles about 30–50 μm in thickness and 60–300 μm in length were mixed in Turbula® T2F (WAB-GROUP, Muttenz, Switzerland) for 20 min in a ratio equal to the composition of Zn-5Mg (5 wt.% Mg). The mixture was subsequently compressed in a mold 20 mm in diameter by loading 80 kN for 2 min. Green compacts were extruded at 200 °C and an extrusion velocity of 5 mm/min. The final diameter of the extruded rod was 6.5 mm, which corresponds to an extrusion ratio equal to 10. All processing steps were performed under an air atmosphere. Pure Zn was prepared in the same way as the reference material.

2.2. Microstructure Characterization

Mechanical grinding and polishing were used for surface pre-treatment. Final polishing was performed using an Eposil F suspension (Metalco Testing s.r.o, Roztoky u Prahy, Czech Republic). The microstructure was characterized by optical microscopy and scanning electron microscopy (TescanVEGA3, TESCAN Brno, s.r.o, Brno, Czech Republic) equipped with Energy-Dispersive X-Ray Spectroscopy-EDS (AZtec, Oxford Instruments, Abingdon, United Kingdom).

2.3. Mechanical Properties

Vickers Hardness HV1 was measured on sample planes perpendicular to the extrusion direction. Compression tests were measured on cylindrical samples 5 mm in diameter and 7 mm in height. Classical dog bone specimens with gauge length equal to 10 mm and diameter equal to 4.5 mm were used for tensile tests. Shoulders on the side of specimens were 15 mm in length and 5.5 mm in diameter. The radius between the shoulders and loaded part was 3.5 mm. Both compressive and tensile tests were performed using a LabTest 5.250SP1-VM universal loading machine (LABORTECH s.r.o., Opava, Czech Republic) at a strain rate of 0.001 s^{-1}. Loading was performed parallel to the extrusion direction. Tensile yield strength (TYS), ultimate tensile strength (UTS), and elongation to fracture were evaluated from engineering stress–strain curves.

2.4. Corrosion Properties

Samples 5 mm in diameter and 7 mm in height were immersed in simulated body fluid (SBF27) prepared according to Müller at al. [11] for 336 h. The ratio between sample surface and volume of solution was 100 mL·cm^{-2}. After immersion testing corrosion products were removed from the samples by repeated immersion of the samples in a solution containing 200 g·L^{-1} CrO_3. This process was performed until the change of the sample weight between steps was less than 0.001 g. Corrosion rates were calculated from the weight changes between the initial sample and dry sample with removed corrosion products and the concentration of released Zn ions into the corrosion media was determined by ICP-MS analyses (Elan DRC-e, Perkin-Elmer, Waltham, MA, USA). Cathodic curves with a general three electrode setup (sample = working electrode, glassy carbon = counter electrode, and Ag/AgCl containing saturated KCl = reference electrode) were measured using Parstat 3000-MC (AMETEK—Measurement, Communications & Testing, Berwyn, Pennsylvania, USA) in SBF. A sample with the same dimensions as for immersion tests mounted through the screw-thread to the Teflon holder was left in the corrosion media for 60 min to stabilize the open circuit potential (OCP); subsequently, polarization started from +0.02 versus OCP and continued to −0.6 versus OCP with a rate of 1 mV·s^{-1}. Before all corrosion tests, samples were ground on SiC papers of up to P2000 and degreased in ethanol.

3. Results and Discussion

3.1. Microstructure Characterization

The microstructure of the extruded materials is depicted in Figure 1. Both Zn (Figure 1a) and Zn-5Mg (Figure 1b,c) can be seen to be characterized by a partially recrystallized bimodal microstructure containing equiaxed grains with grain sizes in the range 10–30 μm and elongated grains 10 μm in thickness and 50–300 μm in length. Oxide shells, which come from the original powder surface, can be observed to surround the deformed particles. No systematic difference between the microstructure of the Zn matrix was observed for pure Zn and Zn-5Mg extruded materials. Magnesium particles (black areas in Figure 1) can be seen to be homogeneously distributed in the Zn matrix. The interface between Zn and magnesium is occupied by an Mg_2Zn_{11} intermetallic phase [21] (Figure 2) which belongs to the gamma brasses type of phase and contains 15.5 wt. % of Mg. This thermodynamically stable intermetallic phase [22] is generally formed during solidification of Zn–Mg alloys as part of the eutectic reaction. The average thickness of the presented phase was 1.2 ± 0.2 μm. Yang et. al. [23] have also studied Zn–Mg composite materials prepared from pure powders by spark plasma sintering (SPS). Final composites were composed of a mixture of stable Mg_2Zn_{11} and metastable Mg_2Zn. In addition, no pure magnesium remained in the microstructure [23].

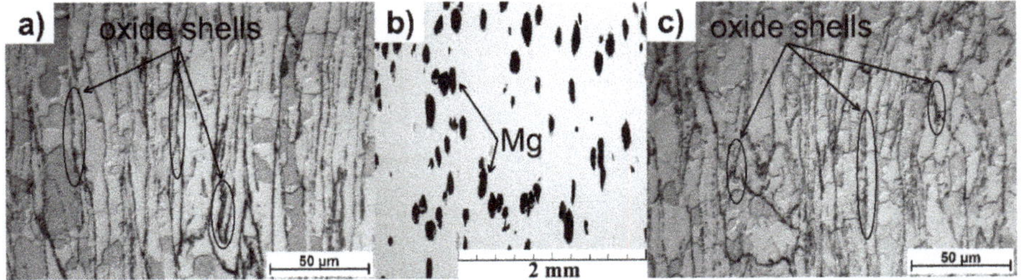

Figure 1. The microstructure of (**a**) Zn; (**b**) and (**c**) general view and detail of Zn-5Mg material prepared by powder metallurgy, respectively.

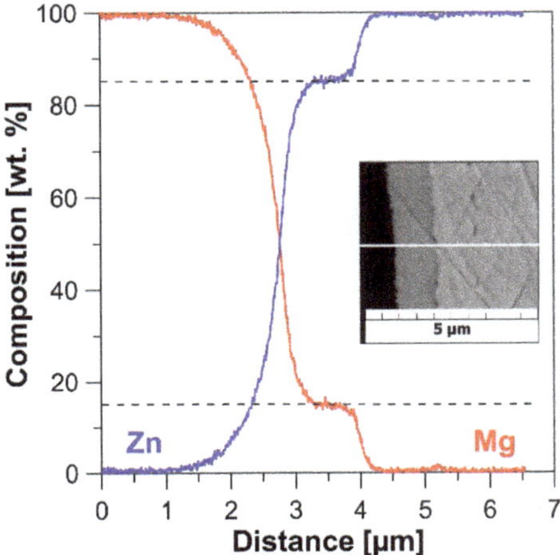

Figure 2. Line-scan of chemical composition at the interface of Zn and Mg.

3.2. Mechanical Properties

Tensile and compressive curves of Zn and Zn-5Mg and evaluated values of mechanical properties are shown in Figure 3 and Table 1, respectively. Due to similarities in the microstructure of the Zn matrix, the difference in mechanical behavior can be attributed to the addition of Mg particles. This improves the yield stress (TYS—tensile yield strength, CYS—compressive yield strength) and ultimate stress (UTS—ultimate tensile strength, UCS—ultimate compressive strength) values under both tension and compression loading. The general increase in presented values is about 40 MPa. Such an improvement is related to the presence of an Mg_2Zn_{11} intermetallic phase formed on the interface between the Mg and Zn particles (Figure 2). On the contrary, the specified phase is brittle and causes a partial decrease in the ductility. Nevertheless, both Zn and Zn-5Mg are still characterized by excellent elongation (35 ± 4% and 16 ± 2%, respectively) compared to various Zn-based alloys [10]. Yang et al. [23] measured CYS for Zn-5Mg (186.63 ± 26.54 MPa) prepared by SPS and indicate the possible use of these materials for some applications without load-bearing requirements. However, the values of 148 ± 6 MPa for TYS and 183 ± 4 MPa for UTS measured for extruded Zn-5Mg are not far from generally required values for fixation medical devices (TYS > 200 MPa, UTS > 270 MPa), and suitable processing of these materials could lead to the desired values. Additionally, the value of modulus of elasticity (E) is slightly lower for Zn-5Mg (81 GPa) as a consequence of the lower value of E for pure Mg (45 GPa) compared to pure Zn (108 GPa). Such behavior is also desirable to prevent the stress shielding effect.

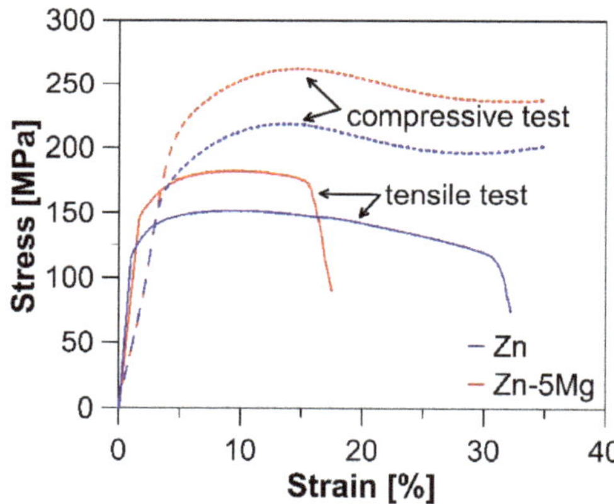

Figure 3. Compressive and tensile stress–strain curves for extruded Zn and Zn-5Mg.

Table 1. Tensile and compressive mechanical properties of Zn and Zn-5Mg materials after compaction by extrusion at 200 °C. Compressive tests were finished at 35% of relative deformation due to the absence of fracture.

	HV1	Tensile Test			Compression Test	
		TYS (0.2 Proof Stress) (MPa)	**UTS (MPa)**	**Elongation (%)**	**CYS (0.2 Proof Stress) (MPa)**	**UCS (MPa)**
Zn	35	114 ± 5	156 ± 5	35 ± 4	170 ± 6	215 ± 4
Zn-5Mg	36 (Zn) 51 (Mg)	148 ± 6	183 ± 4	16 ± 2	209 ± 6	256 ± 6

3.3. Corrosion Properties

Corrosion rates evaluated from weight changes were 0.373 ± 0.004 and 0.043 ± 0.020 mg·cm^{-2}·day^{-1} for Zn and Zn-5Mg respectively. Evaluation from Zn^{2+} concentration led to lower values, namely to 0.082 ± 0.016 and 0.022 ± 0.010 mg·cm^{-2}·day^{-1}, due to the presence of Zn in the solid state of corrosion products. This result reveals a surprising four-time decrease in released Zn^{2+} ions for Zn-5Mg. In fact, during in vitro cytotoxicity testing, even 50% dilution generally highly improves cell viability [15]. Regarding the alloying of Zn-based biodegradable alloys, such a huge change in corrosion rate has never been observed, to the best of our knowledge. To explain this behavior, potentiodynamic curves were measured after 1 and 12 h of immersion in SBF. The curves representing the polarization of pure Zn (blue lines in Figure 4) can be seen to be similar after 1 and 12 h in SBF. Both curves contain a specific linear region which indicates a cathodic reaction controlled by the diffusion of oxygen. The Ecor value of Zn-5Mg (−1.42 V) is due to the dissolution of less noble Mg, producing a more negative value compared to the E_{cor} of pure Zn (−0.98 V). At this potential, the reduction of water (Equation (1)) accompanied by hydrogen release is a dominant cathodic reaction.

$$2H_2O + 2e^- \rightarrow H_2 + 2OH^- \quad (1)$$

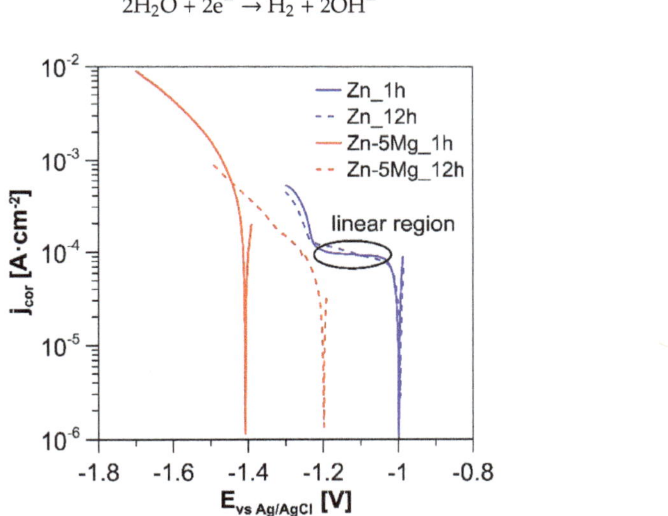

Figure 4. Polarization curves for extruded Zn and Zn-5Mg.

Magnesium works as a sacrificial anode and helps to protect the zinc matrix. Consequently, the decrease in current densities on the measured cathodic curve of sample Zn-5Mg immersed for 12 h was observed. The observed behavior caused the decrease in corrosion rate of the alloy during longer immersion tests and also lowered the amount of released Zn^{2+} ions during the initial stage of the corrosion process, which can particularly suppress the cytotoxic effect and improve the material biocompatibility. In contrary to the present study, Yang et al. [23] observed an increased corrosion rate for Zn-5Mg compared to pure Zn. They claimed that the corrosion rate of Zn-5Mg was increased as a consequence of micro-galvanic cells among $MgZn_2$, Mg_2Zn_{11}, and pure Zn. In this study, the intermetallic phase occupied the only low area at the interface of pure Zn and Mg. Pure magnesium is even less noble than the presented phases, works as strong sacrificial anode, and enables the passivation of Zn. This leads in particular to the lower corrosion rates of the Zn-5Mg composite in this specific case and supports the idea that variation in phase composition of these types of materials can highly affect corrosion behavior.

4. Conclusions

This paper introduces a novel approach for the preparation of Zn-based biodegradable materials using the powder metallurgy technique. In this work, a Zn-5Mg alloy was successfully prepared by extrusion of a green compact containing a mixture of pure Zn and Mg powders. Compared to the product made using Zn powder, the addition of Mg was observed to positively affect TYS and UTS values while maintaining excellent elongation. Additionally, Zn-5Mg was characterized by a decreased value of modulus of elasticity, which is highly desirable for fixation devices intended for traumatology applications, and a lower corrosion rate compared to pure Zn and Zn-based alloys. A lower amount of Zn ions released by the corrosion process can suppress cytotoxicity effects. Generally, the observed behaviour will lead to a significant improvement in the biocompatibility of the material.

Author Contributions: Conceptualization, J.K. and D.V.; methodology, D.D. and J.Č.; investigation, J.K., D.D., and J.P.; writing—original draft, J.K.; writing—review and editing, D.V. and J.Č.; supervision, D.V.

Funding: This research was funded by the Czech Science Foundation (project no. 18-06110S) and specific university research (MSMT No 21-SVV/2019) for financial support of this work.

Conflicts of Interest: The authors declare no conflict of interest.

References

1. Vimalanandan, A.; Bashir, A.; Rohwerder, M. Zn-Mg and Zn-Mg-Al alloys for improved corrosion protection of steel: Some new aspects. *Mater. Corros.* **2014**, *65*, 392–400. [CrossRef]
2. Hausbrand, R.; Stratmann, M.; Rohwerder, M. Corrosion of zinc–magnesium coatings: Mechanism of paint delamination. *Corros. Sci.* **2009**, *51*, 2107–2114. [CrossRef]
3. Prosek, T.; Nazarov, A.; Bexell, U.; Thierry, D.; Serak, J. Corrosion mechanism of model zinc–magnesium alloys in atmospheric conditions. *Corros. Sci.* **2008**, *50*, 2216–2231. [CrossRef]
4. Jin, H.; Zhao, S.; Guillory, R.; Bowen, P.K.; Yin, Z.; Griebel, A.; Schaffer, J.; Earley, E.J.; Goldman, J.; Drelich, J.W. Novel high-strength, low-alloys Zn-Mg (<0.1 wt% Mg) and their arterial biodegradation. *Mater. Sci. Eng. C* **2018**, *84*, 67–79. [CrossRef]
5. Bowen, P.K.; Guillory, R.J.; Shearier, E.R.; Seitz, J.-M.; Drelich, J.; Bocks, M.; Zhao, F.; Goldman, J. Metallic zinc exhibits optimal biocompatibility for bioabsorbable endovascular stents. *Mater. Sci. Eng. C* **2015**, *56*, 467–472. [CrossRef]
6. Yang, H.; Wang, C.; Liu, C.; Chen, H.; Wu, Y.; Han, J.; Jia, Z.; Lin, W.; Zhang, D.; Li, W.; et al. Evolution of the degradation mechanism of pure zinc stent in the one-year study of rabbit abdominal aorta model. *Biomaterials* **2017**, *145*, 92–105. [CrossRef]
7. Zheng, Y.F.; Gu, X.N.; Witte, F. Biodegradable metals. *Mater. Sci. Eng. R Rep.* **2014**, *77*, 1–34. [CrossRef]
8. Vojtech, D.; Kubasek, J.; Serak, J.; Novak, P. Mechanical and corrosion properties of newly developed biodegradable Zn-based alloys for bone fixation. *Acta Biomater.* **2011**, *7*, 3515–3522. [CrossRef] [PubMed]
9. Su, Y.C.; Cockerill, I.; Wang, Y.D.; Qin, Y.X.; Chang, L.Q.; Zheng, Y.F.; Zhu, D.H. Zinc-Based Biomaterials for Regeneration and Therapy. *Trends Biotechnol.* **2019**, *37*, 428–441. [CrossRef]
10. Venezuela, J.; Dargusch, M.S. The influence of alloying and fabrication techniques on the mechanical properties, biodegradability and biocompatibility of zinc: A comprehensive review. *Acta Biomater.* **2019**, *87*, 1–40. [CrossRef]
11. Li, H.F.; Xie, X.H.; Zheng, Y.F.; Cong, Y.; Zhou, F.Y.; Qiu, K.J.; Wang, X.; Chen, S.H.; Huang, L.; Tian, L.; et al. Development of biodegradable Zn-1X binary alloys with nutrient alloying elements Mg, Ca and Sr. *Sci. Rep.* **2015**, *5*, 10719. [CrossRef] [PubMed]
12. Kubasek, J.; Vojtech, D.; Jablonska, E.; Pospisilova, I.; Lipov, J.; Ruml, T. Structure, mechanical characteristics and in vitro degradation, cytotoxicity, genotoxicity and mutagenicity of novel biodegradable Zn-Mg alloys. *Mater. Sci. Eng. C* **2016**, *58*, 24–35. [CrossRef] [PubMed]
13. Mostaed, E.; Sikora-Jasinska, M.; Drelich, J.W.; Vedani, M. Zinc-based alloys for degradable vascular scent applications. *Acta Biomater.* **2018**, *71*, 1–23. [CrossRef]
14. Levy, G.K.; Goldman, J.; Aghion, E. The Prospects of Zinc as a Structural Material for Biodegradable Implants-A Review Paper. *Metals* **2017**, *7*, 18. [CrossRef]
15. Haase, H.; Hebel, S.; Engelhardt, G.; Rink, L. The biochemical effects of extracellular Zn(2+) and other metal ions are severely affected by their speciation in cell culture media. *Metallomics* **2015**, *7*, 102–111. [CrossRef]
16. Gu, X.-N.; Zheng, Y.-F. A review on magnesium alloys as biodegradable materials. *Front. Mater. Sci. China* **2010**, *4*, 111–115. [CrossRef]

17. Zhu, D.; Su, Y.; Young, M.L.; Ma, J.; Zheng, Y.; Tang, L. Biological Responses and Mechanisms of Human Bone Marrow Mesenchymal Stem Cells to Zn and Mg Biomaterials. *ACS Appl. Mater. Interfaces* **2017**, *9*, 27453–27461. [CrossRef]
18. Jablonska, E.; Vojtech, D.; Fousova, M.; Kubasek, J.; Lipov, J.; Fojt, J.; Ruml, T. Influence of surface pre-treatment on the cytocompatibility of a novel biodegradable ZnMg alloy. *Mater. Sci. Eng. C* **2016**, *68*, 198–204. [CrossRef]
19. Krystýnová, M.; Doležal, P.; Fintová, S.; Březina, M.; Zapletal, J.; Wasserbauer, J. Preparation and Characterization of Zinc Materials Prepared by Powder Metallurgy. *Metals* **2017**, *7*, 396. [CrossRef]
20. Sadighikia, S.; Abdolhosseinzadeh, S.; Asgharzadeh, H. Production of high porosity Zn foams by powder metallurgy method. *Powder Metall.* **2015**, *58*, 61–66. [CrossRef]
21. Samson, S. Die kristallstruktur von Mg2Zn11—Isomorphie zwischen Mg2Zn11 und Mg2Cu6Al5. *Acta Chem. Scand.* **1949**, *3*, 835–843. [CrossRef]
22. Smithells, C.J.; Gale, W.F.; Totemeier, T.C. *Smithells Metals Reference Book*, 8th ed.; Elsevier Butterworth-Heinemann: Amsterdam, The Netherlands; Boston, MA, USA, 2004.
23. Yang, H.T.; Qu, X.H.; Lin, W.J.; Chen, D.F.; Zhu, D.H.; Dai, K.R.; Zheng, Y.F. Enhanced Osseointegration of Zn-Mg Composites by Tuning the Release of Zn Ions with Sacrificial Mg-Rich Anode Design. *ACS Biomater. Sci. Eng.* **2019**, *5*, 453–467. [CrossRef]

Article

Impact of the Morphology of Micro- and Nanosized Powder Mixtures on the Microstructure of Mg-Mg_2Si-CNT Composite Sinters

Anita Olszówka-Myalska *, Patryk Wrześniowski, Hanna Myalska, Marcin Godzierz and Dariusz Kuc

Institute of Materials Science, Department of Material Science and Metallurgy, Silesian University of Technology, Krasińskiego 8, 40-019 Katowice, Poland; patryk.wrzesniowski@polsl.pl (P.W.); hanna.myalska@polsl.pl (H.M.); marcin.godzierz@polsl.pl (M.G.); dariusz.kuc@polsl.pl (D.K.)

* Correspondence: anita.olszowka-myalska@polsl.pl; Tel.: +48-32-603-4422

Received: 26 August 2019; Accepted: 2 October 2019; Published: 4 October 2019

Abstract: The problem of preparing a ternary powder mixture, which was meant to fabricate sintered heterophase composite, and consisted of micro- and two nanosized powders, was analyzed. The microsized powder was a pure magnesium, and as nanocomponents, a silicon powder (nSi) and carbon nanotubes (CNTs) with 2% and 1% volume fractions, respectively, were applied. The powder mixtures were prepared using ultrasonic and mechanical mixing in technological fluid, and four mixing variants were applied. The morphology of the powder mixtures was characterized with scanning electron microscopy (SEM), and then, composite sinters were fabricated in a vacuum with hot temperature pressing at 580 °C under 15 MPa pressure, using a Degussa press. The reaction between the nSi and the Mg matrix, which caused the creation of the Mg_2Si phase in the fabricated Mg-Mg_2Si-CNT composite, was confirmed with X-ray diffraction (XRD). The porosity and hardness of the composite sinters were examined, and optical microscopy (OM) and quantitative image analyses were carried out to characterize the microstructure of the composites. In the manufacturing process of the Mg-nSi-CNT mixtures, the best results were the following: first separate de-agglomeration of nanocomponents, then their common mixing, and finally, the deposition of nanocomponents at the surface of the microsized magnesium powder. The applied procedure ensured the uniform layer formation of de-agglomerated nanocomponents on the Mg powder, without re-agglomerated nSi and CNTs. Moreover, this type of powder mixture morphology allows to obtain sinters with lower porosity and higher hardness, which is accompanied by precipitation of a finer Mg_2Si phase. In the Mg-Mg_2Si-CNT composite, the carbon phase was present, and it was located in the magnesium matrix and in silicide.

Keywords: heterophase magnesium matrix composite; Mg_2Si; carbon nanotubes; nanopowders de-agglomeration; sintering

1. Introduction

Development in the fabrication of nanopowder materials, and their increased availability, is gaining momentum around the world. Nanopowder materials, such as carbon nanotubes (CNTs) [1–5], graphene [5–7], and nanosilver [8,9], due to their extraordinary properties, are applied as reinforcing components in composites with polymer [10–12], metal [13–17], and ceramic matrices [18,19]. Research focused on the use of nanocomponents concerns composites with the magnesium matrix, due to the low density of this metal [20–22]. However, proper use of the unique properties of nanocomponents is limited, due to the components' high tendency to agglomerate, which means that the consolidation of agglomerated nanofibers or nanoparticles with the matrix component can cause a decrease in, for example, mechanical properties, as well as thermal or electric conductivity. In the case of metal matrix composites manufactured with powder metallurgy methods, powder mixtures with 0.5–4.0 vol.% of nanocomponents are usually applied, and much attention has focused on the formation of homogenous mixtures. Mixing in the ball mills is a method well-known in the literature for preparing powder mixtures, but this method can cause a loss of nanocomponents, the generation of internal defects and their re-agglomeration. Another solution, which has been applied to Mg-CNTs [23–25] and Mg-nSiO_2 [26,27] mixtures, is ultrasonic mixing in technological fluids.

In the present work, ultrasonic mixing was applied to develop a beneficial procedure for preparing ternary mixtures, for a system that included a microsized powder and two nanopowders. That issue is crucial in the design of a new multiphase material technology. In the experiments, carbon nanotubes and silicon nanopowder (nSi) were used. During sintering, the silicon reacts with magnesium, and an intermetallic Mg_2Si phase is formed [28–32]. Thus, from the Mg-nSi-CNT powder mixture, a composite with Mg_2Si particles and nanotubes will be created. Due to the mechanical properties, magnesium silicide is often used as a reinforcement phase in ex situ and in situ composites with aluminum [33–36] and the magnesium [26–30] matrix. The intermetallic phase also exhibits good thermoelectric properties, and many studies have focused on that aspect of application [29,30]. However, the structural effect of the formation of Mg_2Si in an environment enriched with CNTs is not known in the literature. Some movement of CNTs induced by synthesis reaction can be expected in the composite. The thermodynamic data indicate the reaction $2Mg + Si \rightarrow Mg_2Si$, and Gibbs free energy ΔG at 580 °C (the calculations were carried out using HSC Chemistry 4) reaches the value of −71.76 kJ. However, transformation of the CNTs into SiC is also possible, because ΔG for the reaction $Si + C \rightarrow SiC$ is also negative, and equals −64.96 kJ.

In the here presented studies, during the preparation of the ternary powder mixtures, it was assumed that de-agglomeration of the main nanocomponents would occur due to the ultrasound action in technological liquid (alcohol) [37] supported by mechanical mixing. The issue that requires detailed experiments and analyses is the correct order of the mixing steps which may limit uncontrolled re-agglomeration of the nanocomponents. Four variants of the mixing procedure were applied, and the effectiveness of the de-agglomeration was evaluated based on the powder mixture morphology. The porosity, hardness and microstructure of the composite sinters were also examined. Special attention was paid to Mg_2Si phase dispersion, because the presence of nSi clusters in a ternary powder mixture induces relatively massive silicides in the composite, and on this basis, the effectiveness of the de-agglomeration of the nanocomponents can be characterized. The work concerns the general problem as the simultaneous use of fibrous and particulate nanocomponents in the conventional powder metallurgy processes, where a microsized powder is the main ingredient. This issue will occur in materials design regardless of the components chemical and phase composition. The powder ultrasonic mixing used in technological procedure allows to de-agglomerate primary nanofibers and nanoparticles agglomerates, and then their mixture with microsized component formation. The morphology of the mixture is controlled by interaction processes occurring in systems: nanofiber-microparticle, nanoparticle-microparticle, and nanofiber-nanoparticle. This type of powder mixture can be potentially applied to fabricate sintered composites or semi-products, in form of cold-pressed or hot-pressed moulds. In the literature, an application of multiphase moulds obtained by powder metallurgy was proposed for metal matrix composite fabrication with the different methods of plastic working processes [38–44] or casting methods [45–47]. The goal of such technological solutions is an increase of the reinforcing phases homogeneity, and its dispersion, in the case of in situ formed phases.

2. Materials and Methods

As a raw material in the experiments, magnesium powder (63–250 μm, Sigma Aldrich, 13112, Saint Louis, MO, USA), silicon nanopowder (average particle size 50 nm, Sigma Aldrich, Saint Louis, MO, USA) and carbon nanotubes (diameter 50–85 nm, length 10–15 μm, Multi-Walled Carbon Nanotube Powder, GRAPHENE SUPERMARKET, Ronkonkoma, State NY, US) were applied. Micrographs of the fabricated Mg, nSi and CNT powders observed with a scanning electron microscope (SEM) are presented in Figures 1–3. The surface of the magnesium granules (Figure 1) is irregular, which might be a beneficial feature during the mixing process and the deposition of the nanocomposites on a single granule. The SEM micrographs of nSi (Figure 2) and the CNTs (Figure 3) (raw materials) revealed a few or dozen micrometer-sized clusters, and that disadvantage indicates the necessity of de-agglomeration in processing ternary powder mixtures. Otherwise, correct exploitation of unique nanoaddition properties will be ineffective, due to the submicro- and nanopores between single particles or fibers consolidated with the metal matrix, independent of the technology.

In the experiments, the same composition of the powder mixture, 97% Mg; 2% nSi; 1% CNT (vol.%), was chosen. Based on previous experiences regarding preparation of mixtures with nanocomponents [26,27,33], four variants of mixture processing consisting of ultrasonic and mechanical mixing in alcohol were applied (Table 1). The main differences were in the sequence of de-agglomeration of the nanocomponents and then mixing with the microsized powder. When the mixing procedure was completed, the liquid alcohol was removed from the suspension, and then the powder mixture was dried for 18 h at 60 °C.

The microstructure of the ternary powder mixtures was examined with a field emission (FE)SEM (FE Hitachi S-4200, Hitachi Group, Tokyo, Japan), and the distribution of the components was analyzed. Then the Mg-nSi-CNT powder mixtures and pure Mg powder as a reference material were sintered in a Degussa press, in a vacuum atmosphere, at 580 °C under 15 MPa pressure. Sinters 20 mm in diameter and 10 mm high were obtained.

The porosity measurements of the fabricated composite samples were carried out with the Archimedes method, and Vickers hardness HV0.2 was determined with the Zwick 110 hardness tester (Zwick, Ulm, Germany). For the phase composition examination of the sinters, X-ray diffraction (XRD) was applied (X'Pert 3 Powder X- ray diffractometer, Malvern Panalytical Ltd., Royston, UK). For microstructure characterization, polished samples without additional etching were prepared. The observations were carried out with an optical microscope (OM; GX71 Olympus, Tokyo, Japan) and scanning electron microscope (SEM, Hitachi 3400N, Tokyo, Japan) equipped with wavelength-dispersive X-ray spectroscope (WDS Thermo Scientific Magna Ray, Waltham, Massachusetts, US). By WDS method, the elemental mapping of magnesium (TAP crystal), silicon (TAP crystal), carbon (NiC80 crystal), and oxygen (NiC80 crystal) were obtained. The quantitative metallography examinations were focused on Mg_2Si phase characterization, the volume fraction evaluated with the area fraction and dispersion evaluated with the particle cross-section area. The examinations were performed using Met-Ilo software (J. Szala, Silesian University of Technology, Katowice, Poland), to analyze the influence of the morphology of the Mg-nSi-CNT powder mixture on the microstructure of the composite sinter. The stereological parameters were determined using OM images, although the resolution of method excludes identification of a single CNT, and allows observation only of bigger objects, due to the characteristic blue of the Mg_2Si phase, identifying this silicide was very simple, in contrast to the SEM observation. The size of the Mg_2Si phase particles and the quantity were measured, and then the particles divided into five size classes (left closed intervals: 10–100, 100–1000, 1000–10,000, 10,000–100,000 and > 100,000 μm^2). The measurement procedure was conducted in a relatively large area (15 different areas were chosen for investigation, and they were conducted at 50X magnification). The image analyses allowed to indicate the most effective mixing method for preventing the re-agglomeration phenomenon. In the case of this phenomenon, the methodology is more accurate in comparison with thin foils, for example, where only a small area of the material can be analyzed. Additionally, the silicide particles detected with the OM, and characterized with quantitative metallography, were only a portion of the nSi and Mg reaction product. The other particles were smaller, even nanosized, and more difficult to distinguish.

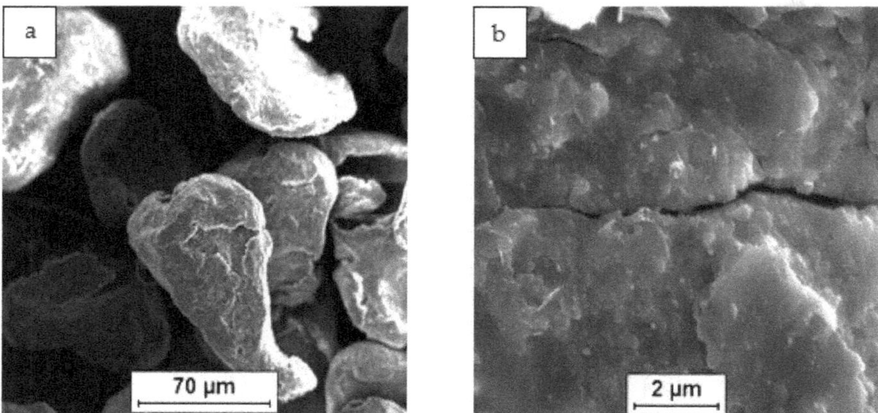

Figure 1. SEM micrographs of Mg powder as-fabricated: (**a**) particles, (**b**) surface of single particle.

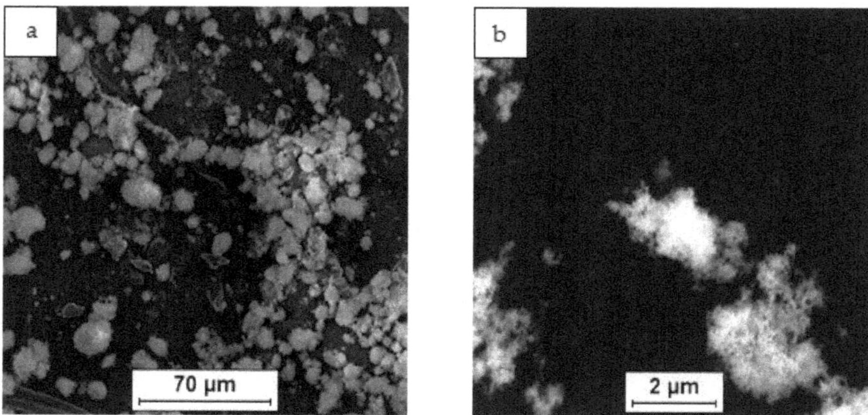

Figure 2. SEM micrographs of silicon nanopowder (nSi) as-fabricated: (**a**) agglomerates of different size, (**b**) morphology of single agglomerate.

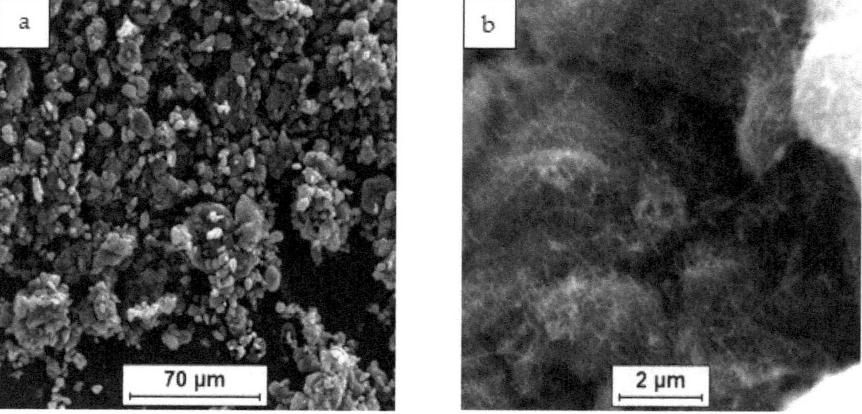

Figure 3. SEM micrographs of as-fabricated carbon nanotubes (CNT): (**a**) agglomerates of different size, (**b**) morphology of single agglomerate.

Table 1. Applied procedure of ternary powder mixture preparation and condition of their sintering process (subscript D means the ultrasonic de-agglomeration in alcohol).

Mixture Signature	Mixture Composition [vol.%]	Mixing Procedure	Sintering Conditions
Mg	100 Mg	Reference sample	580 °C, 15 MPa, vacuum
Mg+(nSi_D + CNT_D)	97% Mg; 2% nSi; 1% CNT	- separate de-agglomeration nSi and CNT - preparation of $(nSi)_D$ + $(CNT)_D$ suspension - addition of Mg to $(nSi)_D$ + $(CNT)_D$ suspension	
Mg + $(nSi + CNT)_D$		- common de-agglomeration of nSi and CNT - addition of Mg+$(nSi+CNT)_D$ suspension	
(Mg + nSi_D) + CNT_D		- separate de-agglomeration of nSi and CNT - preparation of $(nSi)_D$ + Mg suspension - addition of $(nSi)_D$ to Mg + $(CNT)_D$ suspension	
(Mg + CNT_D) + nSi_D		- separate de-agglomeration of nSi and CNT - preparation of $(CNT)_D$ + Mg suspension - addition of $(CNT)_D$ to Mg + $(nSi)_D$ suspension	

3. Results

3.1. Characterization of the Powder Mixtures

Examinations of the Mg-nSi-CNT powder mixtures with SEM revealed that the main effect of the mixing procedure was the deposition of the nanopowders on the Mg powder. This phenomenon is known from the literature for a binary mixture of microsized and nanosized components and occurs due to adhesion forces [23–28]. Moreover, the macroscopic observations of the mixtures did not reveal a significant residue of nanocomponents on the mixer walls after mixing, and generally, the metallic color of the magnesium granules disappeared. Additionally, the technological liquid was transparent after the process.

The microscopic examination of the dried ternary powder mixtures (Figures 4–7) showed differences in the distribution of the CNTs and nSi, which was related to the mixing procedure. Magnesium powder was coated with nanocomponents, but an unfavorable phenomenon was observed in the case of the (Mg + nSi_D) + CNT_D and (Mg + CNT_D) + nSi_D samples (Figures 6a and 7a). Agglomerates of CNTs and nSi, a few micrometers and larger, were detected, and they were poorly connected to the Mg grains, and even separate from the metallic granules. This effect was more intense for the $(Mg + nSi)_D$ + CNT_D mixture, but in general, the tendency to re-agglomeration of two different nanocomponents in the ultrasonic mixed suspension was noticed. It occurred when the microsized component (Mg) was not mixed at the same time with the two previously de-agglomerated nanocomponents. The main reason for that type of re-agglomeration likely is the fibrous morphology of the CNTs. In the procedure for mixing the (Mg + nSi_D) + CNT_D powder, the nSi particles initially deposited at magnesium were caught by the de-agglomerated fibers, while in the processes for preparing (Mg + CNT_D) + nSi_D, the nSi accumulated around the CNTs.

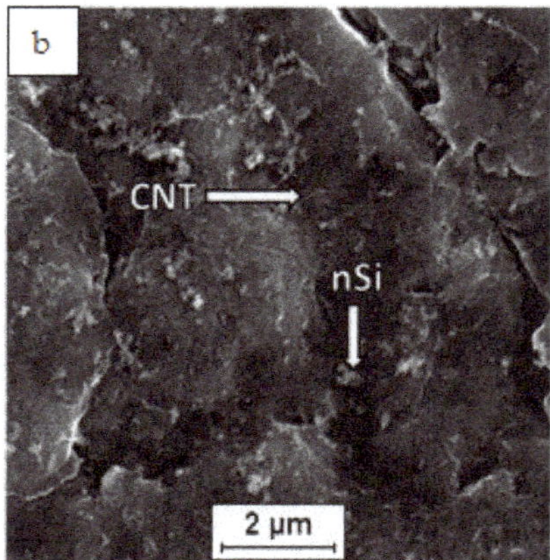

Figure 4. SEM micrographs of Mg+(nSi_D+CNT_D) powder mixture: (**a**) morphology of powder mixture, (**b**) surface of microsized Mg coated with CNT and nSi.

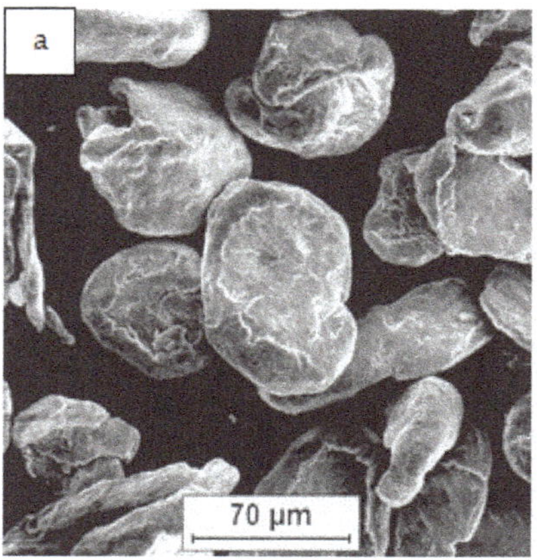

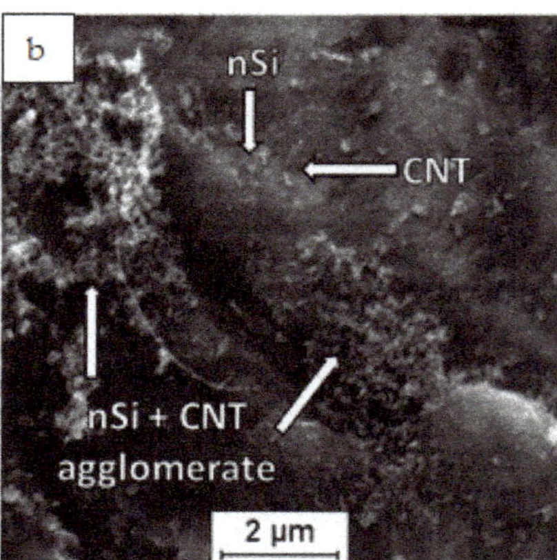

Figure 5. SEM micrographs of Mg + $(nSi + CNT)_D$ powder mixture: (**a**) morphology of powder mixture, (**b**) surface of microsized Mg coated with CNT and nSi.

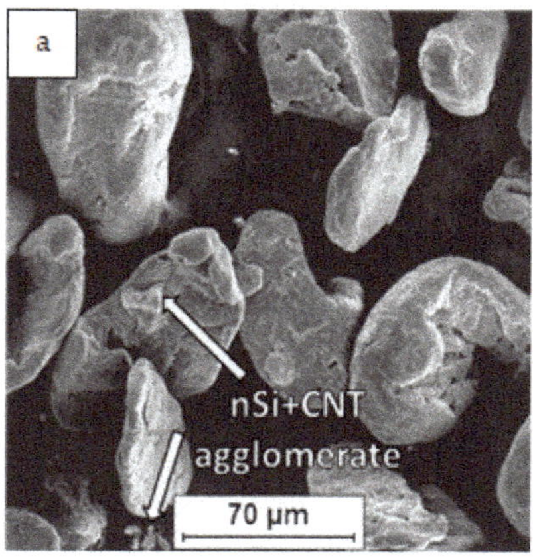

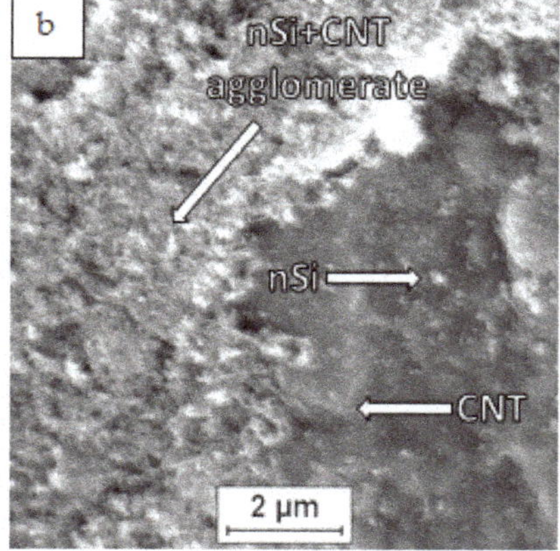

Figure 6. SEM micrographs of (Mg + CNT_D) + nSi_D powder mixture: (**a**) morphology of powder mixture, (**b**) surface of microsized Mg coated with CNT and nSi.

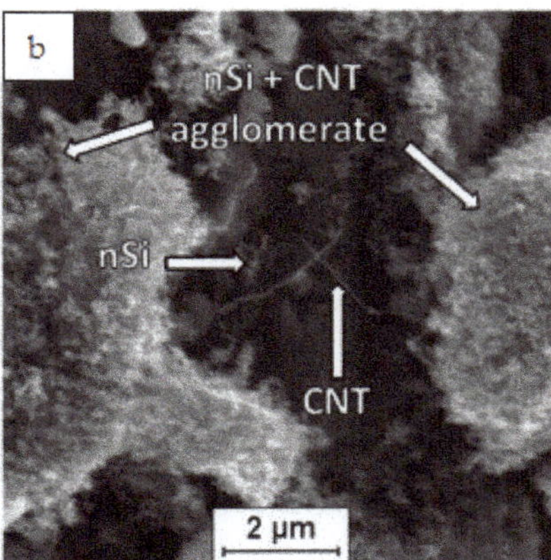

Figure 7. SEM micrographs of (Mg + nS_D) + CNT_D powder mixture: (**a**) morphology of powder mixture, (**b**) surface of microsized Mg coated with CNT and nSi.

Moreover, the microstructure observations indicated that in the proposed mixing procedures, for one micro- and two nanosized components, selective deposition of previously de-agglomerated nanocomponents could not be expected. This means that the formation of a layered structure, where first, nSi covers the Mg particles, and then the CNTs create another layer, or vice versa, cannot be achieved. That excludes formation of the CNT nanozone at the Mg-nSi interface, and limits further de-agglomeration of the fibrous component, induced in conditions where the silicide forms not by reactive diffusion, but by self-propagating high-temperature synthesis (SHS).

In comparison with the powder mixtures described previously, the results of the FE-SEM examinations for the two other mixtures, Mg + $(nSi + CNT)_D$ and Mg + (nSi_D + CNT_D), revealed some differences in the microstructure. Generally, the Mg surface was coated with distinctly separate CNT and nSi grains, and the layer of nanocomponents was uniform and relatively thin (Figure 4b). However, when the CNT and nSi were de-agglomerated together before being introduced in the Mg-alcohol suspension (the Mg + $(nSi + CNT)_D$ mixture), the agglomerates of CNT + nSi at the Mg surface were revealed. That effect may confirm a tendency of CNTs and nSi to mutual agglomeration, induced by the fibrous component.

3.2. *Characterization of Composite Sinters*

An example of the XRD pattern obtained for the composites is shown in Figure 8, and it confirms the presence of αMg and a new Mg_2Si phase, formed in situ as a result of the nSi reaction with the Mg matrix. A weak signal coming from MgO phase was identified as well. The presence of a carbon component, or SiC, was not detected due to the low carbon component content and insufficient method sensitivity. Therefore, in a future investigation, the identification of nanostructural phases in composite will be performed, and the high-resolution transmission electron microscopy and selective area electron diffraction methods will be conducted.

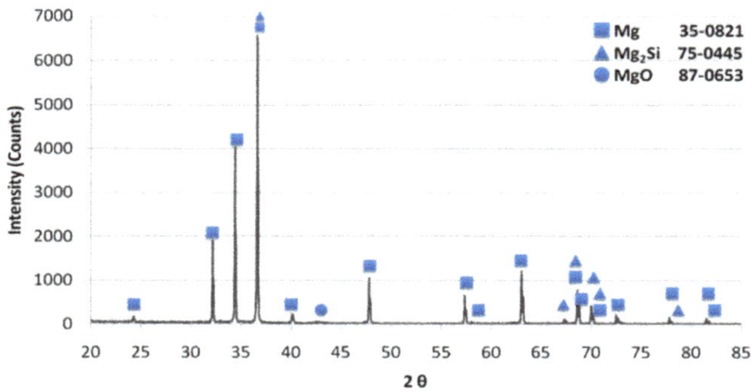

Figure 8. X-Ray Diffraction Pattern of Diffraction Mg + nSi_D + CNT_D composite.

Results of composite microstructure examinations with the OM without additional etching are shown in Figures 9–12. Two main structural elements were revealed: rounded magnesium areas (light) with characteristic MgO boundaries derived from the initial magnesium powder, and an irregular dark blue phase of similar or greater size (Figures 9a, 10a, 11a and 12a). At higher magnification (Figures 9b, 10b, 11b and 12b), it is visible that the irregular dark blue phase contains very fine black and elongated phases, and this suggests that the particles are a mixture of Mg_2Si and CNTs. Similar black elongated phases were detected in the initial magnesium powder grains except the oxides (beige), and inside the Mg grains as well. Within the Mg grains, light blue, irregular and very fine precipitations of Mg_2Si can be observed, and that suggests the SHS reaction. Moreover, that explains the movement of CNTs, previously deposited on the microsized Mg powder, in the magnesium matrix. The comparison of the microstructure of all composites indicates that in the material obtained from the Mg + (nSi_D + CNT_D) mixture, the Mg_2Si phase is the finest (Figure 9), and more beneficial conditions for matrix reinforcement formation are created.

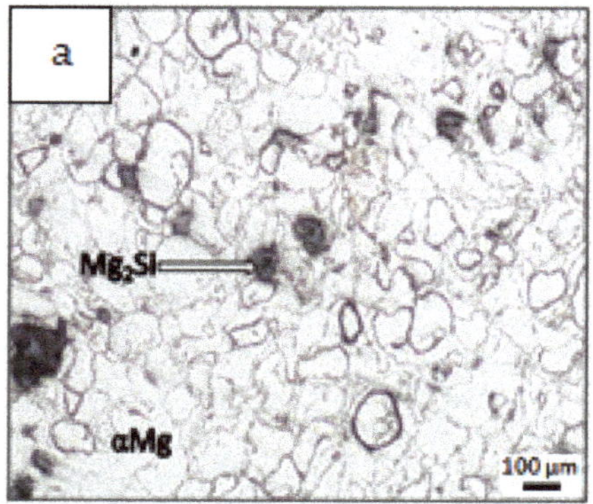

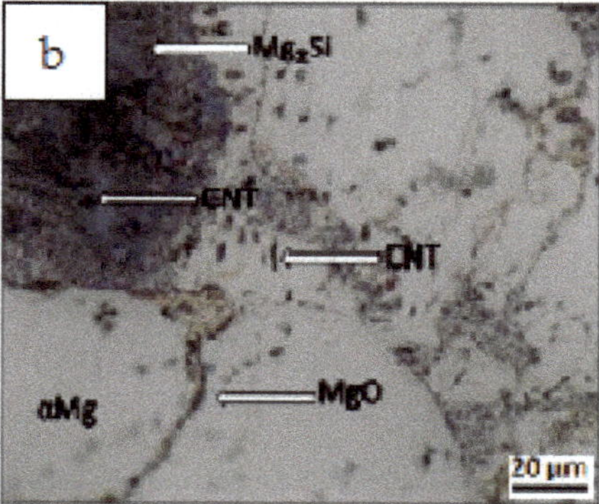

Figure 9. OM micrographs of Mg-Mg_2Si-CNT composite fabricated from Mg+(nSi_D+CNT_D) powder mixture: (**a**) Mg_2Si agglomerates in Mg matrix, (**b**) phases of different size and morphology.

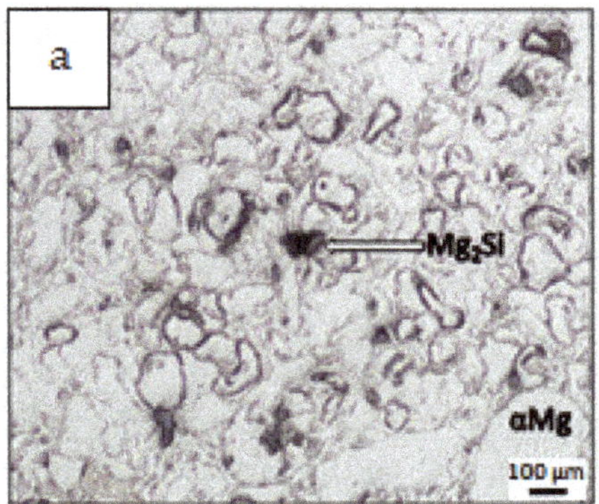

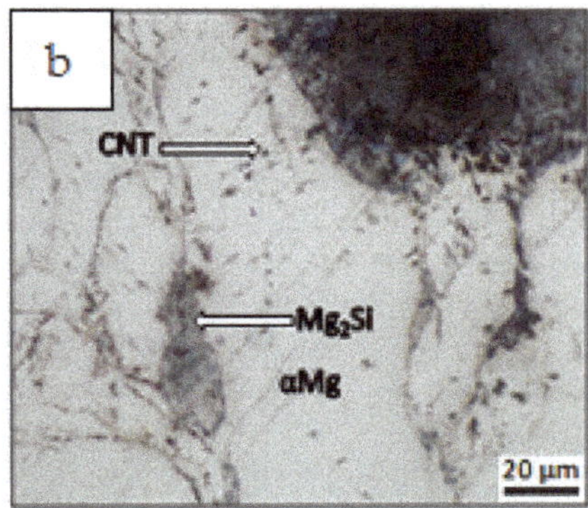

Figure 10. OM micrographs of Mg-Mg_2Si-CNT composite fabricated from Mg + (nSi + CNT)$_D$ powder mixture: (**a**) Mg_2Si agglomerates in Mg matrix, (**b**) phases of different size and morphology.

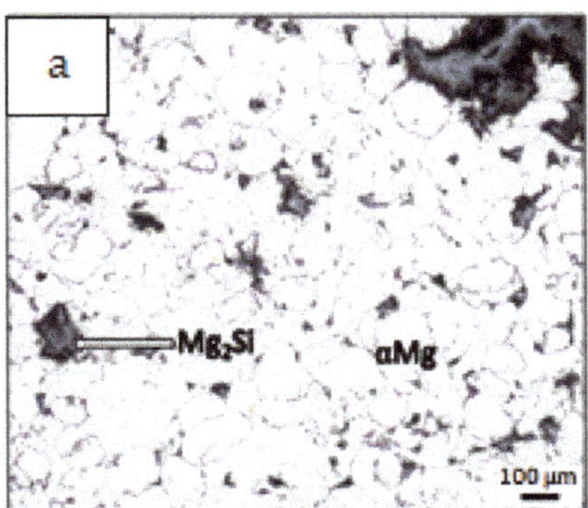

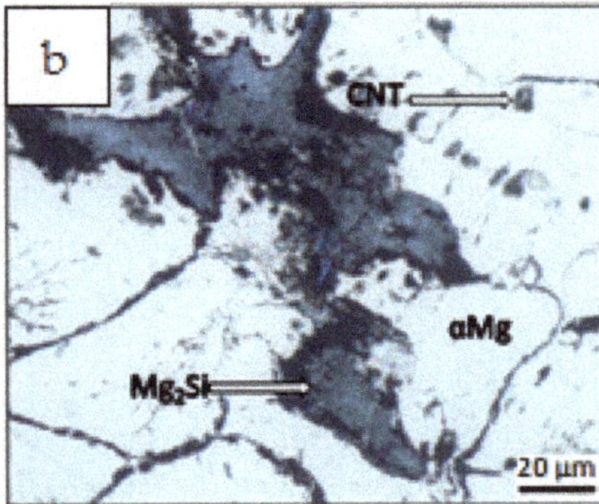

Figure 11. OM micrographs of Mg-Mg_2Si-CNT composite fabricated from (Mg + nSi$_D$) + CNT$_D$ powder mixture: (**a**) Mg_2Si agglomerates in Mg matrix, (**b**) phases of different size and morphology.

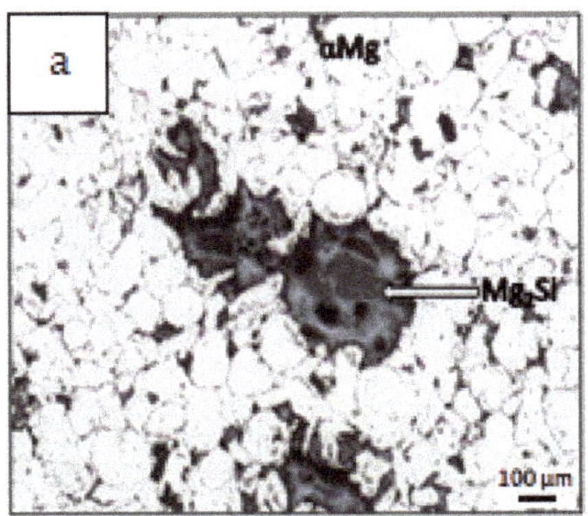

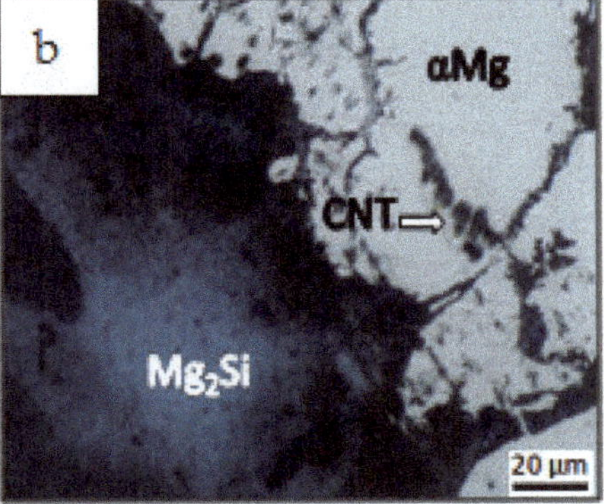

Figure 12. OM micrographs of Mg-Mg_2Si-CNT composite fabricated from (Mg + CNT$_D$) + nSi$_D$ powder mixture: (**a**) Mg_2Si agglomerates in Mg matrix, (**b**) phases of different size and morphology.

Results of composite microstructure examinations with the SEM and WDS are presented in Figure 13. The results revealed distribution of Mg, Si, O and C. The overlapping areas, enriched in Mg and Si, confirmed in situ formation of Mg_2Si particles in the composite. Moreover, in the regions containing Mg_2Si particles, an increase of carbon concentration was observed on the carbon elemental mapping (Figure 13e), which may suggest the presence of CNT's. Furthermore, the microareas with higher concentration of oxygen were found in the magnesium matrix. The oxygen distribution indicates oxides presence originating from the initial Mg powder surface. Obtained WDS elemental mapping results are consistent with the OM and XRD results.

Results for the porosity and microhardness measurements are presented in Table 2, and the differences depending on the mixing procedure and the effectiveness of the de-agglomeration of the nanocomponents are demonstrated. The lower porosity of the composite obtained from the initial powder mixture with the uniform nanocomponents distribution, that is, Mg + (nSi_D + CNT_D), indicated that this type of powder mixture morphology is the most effective in material compaction, and it influences the composite microhardness, which is a little bit higher for a sinter obtained from this mixture. The highest porosity was obtained for the reference Mg sinter, in comparison with the composites obtained. This result can be explained by the impact of an exothermal reaction between Mg and nSi, which induced a local increase in the temperature in the composite sinter, and better compaction under the same pressure applied.

Table 2. Hardness and open porosity of composite sinters obtained from Mg-nSi-CNT mixtures of the same composition and prepared by different way.

Material	Hardness, HV 0.2	Open Porosity, %
Mg	41.98 ± 2.8	1.22
Mg + (nSi_D + CNT_D)	46.5 ± 3.0	0.71
Mg + (nSi + CNT)$_D$	44.5 ± 2.7	1.05
(Mg + nSi_D) + CNT_D	43.6 ± 4.7	0.88
(Mg + CNT_D) + nSi_D	44.6 ± 3.9	0.76

An example of the image transformation procedure applied for the quantitative metallography examination of the Mg_2Si phase with Met-Ilo software is presented in Figure 14. For each sample, 15 areas were measured, and the results are presented in Table 3 and in Figure 15.

Differences in Mg_2Si synthesis depending on the preparation of the powder mixture were revealed. The main difference was in the value of the silicide area fraction A_A, which is two times lower for the composites obtained from Mg + (nSi + CNT)$_D$ and Mg + (nSi_D + CNT_D) mixtures in comparison to (Mg + nSi_D) + CNT_D and (Mg + CNT_D) + nSi_D. That effect was obtained for the same initial powder composition and sintering parameters, which suggests that very fine Mg_2Si particles smaller than 10 μm^2 were formed, and many below the OM resolution. The analysis of the detected number of particles in size classes also exhibited an evident difference for the composites from the (Mg + nSi_D) + CNT_D and (Mg + CNT_D) + nSi_D mixtures. For those composites, the number of particles was greater compared to the two other samples. This effect was directly connected to the morphology of the mixture of the initial powders, and it can be explained as a result of the Mg reaction with few micrometer-sized nSi/CNT agglomerates, located either on the Mg microsized powder surface or separately from the metallic particles (Figures 6 and 7).

The preliminary research on the hybrid composite obtained with powder metallurgy showed that the parameters applied for Mg-nSi-CNT sintering and the mixture composition required optimization.

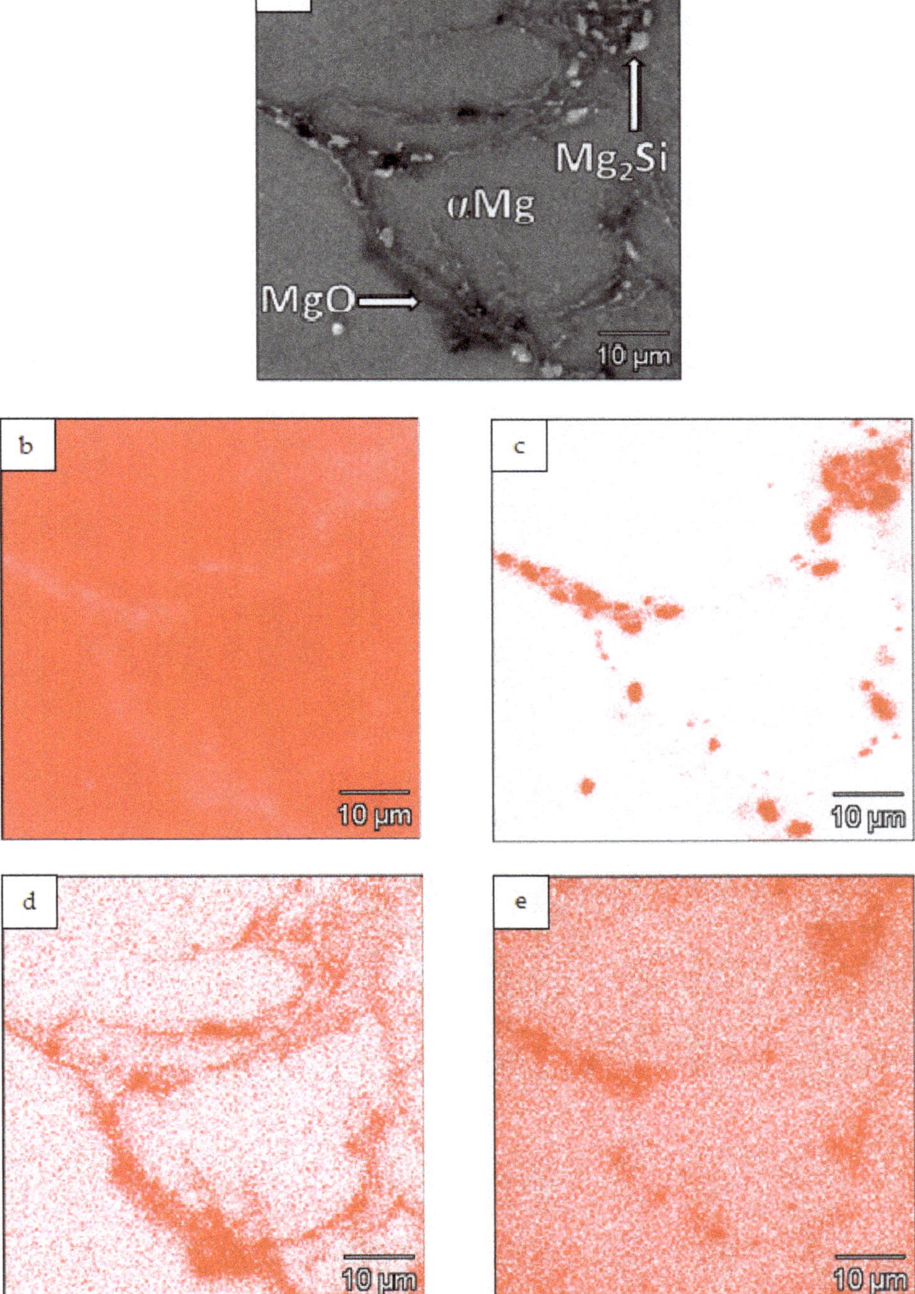

Figure 13. SEM micrographs of Mg-Mg_2Si-CNT composite fabricated from Mg+(nSi_D+CNT_D) powder in composition mode (**a**) with WDS elemental mapping of Mg (**b**), Si (**c**), O (**d**) and C (**e**).

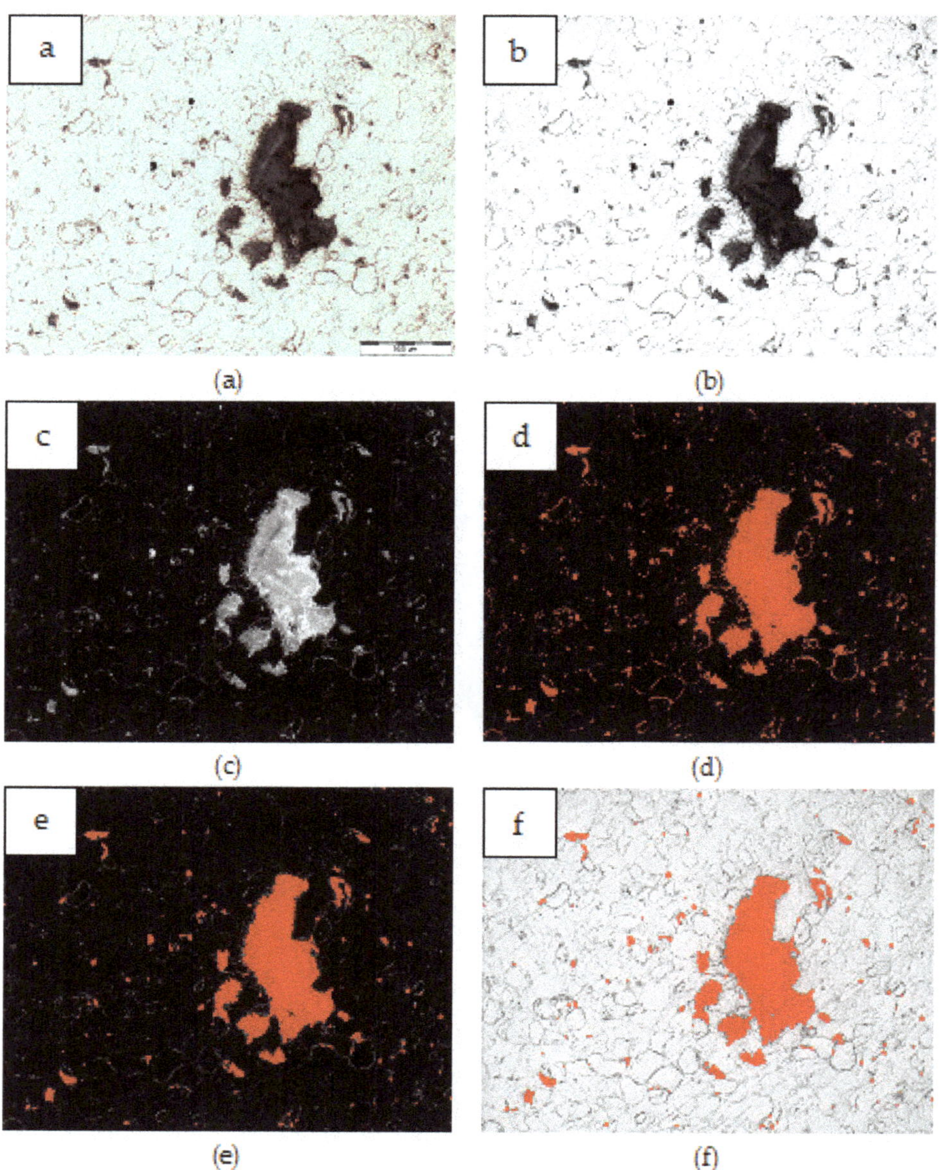

Figure 14. Procedure applied for Mg_2Si particles detection in Mg-Mg_2Si-CNT composite sinters: (**a**) initial OM image, (**b**) grey image-histogram equalization, (**c**) numerical inversion, (**d**) automatic binarization (k-means; white phase), (**e**) opening 2 and (**f**) complete image - initial + binary.

Table 3. Quantitative characteristics of Mg_2Si particles sized of more than 10 μm^2 formed in $Mg-Mg_2Si-CNT$ composite sinters.

Material	Area fraction, A_A [%]	Area [μm^2]	Count	Count
Mg + (nSi_D + CNT_D)	5.38 ± 2.66	10–100	747	2367
		100–1000	1295	
		1000–10,000	272	
		10,000–100,000	49	
		> 100,000	4	
Mg + (nSi + CNT)$_D$	6.69 ± 4.11	10–100	441	1950
		100–1000	1198	
		1000–10,000	246	
		10,000–10,0000	52	
		> 10,0000	13	
(Mg + nSi_D) + CNT_D	18.75 ± 9.88	10–100	1708	8018
		100–1000	5208	
		1000–10,000	994	
		10,000–10,0000	97	
		> 100,000	11	
(Mg + CNT_D) + nSi_D	17.50 ± 2.59	10–100	1917	10814
		100–1000	6935	
		1000–10,000	1833	
		10,000–100,000	122	
		> 100,000	7	

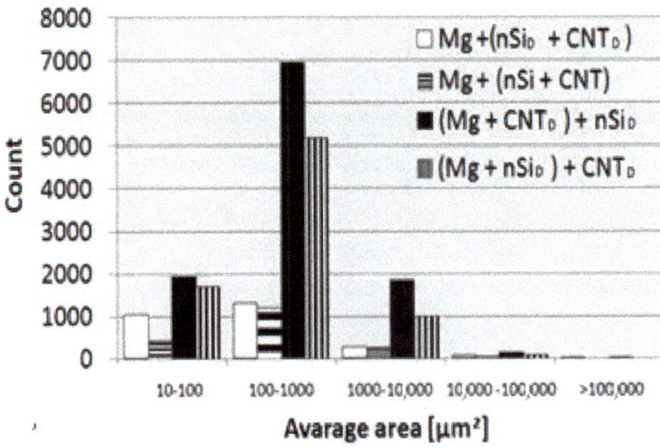

Figure 15. Quantity of Mg_2Si particles in composite sinters divided into size classes.

From the literature [13,14,21–23] it is known, that due to CNT's tendency for agglomeration, an application of more than 1 vol.% of CNT's may cause a decrease of mechanical and electrical properties in metal matrix composite. Thus, in the present work, the highest effective reinforcing amount of fibrous carbon nanocomponent was used. In our previous experiments, the procedure of Mg-nSi-CNT mixture preparing with proposed method was successfully examined to volume fraction less than 2% of nSi and 1% of CNT. However, the effects related with Mg_2Si formation can be useful in CNT de-agglomeration processes. Therefore, in further experiments two main aspects should be considered. The first aspect is the ratio of nanosilicon to CNT in the powder mixture, and the maximum number of nanocomponents, which can introduced into the Mg-base powder mixture. The second important issue is the adjustment of sintering parameters such as heating speed, sintering temperature and pressure.

Generally, the experiment revealed the possibilities of a design for micro- and nanopowder mixture morphology, and its influence on the final product microstructure and properties, including ex situ and in situ nano-reinforcement in composites.

4. Conclusions

A novel approach for processing ternary powder mixtures intended for fabricating hybrid composites, which consisted of a microsized component and two nanosized components, was presented in this work. The main conclusions are drawn as follows:

1. The fabrication of the microsized powder, nanopowder and nanotube mixture was successfully tested for the Mg-nSi-CNT system, and the ultrasonic method for de-agglomerating the nanocomponents in liquid base suspension proved to be useful if the appropriate order of technological procedures has been preserved.
2. The final ternary powder mixture consisted of a microsized powder coated with two nanocomponents, and the components were uniformly distributed when the nanocomponents were first de-agglomerated separately, and then mixed together and deposited at the microsized powder surface.
3. Differences in the microstructure of the Mg-Mg_2Si-CNT composite depending on the initial morphology of the powder mixture were observed. The most noticeable changes were in the size of the Mg_2Si particles, where the values could be very high (a few dozen micrometers in diameter), when nSi re-agglomeration occurred. Simultaneously, the size of the CNT agglomerates, detected in the composite, often surrounded by the Mg_2Si phase, was also greater.
4. The procedure for preparing the Mg-nSi-CNT mixture, which was proposed as the most effective, ensured the lowest porosity and the highest hardness of the Mg-Mg_2Si-CNT composite obtained by sintering under pressure.

Author Contributions: Conceptualization, A.O.-M.; methodology, A.O.-M. and H.M.; validation, H.M., P.W. and M.G.; formal analysis, A.O.-M. and P.W.; investigation, H.M., P.W. and M.G.; resources, D.K.; data curation, H.M., P.W., M.G. and D.K.; writing—original draft preparation, A.O.-M.; writing—review and editing, A.O.-M., H.M. and P.W.; visualization, P.W.; supervision, A.O.-M.

Funding: The study was carried out as a part of statutory research of the Institute of Materials Science at the Silesian University of Technology for 2019 (BK-205/ RM0/2019).

Conflicts of Interest: The authors declare no conflict of interest.

References

1. Tiwari, S.K.; Kumar, V.; Huczko, A.; Oraon, R.; Adhikari, A.D.; Nayak, G.C. Magical Allotropes of Carbon: Prospects and Applications. *Crit. Rev. Solid State Mater. Sci.* **2016**, *41*, 257–317. [CrossRef]
2. De Volder, F.L.M.; Tawfick, S.H.; Baughman, R.H.; Hart, A.J. Carbon Nanotubes: Present and Future Commercial Applications. *Science* **2013**, *399*, 535–539. [CrossRef] [PubMed]
3. Begtrup, G.E.; Ray, K.G.; Kessler, B.M.; Yuzvinsky, T.D.; Garcia, H.; Zettl, A. Extreme thermal stability of carbon nanotubes. *Phys. Status Solidi* **2007**, *11*, 3960–3963. [CrossRef]
4. Mahajan, A.; Kingon, A.; Kukovecz, A.; Konya, Z.; Vilarinho, P. Studies on the thermal decomposition ofmultiwall carbon nanotubes under different atmospheres. *Mater. Lett.* **2013**, *90*, 165–168. [CrossRef]
5. Kozioł, M.; Jesionek, M.; Szperlich, P. Addition of a small amount of multiwalled carbon nanotubes and flaked grapheme to epoxy resin. *J. Reinf. Plast. Comp.* **2017**, *36*, 640–654. [CrossRef]

6. Rutkowski, P.; Klimczyk, P.; Jaworska, L.; Stobierski, L.; Dubiel, A. Thermal properties of pressure sintered alumina–grapheme. *J. Therm. Anal. Calorim.* **2015**, *122*, 105–114. [CrossRef]
7. Sobczak, N.; Sobczak, J.J.; Kudyba, A.; Homa, M.; Bruzda, G.; Grobelny, M.; Kalisz, M.; Strobl, K.; Singhal, R.; Monville, M. Wetting transparency of graphene deposited on copper in contact with liquid tin. *Transact. Foundry Res. Inst.* **2014**, *54*, 3–11.
8. Wijnhoven, S.; Peijnenburg, W.; Herberts, C.; Hagens, W.; Oomen, A.; Heugens, E.; Roszek, B.; Bisschops, J.; Gosens, I.; Van De Meent, D.; et al. Nano–silver—A review of available data and knowledge gaps in human. *Nanotoxicology* **2009**, *3*, 109–138. [CrossRef]
9. Echegoyen, Y.; Nerín, C. Nanoparticle release from nano–silver antimicrobial food containers. *Food Chem. Toxicol.* **2013**, *62*, 16–22. [CrossRef]
10. Latko–Durałek, P.; Dydek, K.; Sobczak, M.; Boczkowska, A. Processing and characterization of thermoplastic nanocomposite fibers of hot melt copolyamide and carbon nanotubes. *J. Thermoplast. Compos.* **2018**, *31*, 1–17. [CrossRef]
11. Gupta, A.K.; Harsha, S.P. Analysis of mechanical properties of carbon nanotube reinforced polymer composites using continuum mechanics approach. *Procedia Mater. Sci.* **2014**, *6*, 18–25. [CrossRef]
12. Arash, B.; Wang, Q.; Varadan, V.K. Mechanical properties of carbon nanotube/polymer composites. *Sci. Rep.* **2014**, *4*, 1–8. [CrossRef] [PubMed]
13. Shimizu, Y.; Miki, S.; Soga, T.; Itoh, I.; Todoroki, H.; Hosono, T.; Sakaki, K.; Hayashi, T.; Kim, Y.A.; Endo, M.; et al. Multi-walled carbon nanotube-reinforced magnesium alloy composites. *Scr. Mater.* **2008**, *58*, 267–270. [CrossRef]
14. Han, G.; Wang, Z.; Liu, K.; Li, S.; Du, X.; Du, W. Synthesis of CNT reinforced AZ31 magnesium alloy composites with uniformly distributed CNTs. *Mater. Sci. Eng. A* **2015**, *628*, 350–357. [CrossRef]
15. Hekner, B.; Myalski, J.; Pawlik, T.; Sopicka–Lizer, M. Effect of Carbon in Fabrication Al–SiC Nanocomposites for Tribological Application. *Materials* **2017**, *10*, 679. [CrossRef] [PubMed]
16. Myalska, H.; Swadźba, R.; Rozmus, R.; Moskal, G.; Wiedermann, J.; Szymański, K. STEM analysis of WC–Co coatings modified by nano–sized TiC and nano–sized WC addition. *Surf. Coat. Tech.* **2017**, *318*, 279–287. [CrossRef]
17. Zhang, L.; Wang, Q.; Liao, W.; Guo, W.; Li, W.; Jiang, H.; Ding, W. Microstructure and mechanical properties of the carbon nanotubes reinforced AZ91D magnesium matrix composites processed by cyclic extrusion and compression. *Mater. Sci. Eng. A* **2017**, *689*, 427–434. [CrossRef]
18. Kostecki, M.; Grybczyk, M.; Klimczyk, P.; Cygan, T.; Woźniak, J.; Wejrzanowski, T.; Jaworska, L.; Morgiel, J.; Olszyna, A. Structural and mechanical aspects of multilayer graphene addition in alumina matrix composites–validation of computer simulation model. *J. Eur. Ceram. Soc.* **2016**, *36*, 4171–4179. [CrossRef]
19. Stankiewicz, N.; Lelusz, M. Nanotechnologia w budownictwie–przegląd zastosowań. *Civ. Environ. Eng.* **2014**, *5*, 101–112.
20. Li, H.; Cheng, L.; Sun, X.; Li, Y.; Li, B.; Liang, C.; Wang, H.; Fan, J. Fabrication and properties of magnesium matrix composite reinforced by urchin–like carbon nanotube–alumina in situ composite structure. *J. Alloy Compd.* **2018**, *746*, 320–327. [CrossRef]
21. Li, Q.; Viereckl, A.; Rottmair, C.A.; Singer, R.F. Improved processing of carbon nanotube/magnesium alloy composites. *Compos. Sci. Technol.* **2009**, *69*, 1193–1199. [CrossRef]
22. Kondoh, K.; Fukuda, H.; Umeda, J.; Imai, H.; Fugetsu, B. Microstructural and mechanical behavior of multi–walled carbon nanotubes reinforced Al–Mg–Si alloy composites in aging treatment. *Carbon* **2014**, *72*, 15–21. [CrossRef]
23. Kondoh, H.; Fukuda, H.; Umeda, J.; Imai, H.; Fugetsu, B.; Endo, M. Microstructural and mechanical analysis of carbon nanotube reinforced magnesium alloy powder composites. *Mater. Sci. Eng. A* **2010**, *527*, 4103–4108. [CrossRef]
24. Murugan, S.; Nguyen, Q.B.; Gupta, M. Synthesis of Magnesium Based Nano–composites. *IntechOpen* **2019**, 1–19.
25. Malaki, M.; Xu, W.; Ashish, K.; Kasar, A.K.; Menezes, P.L.; Dieringa, H.; Varma, R.S.; Gupta, M. Advanced Metal Matrix Nanocomposites. *Metals* **2019**, *9*, 330. [CrossRef]
26. Olszówka–Myalska, A.; McDonald, S.A.; Withers, P.J.; Myalska, H.; Moskal, G. Microstructure of in situ Mg metal matrix composites based on silica nanoparticles. *Solid State Phenom.* **2012**, *191*, 189–198. [CrossRef]
27. Olszówka–Myalska, A. Sintered in situ magnesium matrix composites. *Mechanik* **2016**, *89*, 504–505. [CrossRef]

28. Umeda, J.; Kondoh, K.; Kawakami, M.; Imai, H. Powder metallurgy magnesium composite with magnesium silicide in using rice husk silica particles. *Powder Technol.* **2009**, *189*, 399–403. [CrossRef]
29. Nazer, N.S.; Denys, R.V.; Andersen, H.F.; Arnberg, L.; Yarty, V.A. Nanostructured magnesium silicide Mg_2Si and its electrochemical performance as an anode of a lithium ion battery. *J. Alloy Compd.* **2017**, *718*, 478–491. [CrossRef]
30. Nieroda, P.; Zybała, R.; Wojciechowski, K.T. Właściwości termoelektryczne Mg_2Si otrzymywanego techniką SPS. *Ceram. Mater.* **2012**, *64*, 490–493.
31. Zhang, S.; Chen, T.; Cheng, F.; Li, P. A Comparative Characterization of the Microstructures and Tensile Properties of As–Cast and Thixoforged in situ AM60B–10 vol% Mg_2Sip Composite and Thixoforged AM60B. *Metals* **2015**, *5*, 457–470. [CrossRef]
32. Chegini, M.; Shaeri, M.H.; Taghiabadi, R.; Chegini, S.; Djavanroodi, F. The Correlation of Microstructure and Mechanical Properties of In–Situ Al–Mg_2Si Cast Composite Processed by Equal Channel Angular Pressing. *Materials* **2019**, *12*, 1553. [CrossRef] [PubMed]
33. Zainon, F.; Rafezi Ahmad, K.R.; Daud, R. The effects of Mg_2Si(p) on microstructure and mechanical properties of AA332 composite. *Adv. Mat. Res.* **2016**, *5*, 55–66. [CrossRef]
34. Ellendt, N.; Uhlenwinkel, V.; Stelling, O.; Irretier, A.; Kessler, O. Spray Forming of Mg_2Si Rich Aluminum Alloys. *Mater. Sci. Forum* **2007**, *534*, 437–440. [CrossRef]
35. Sun, Y.; Johnson, D.R.; Trumble, K.P.; Priya, P.; Krane, M.J. Krane1 Effect of Mg_2Si Phase on Extrusion of AA6005 Aluminum Alloy. *Light Metals* **2014**, 429–433.
36. Prusova, E.; Deevb, V.; Rakhubab, E. Aluminum Matrix In–Situ Composites Reinforced with Mg_2Si and Al_3Ti. *Mater. Today Proc.* **2019**, *11*, 386–391. [CrossRef]
37. Olszówka–Myalska, A.; Myalska, H. Method For Initial Consolidation of Nanograin Powders with Micrograin Powders; Nanograin Powders with Submicrograin Powders in Order to Obtain the In Situ Type Composites. Polish Patent Application No. P.225001, 22 December 2014.
38. Kwon, H.; Estili, M.; Takagi, K.; Miyazaki, T.; Kawasaki, A. Combination of hot extrusion and spark plasma sintering for producing carbon nanotube reinforced aluminum matrix composites. *Carbon* **2009**, *47*, 570–577. [CrossRef]
39. Chen, B.; Kondoh, K.; Imai, H.; Umeda, J.; Takahashi, M. Simultaneously enhancing strength and ductility of carbon nanotube/aluminum composites by improving bonding conditions. *Scr. Mater.* **2016**, *113*, 158–162. [CrossRef]
40. Chen, B.; Kondoh, K. Sintering behaviors of carbon nanotubes–aluminum composite powders. *Metals* **2016**, *6*, 213. [CrossRef]
41. Liu, Z.Y.; Zhao, K.; Xiao, B.L.; Wang, W.G.; Ma, Z.Y. Fabrication of CNT/Al composites with low damage to CNTs by a novel solution–assisted wet mixing combined with powder metallurgy processing. *Mater. Design* **2016**, *97*, 424–430. [CrossRef]
42. Seo, H.Y.; Jiang, L.R.; Kang, C.G.; Jin, C.K. A hot extrusion process without sintering by applying MWCNTs/Al6061 composites. *Metals* **2018**, *8*, 184. [CrossRef]
43. Seo, H.Y.; Jiang, L.R.; Kang, C.G.; Jin, C.K. Effect of compression process of MWCNT–reinforced AL6061 powder on densification characteristics and its mechanical properties. *Metals* **2017**, *7*, 437. [CrossRef]
44. Wang, J.T.; Figueiredo, R.B.; Langdon, T.G. Mechanical properties and microstructures of Severely Plastic Deformed pure titanium by mechanical milling and spark plasma sintering. *Mater. Sci. Forum* **2010**, *667*, 559–564.
45. Lakshmikanthan, P.; Prabu, B. Mechanical and tribological behavior of aluminium Al6061–coconut shell ash composite obtained using stir casting pellet method. *J. Balk. Tribol. Assoc.* **2016**, *22*, 4008–4018.
46. Sujith, S.V.; Mahapatra, M.M.; Mulik, R. An investigation into fabrication and characterization of direct reaction synthesized Al–7079–TiC in situ metal matrix composites. *Arch. Civ. Mech. Eng.* **2018**, *19*, 63–78. [CrossRef]
47. Olejnik, E.; Szymański, Ł.; Tokarski, T.; Tumidajewicz, M. TiC—Based local composite reinforcement obtained in situ in ductile iron based castings with use of rode preform. *Mater. Lett.* **2018**, *222*, 192–195. [CrossRef]

Review

The Critical Raw Materials in Cutting Tools for Machining Applications: A Review

Antonella Rizzo [1,*], Saurav Goel [2,3], Maria Luisa Grilli [4], Roberto Iglesias [5], Lucyna Jaworska [6,7], Vjaceslavs Lapkovskis [8], Pavel Novak [9], Bogdan O. Postolnyi [10,11] and Daniele Valerini [1]

1. ENEA–Italian National Agency for New Technologies, Energy and Sustainable Economic Development, Brindisi Research Centre, S.S. 7 Appia–km 706, 72100 Brindisi, Italy; daniele.valerini@enea.it
2. School of Engineering, London South Bank University, 103 Borough Road, London SE1 0AA, UK; saurav.goel@cranfield.ac.uk
3. School of Aerospace, Transport and Manufacturing, Cranfield University, Cranfield MK4 30AL, UK
4. ENEA–Italian National Agency for New Technologies, Energy and Sustainable Economic Development, Casaccia Research Centre, Via Anguillarese 301, 00123 Rome, Italy; marialuisa.grilli@enea.it
5. Department of Physics, University of Oviedo, Federico Garcia Lorca 18, ES-33007 Oviedo, Spain; roberto@uniovi.es
6. Łukasiewicz Research Network, Institute of Advanced Manufacturing Technology, 30-011 Krakow, Poland; lucyna.jaworska@ios.krakow.pl
7. Faculty of Non-Ferrous Metals, AGH University of Science and Technology, 30-059 Krakow, Poland
8. Faculty of Civil Engineering, Scientific Laboratory of Powder Materials/Faculty of Mechanical Engineering, Institute of Aeronautics, 6A Kipsalas str, lab. 110, LV-1048 Riga, Latvia; lap911@latnet.lv
9. Department of Metals and Corrosion Engineering, University of Chemistry and Technology, Prague, Technická 5, 166 28 Prague 6, Czech Republic; Paja.Novak@vscht.cz
10. IFIMUP—Institute of Physics for Advanced Materials, Nanotechnology and Photonics, Department of Physics and Astronomy, Faculty of Sciences of the University of Porto, 687 Rua do Campo Alegre, 4169-007 Porto, Portugal; b.postolnyi@gmail.com
11. Department of Nanoelectronics, Sumy State University, 2 Rymskogo-Korsakova st., 40007 Sumy, Ukraine

* Correspondence: antonella.rizzo@enea.it; Tel.: +39-0831201428

Received: 20 December 2019; Accepted: 2 March 2020; Published: 18 March 2020

Abstract: A variety of cutting tool materials are used for the contact mode mechanical machining of components under extreme conditions of stress, temperature and/or corrosion, including operations such as drilling, milling turning and so on. These demanding conditions impose a seriously high strain rate (an order of magnitude higher than forming), and this limits the useful life of cutting tools, especially single-point cutting tools. Tungsten carbide is the most popularly used cutting tool material, and unfortunately its main ingredients of W and Co are at high risk in terms of material supply and are listed among critical raw materials (CRMs) for EU, for which sustainable use should be addressed. This paper highlights the evolution and the trend of use of CRMs) in cutting tools for mechanical machining through a timely review. The focus of this review and its motivation was driven by the four following themes: (i) the discussion of newly emerging hybrid machining processes offering performance enhancements and longevity in terms of tool life (laser and cryogenic incorporation); (ii) the development and synthesis of new CRM substitutes to minimise the use of tungsten; (iii) the improvement of the recycling of worn tools; and (iv) the accelerated use of modelling and simulation to design long-lasting tools in the Industry-4.0 framework, circular economy and cyber secure manufacturing. It may be noted that the scope of this paper is not to represent a completely exhaustive document concerning cutting tools for mechanical processing, but to raise awareness and pave the way for innovative thinking on the use of critical materials in mechanical processing tools with the aim of developing smart, timely control strategies and mitigation measures to suppress the use of CRMs.

Keywords: critical raw materials; cutting tools; new materials; new machining methods; modelling and simulation

1. Introduction

In the era of globalisation and high competitiveness, it is of utmost importance for industries to work on reducing manufacturing costs and simultaneously providing added value in terms of increased life of the product when put into service. It is also imperative for them to have a flawless supply chain which contains a key component of sourcing critical raw materials (CRMs) [1] for them to remain sustainable.

In the EU, the excess of imports of CRMs, especially in the mechanical manufacturing industry—here, we refer to metals that are vital to EU industries such as tungsten, chromium, and niobium to name but a few—has reached an alarming stage, bringing a great degree of dependency on countries that are monopolistic suppliers of these CRMs. To mitigate this foreseeable problem, the EU has already launched several campaigns, including the "Raw Materials Initiative" (RMI) [2].

The issue of CRMs must be tackled with scientific rigour by pursuing different parallel actions; in particular, by (1) improving the production processes of CRMs (increasing sustainable mining, reducing extraction costs, increasing the efficiency of materials, increasing security, etc.); (2) finding suitable candidates to partially or totally substitute the CRMs; and (3) increasing their recycling.

In this context, the development of new materials with superior characteristics or better performance than existing materials is desirable to lead to a longer product life and therefore to reduce the cost of the product. The cost of cutting tools used in the manufacturing industry at present, particularly to machine high-value components such as turbine blades, automotive and aerospace parts, machine parts and biomedical implants, is significant and dictates the total manufacturing cost and thus the final price of the product [3]. The tooling cost is not only associated with the cost incurred on relapping the tools, but it also involves the cost incurred due to the increased cycle time owing to the unloading and reloading of the new tool. The main focus in manufacturing research currently is to improve productivity, and thus, there is a focus on developing materials (especially for cutting tools) that can withstand higher cutting loads than the current attainable machining limits. The cutting tools discussed here are meant for all mechanical-based contact loading processes, including, for example, milling, drilling, turning, honing, chamfering, hobbing, knurling, parting-off, and so on. The current development of new materials specifically for cutting tools has remained a slow process, particularly due to the high initial investment, but interest has been growing recently towards developing more sustainable ways to de-risk the supply chain.

Cemented carbide is one of the most popular cutting tool materials. Typically, a carbide cutting tool is manufactured with a mixture of tungsten and cobalt (the binder that holds tungsten carbide together), with a wealth of variations in carbide grain size and the ratio of carbide to binder. Preferred blends have been developed over time to achieve effectiveness with different cutting depths and widths, as well as workpiece materials. The distribution of cutting tools in the global market in 2018 by the cutting method and workpiece material is shown in Figure 1a,b, respectively. It can be seen that milling, turning and drilling are the processes mostly used in primary machining operations, and the tools for these operations capture almost 87% of the total tooling market. In terms of the type of cutting material, cemented carbides capture one half of the market, followed by high speed steel. Ceramics, cermets and superhard materials such as polycrystalline diamond (PCD) and polycrystalline cubic boron nitride (PCBN) capture the remaining share of the tool material market. The two important ingredients of the carbide tools mentioned above—namely tungsten and cobalt—were earlier identified (in 2011) as being among the list of the 14 critical raw materials (see Figure 2) vital to EU industries [4]. More recently, in 2014 and in 2017, they have continued to remain on this list on account of their economic importance and risk of supply interruption [5,6]. Another driving force for Co substitution lies in its well-known genotoxic and carcinogenic activities. A pressing need has thus emerged for the

EU to develop alternative solutions to the existing cutting tool materials to avoid a complete closure of machining businesses in the absence of the raw material of the cutting tools.

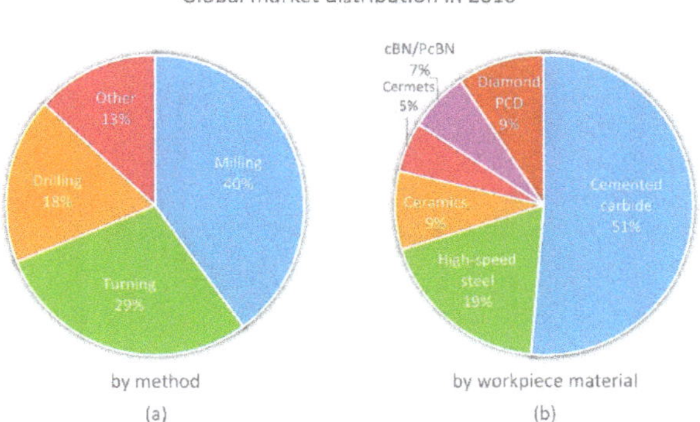

Figure 1. Cutting tool global market distribution by cutting technology (**a**) and by workpiece material (**b**) as presented by Dedalus Consulting. Data taken from [7].

Critical Raw Materials for the EU

(reported by the European Commission)

Listed as CRMs in 2011
Listed as CRMs in 2014
Listed as CRMs in 2017

	1	2	3	4	5	6	7	8	9	10	11	12	13	14	15	16	17	18
1	H																	He
2	Li	Be											B	C	N	O	F	Ne
3	Na	Mg											Al	Si	P	S	Cl	Ar
4	K	Ca	Sc	Ti	V	Cr	Mn	Fe	Co	Ni	Cu	Zn	Ga	Ge	As	Se	Br	Kr
5	Rb	Sr	Y	Zr	Nb	Mo	Tc	Ru	Rh	Pd	Ag	Cd	In	Sn	Sb	Te	I	Xe
6	Cs	Ba	La	Hf	Ta	W	Re	Os	Ir	Pt	Au	Hg	Ti	Pb	Bi	Po	At	Rn
7	Fr	Ra	Ac	Rf	Db	Sg	Bh	Hs	Mt	Ds	Rg	Cn	Nh	Fl	Mc	Lv	Ts	Og
			La	Ce	Pr	Nd	Pm	Sm	Eu	Gd	Tb	Dy	Ho	Er	Tm	Yb	Lu	
			Ac	Th	Pa	U	Np	Pu	Am	Cm	Bk	Cf	Es	Fm	Md	No	Lr	

Fluorspar CaF_2	Natural Graphite	Baryte	Borate	Coking coal	Natural rubber	Phosphate rock	Magnesite

Figure 2. Critical raw materials list for 2011–2017 overlaid on the periodic table of the elements [5,6].

Perhaps, the most direct impact of the new era of digital manufacturing [8] on future cutting tools is the potential use of additive manufacturing technologies for the generation of quasi-network-shaped tools with geometric structures and challenging sub-structures and the potential to produce under-functionally different structures and optional materials with graded properties (4D printing). One of the four themes defined by the Roadmap of High Performance Cutting (HPC) [9] points at the use of integrated tooling based on cycle time reduction and quality improvement in high-end batch production. This theme was further developed, and the term "smart tooling" has been coined [9], which refers to the insertion of sensor actuators [10] within the cutting tool as a step forward in the process monitoring of the tool, allowing users to collect the necessary data for the creation of more accurate digital twins for the machining processes (Figure 3).

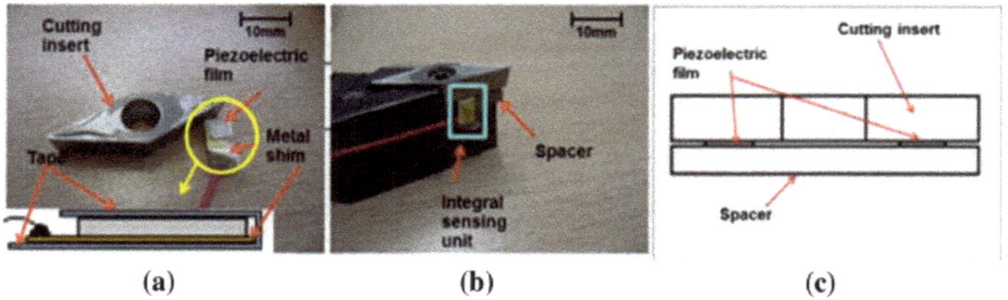

(a) (b) (c)

Figure 3. (**a**) Composition of the smart cutting tool, (**b**) assembly of the tool, (**c**) cross-section view of the tool, as taken from [10].

An example of the additional functions which are commercially available in this regard is the use of low-frequency vibration-assisted drilling (LFVAD) [11] for improving the quality of the machining of brittle materials such as carbon fibre reinforced composite (CFRP) via an interrupted cutting mechanism.

Besides such alterations in machining mechanisms, there is remaining interest regarding the successful realisation of novel tool coating materials as well as the development of enhanced cutting technologies, including laser-assisted cutting and cutting with cryogenic assistance. The idea behind the use of these methods depends on whether the main wear mechanism of the tool is governed by the high physical hardness of the workpiece or the high chemical affinity. In the former case, laser heating ahead of the cutting tool can reduce the cutting resistance, making the workpiece more compliant to cutting, while in the latter case, the cryogenic environment helps delay the kinetics of the chemical reactions, thus inhibiting the occurrence of diffusion-induced chemical wear.

As an additional strategy, there are many existing coatings that protect cutting tools working under extreme conditions. The most desired attribute of tool coatings is to provide high hardness and toughness for high wear performance and thermal and chemical stability to withstand extreme cutting environments, especially during high-speed machining. Most of the widely used coatings do not contain CRMs, but multi-elemental and high-entropy coatings designed for specific extreme conditions may include them [12]. Transition metals are one of the biggest groups of chemical elements in the periodic table and, as can be seen in Figure 2, half of the refractory metals group are marked as CRMs. However, the amount of material needed for a coating is significantly lower than for analogous bulk CRMs used in cutting tools. Moreover, the nano and microstructural design of protective coatings, such as nanocomposites or multilayers, suppresses the usage of certain CRMs in coatings.

Several recent studies have addressed technical–scientific actions with different methods that can be pursued to address the problems of CRMs in cutting tools; in particular, to achieve the following:

1. Increase the life of the tools by enhanced material removal techniques including laser assistance, cryogenic assistance, vibration assistance and use of protective coatings;
2. Develop and synthesise newer materials that can adequately partially or totally replace the CRMs used in the tools;
3. Rigorously involve modelling and simulation in the tool design and deploy digital twins to make improved predictions in the present era of digital manufacturing;
4. Improve the recycling of worn-out tools.

In this paper, we summarise the current understanding regarding the use of CRMs in the tooling industry, considering the four aspects listed above, without any intentional effort to propose an immediate solution to this problem, but with the aim of stimulating the development of effective strategies in this field.

2. Attempts to Expand the Life of Tungsten Carbide (WC)–Co Based Tools

There are several methods to increase the lifetime and efficiency of tools in machining applications, such as the proper modification of the base tungsten carbide (WC)–Co material of the tool, the use of advanced processing techniques, and the application of protective coatings. These methods are reviewed in the following subparagraphs.

2.1. WC–Co Based Materials

Cemented carbides belong to the most common and the longest-used tool materials produced by powder metallurgy methods. Sintered carbides are characterized by their high strength and abrasion resistance and include one or more high-melting metal carbides constituting the basic component together with the metallic binding phase [13]. According to this standard, sintered carbides are divided into groups: those used for the production of tools for metal machining, plastic forming and for the use of mining tools. The basic component of cemented carbides is WC, which, depending on the manufacturer and group of material applications, can constitute from 50% to 90% by weight of the sintered content. The other ingredients are carbides of titanium (TiC), tantalum (TaC) and niobium (NbC), the content of which can be from 0% to 35% by weight. These carbides dissolve each other and can also dissolve a large amount of tungsten carbide. The rest of the composition is usually cobalt. Cobalt is characterized by very good wettability with most materials that are components of sintered tool materials and a fairly high melting point, which is 1493 °C. Such a microstructure, which is characteristic for sintered carbides, allows for the presence of a ductile binding phase; additionally, the hard and brittle carbide phase allows the bonding of opposing features in one material, such as high abrasion resistance and hardness with high strength and fairly good ductility. Many carbide applications feature very good abrasion resistance, which depends on the chemical composition.

Two-component cemented carbides of the WC–Co type with a low content of cobalt are characterized by the highest abrasion resistance. These grades can be used if there are no impacts during operation, and abrasion is the main wear mechanism.

The universality of WC–Co results from its very good mechanical and tribological properties. The technological process of obtaining cemented carbides is characterized by relatively easy formation, sintering temperatures lower than ceramic materials and electrical conductivity, and these factors have a positive effect on the possibility of shaping products with complicated geometry using erosive treatment as well as on the ease of applying anti-wear and anti-corrosion coatings. The disadvantage of this material is that cutting tools made of cemented carbides work at relatively low cutting speeds, and these materials tend to oxidize already at temperatures above 400 °C. The most commonly used carbides are tungsten, titanium, tantalum or other high-melting metal elements in an amount of 75 to 94 wt.% and cobalt, nickel or molybdenum, and sometimes other metals are usually used as the binding matrix. The toughness of WC–Co can be increased by increasing the cobalt content, whilst the wear resistance is increased by decreasing the cobalt content and decreasing the carbide grain size. The substitution of cubic carbides (TiC, NbC and TaC) for WC leads to improvements in wear resistance and resistance to plastic deformation. The microstructure of cemented carbide WC–Co consists of tungsten carbide particles combined with cobalt, which are obtained in the sintering process with the participation of the liquid phase. When machining a metal with high plasticity, such as pure iron, using conventional WC–Co cemented carbide, a chip tends to adhere on the rake face of the cutting tool, resulting in serious adhesive wear due to the existence of cobalt, which has a lower melting point compared to WC. The presence of cobalt in cemented carbides during the machining of steel, for example, causes the chip to stick to the cutting blade. Crater wear is caused by the chemical interaction between the rake face and hot chip. Wear occurs by the diffusion of the tool material into the chip or by the adhesion between the chip and tool followed by a fracture below the adhered interface within the tool material. Crater wear can be reduced by increasing the chemical stability of the tool material, decreasing the solubility of the workpiece or barrier protection by substrate alloying or coating [14]. Thermal shock cracks are caused by large temperature gradients at the cutting edge. There is a large difference in the coefficients of thermal expansion between cobalt and tungsten carbide, which is why cracks may appear on the blade during its operation. Therefore, cemented carbides are often used as substrates for coatings.

However, studies are being carried out on uncoated WC–Co intended for machining materials. An example consistent with the idea of reducing the use of critical materials in cemented carbides is cobalt-free carbide. The first cobalt-free carbides were obtained by conventional methods from micropowders [15]. These materials were characterized by high hardness and abrasion resistance, but exhibited greater brittleness compared to cemented carbide WC–Co. For binderless tungsten carbide, the sintering took place in a solid phase, which for conventional, pressure-free sintering means higher sintering temperatures and/or longer sintering times, which in turn cause grain growth in the polycrystalline material. The use of nano-powders and pressure sintering methods (spark plasma sintering, hot pressing sintering) improved the properties of the cobalt-free cemented carbides [16–19].

The addition of free carbon to the binderless WC reduces the amount of the brittle phase of W_2C, Co_3W_3C and oxide creation [20–22].

Spark plasma sintering (SPS, also called "Field Assisted Sintering Technology") and hot pressing sintering, thanks to the shorter sintering time and pressure, allow the reduction of grain growth and of the porosity of materials sintered in the solid phase [23].

Composite materials with the addition of ceramic powders are another idea which has been developed in relation to cemented carbides. In the case of ZrO_2, additional reinforcement is used, which is the result of stresses created during the ZrO_2 phase transformation. They are usually added in an amount of 5% to 15 wt.% in the form of oxides, carbides and nitrides [24,25].

This idea is consistent with the substitution of the critical materials of W and Co. Composite materials with a WC or WC–Co matrix have recently been developed very intensively thanks to the SPS method [26]. There are already individual examples of the industrial application of such solutions.

Research related to the reduction of cemented carbide consumption is also associated with attempts to increase their durability. Superhard materials such as diamond and cubic boron nitride (cBN) are introduced into the WC–Co [27]. Composites containing ultrafine tungsten carbide/cobalt (WC–Co) cemented carbides and 30 vol.% cBN were fabricated mainly by FAST (field-assisted sintering technology) methods. WC–Co/cBN composites have been considered as a next-generation material for use in cutting-tool edges and are characterized by an optimal combination of hardness and toughness. The major challenge in sintering these composites is to produce a well-bonded interface between the WC–Co matrix and cBN particles [28–31].

Whisker toughening is mainly used for binderless tungsten carbide. There are known studies in which SiC_W, Si_3N_{4W} and Al_2O_3 whiskers were used. The whisker participation was up to 10 vol.% [32,33].

The literature describes research related to toughening using nanotubes and graphene to improve the thermal conductivity of cemented carbides. The study found that thermal stress is the main reason for the failure of cemented carbide shield tunnelling tools when shield tunnelling is carried out in uneven soft and hard soil [34].

2.2. *Advanced Machining Techniques*

The machining of "difficult to cut materials" such as the nickel-based superalloys used in turbine blades requires an extended tool life for unhindered machining [35]. Traditional carbide tools are limited to working in the range of 30 m/min to 70 m/min because of their poor thermochemical stability; however, they can be used at high feeds due to their high toughness. Improvements in the poor thermochemical stability of cemented carbide tools could be aided by advanced modelling. Additive manufacturing technology [36] and new machining techniques have now opened newer possibilities of making complex tool shapes. This section presents a brief review of recent attempts made to expand tool life either by reducing the load on the cutting tool or by deferring the known pathway of the accelerated wear mechanism. In either case, the end result is the improved quality of the machined part.

2.2.1. Machining by Laser Assistance (Thermal-Assisted Machining)

In pursuit of overcoming the challenges of difficult-to-machine materials, laser assistance has been adopted worldwide for incorporation with mechanical micromachining. To date, seven major patents (in the years 1982 [37,38], 2006 [39], 2013 [40], 2011 [41], 2014 [42], 2016 [43], and 2017 [44]) have been granted in the US concerning the use of laser assistance during mechanical micromachining. All these patents can be connected by a single chain of a process which is now well known as thermal-assisted machining (TAM) [45]. The concept of TAM is schematically shown in Figure 4a and relies on pre-heating the cutting zone of the material being cut ahead of the cutting tool. This methodology reduces the physical hardness of the workpiece, making it more compliant to cutting by reducing the specific cutting energy (i.e., the work done by the tool in removing a unit volume of the material).

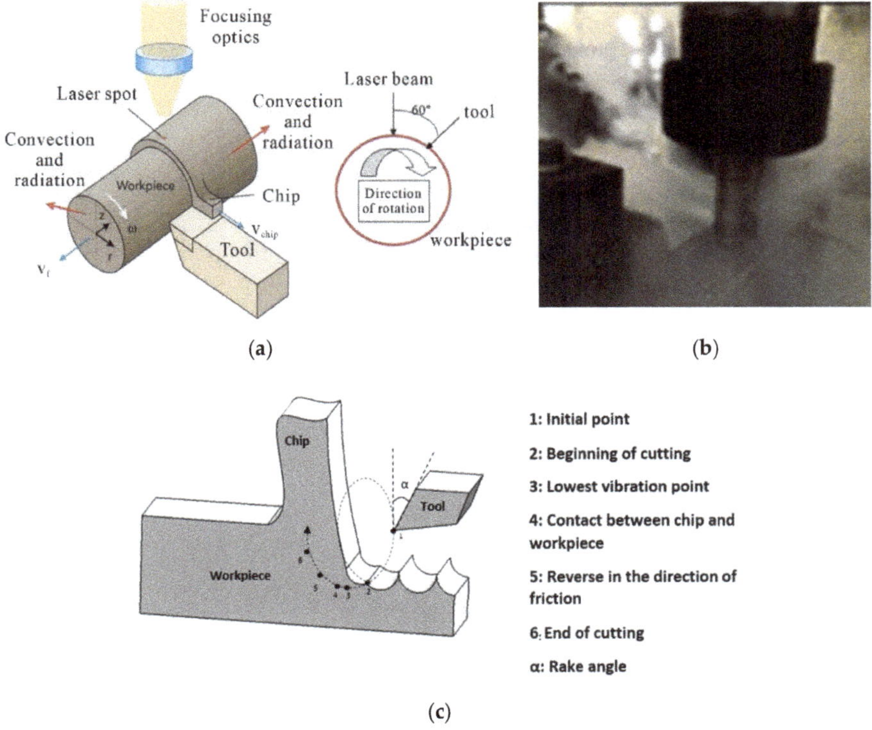

Figure 4. Important advancements made in machining technology. (**a**) Thermal-assisted machining [46]; (**b**) cryogenic machining [47]; (**c**) schematic diagram to illustrate the mechanism of vibration-assisted machining [48].

2.2.2. Cryogenic Machining

As shown in Figure 4b, cryogenic machining relies on freezing the cutting tool to extreme temperatures of about −196 °C by means of either liquid nitrogen or liquid carbon dioxide [49]. Cryogenic machining helps to serve two purposes: (i) it elevates the relative hardness gradient between the tool and the workpiece; and (ii) it delays the kinetics of any chemical diffusion that may likely trigger the tribo-chemical wear of the tool. S. Rakesh highlighted how cryogenic technology is ecological, non-toxic and non-explosive [50].

Cryogenic cooling has been executed in cutting operations in different ways by using liquid nitrogen for precooling the workpiece, cooling the chip, and cooling the cutting tool and cutting zone [51]. Numerous studies have compared conventional cutting strategies and cryogenic cooling methods. However, studies conducted in an attempt to determine the best technique found many contradictions, as the conclusions described above could change in relation to tool–work pairs, to the cutting conditions and to the general evaluation parameters.

The literature dealing with tungsten carbide cutting tools in cryogenic processing is significantly less numerous compared to the literature on steel tools, and only a few papers consider an evaluation parameter of the tool's lifetime [52]. In this context, all these studies show that the cryogenic treatment has positive effects both on the lifetime of the tools and on the surface finishing of the product. The cryogenic treatment of the materials showed significant positive effects such as increased wear resistance, reduction of residual stresses, increased hardness and fatigue resistance.

Seah et al. [53] carried out a series of experiments with the aim to study the aspects of cold and cryogenic treatments on uncoated WC inserts for carbon steel ASSAB 760. They showed that at different cutting speeds the cryo-treated inserts exhibited greater resistance to wear than the untreated and recovered counterpart. In addition, they found that the cold and cryogenic treatment significantly increased the resistance of the cutting insert to chip removal, which became increasingly important as the cutting speed increased.

Yong et al. [54] subjected uncoated WC inserts to a cooling treatment down to −184.5 °C for 24 h and then heated the insert to room temperature, keeping the rate of 0.28 °C/min unchanged both when increasing and decreasing the temperature. They developed a series of face milling operations using different cutting speeds but kept all the other processing parameters constant using untreated and cryo-processed inserts. Two pieces of information were highlighted: the first concerned the cryo-processed inserts, which generally performed better than their untreated counterparts; and the second focused on the increase in tool life of 28%–38% during the cryogenic treatment in wet machining compared to dry machining.

Sreeramareddy et al. [55] studied the tool wear, cutting forces and the surface finish of parts worked using a WC insert coated with a multilayer and subjected to cryogenic treatment. They showed that cryogenic processing reduced the wear on the side of the inserts as well as the cutting forces and the surface roughness of the workpiece machined with untreated inserts.

These studies suggest that the cryogenic treatment of carbide tools is capable of improving the productivity and quality of the final product because it guarantees greater resistance to wear and surface finish. Improvements have been reported in the red-hardness of cryogenically treated inserts, which resulted in low flank wear [56].

Bryson [57] claimed that the increase in wear resistance, with a consequent increase in the lifetime of the carbide tool, was due to the greater strength of the binder after cryogenic treatment.

Thakur et al. [58] highlighted that WC tools undergo a less-strong microstructural modification under cryogenic treatment compared to that detected with conventional heat treatments; some physical transformations actually occur concerning the densification of cobalt, which induces an increase in the gripping of carbide particles and an improvement in the tool life amounting to a wear resistance increase of 27%. The cryogenic treatment of the tool is one approach to enhance its properties by introducing microstructural changes. The formation of complex compounds such as Co_6W_6C or Co_3W_3C might have increased the hardness in the samples due to forced air cooling and oil quenching.

In another work, the effect of cryogenic treatment on the tool was shown to include an increase in tool life, lower cutting force and better surface finish compared to the untreated condition [59].

Reddy et al. [60] examined the workability of C45 steel with untreated and treated (−110 °C for 24 h) ISO P-30 tungsten carbide inserts by measuring the flank wear, main cutting force and surface finish. They concluded that the best machinability found was caused by the increase of the thermal conductivity of tungsten carbide induced by cryogenic treatment. The surface roughness of the workpiece was lower by approximately 20% when the workpiece was machined with deep cryogenically treated tungsten carbide tool inserts in comparison with untreated inserts for cutting speeds in the range between 200 and 350 m/min. Vadivel et al. [61] studied the microstructure of cryogenically treated (TiCN + Al_2O_3) coated and untreated inserts in turning nodular cast iron. Their results highlighted that coated and treated tools have better properties that help the cutting tool to operate under hostile conditions for a longer time.

A. Swamini et al. [62] drew up an overview of the metallurgy behind the cryogenic treatment of cutting tools; their results regarding WC–Co tools can be summarized as follows:

- Cryogenic treatment induces a structural variation with the formation of carbides of the eta phase and the redistribution or densification of Co, which increases its hardness;
- The micro-hardness of the treated tools is greater than that of untreated tools;
- The cryogenic treatment of tungsten carbide inserts increases the tool's lifetime during the processing of austenitic AISI 316 steel;
- Wear patterns are smoother and more regular;
- Cryogenic treatment increases chipping resistance;
- The radius of the chip coil as well as the thickness of the chip itself is smaller after processing with cryogenically treated inserts or in cryo-processing conditions.

M.I. Ahmed et al. [63] used a modified tool holder for the efficient use of cryogenic cooling for machine cutting. The modified tool holder uses the direct continuous contact of liquid nitrogen with the cutting insert for the perfect cooling of the cutting tool. The results showed an average 30-fold increase in the lifetime of the carbide tools.

M. Dhananchezian [64] described the machinability characteristics of turning Hastelloy C-276 with a Physical Vapour Deposition (PVD) coated nano-multilayer TiAlN carbide cutting tool using dry turning and liquid nitrogen cooling methods. The improvement of cutting tool performance was achieved under liquid nitrogen cooling by the control of the wear mechanisms, which in turn reduced the wear rate. Yildirim [65] examined the effect of some cooling conditions on machinability which may be an alternative to conventional cooling. The results showed that the 0.5 vol.% hBN cooling method used in conjunction with liquid nitrogen gave the best results in terms of the machining performance and lifetime of carbide tools.

Particularly noteworthy is the recent work of Biswal et al. [66] which, based on experimental results, highlighted how uncoated cryogenically treated tempered cermet inserts perform better than other cermet inserts thanks to their better wear resistance, micro-hardness and toughness. Despite all this, the cryogenic processing technique has not yet displaced conventional industrial processing. Further studies on cryogenic processing are necessary in order to highlight the increase in the lifetime of WC tools with the consequent saving of critical materials such as W and Co.

2.2.3. Vibration-Assisted Machining

The literature refers to vibration-assisted machining (VAM) as the process of intermittent cutting. The reported benefits include the low wear rate of the tools, reduced burr formation on the workpiece and higher cutting depths being achieved.

The attainable cutting speeds during VAM are limited by the hardware and system, and hence ultrasonic-assisted machining methods are classed as low-speed machining techniques [67]. Also, the process of vibration assistance can be implemented in two ways:

(i) uniaxial tool movement (1D VAM), where the tool vibrates in a plane parallel to the surface of the workpiece; and
(ii) elliptical tool movement [68] (EVAM) where the tool vibrates with an elliptical motion. Both methods can be both resonance-based and non-resonance-based.

The resonant system operates at discrete frequencies, normally higher than 20 kHz and at amplitudes of less than 6 μm, whereas the non-resonant system operates at frequencies between 1 to 40 kHz and with amplitudes 10 times higher than the resonant system. As shown in Figure 4c, the tool is prescribed an oscillatory motion by a wiggle function or the tool is vibrated at a high frequency. Some of the reported benefits are highlighted in Table 1.

Table 1. Tabulated summary of recent efforts made in the vibration-assisted machining (VAM) of titanium alloys and steel.

Work Material	Cutting Parameters Used	Oscillation Parameters (Frequency (f), Amplitude (a))	Cutting Force Comparison with Conventional Turning	Additional Conclusions
Ti6Al2Sn4Zr6Mo (α + β Ti alloy) [69]	fr = 0.1 mm/rev; v = 10–60 m/min; d = 0.2 mm	f = 20 kHz; a = 10 μm	Reduction by 74%	Surface roughness improved by 50%
Ti-15333 (β alloy) [70]	fr = 100 μm/rev; v = 10 m/min; d = 100–500 μm	F = 20 kHz a = 8 μm	Reduction by 80%–85%	Surface roughness improved by 50% while heat was applied during ultrasonic assisted machining
Ti6Al4V [71]	fr = 0.1 mm/rev; v = 10–300 m/min d = 0.1mm	f = 20 kHz a = 20 μm	Reduction by 40%–45%	Surface roughness improved by 40%
Ti 15-3-3-3 (β Ti-alloy) [72]	Fr = 0.1 mm/rev; v = 10–70 m/min; d = 50–500 μm	f = 17.9 kHz a = 10 μm	Reduction by 71%–88%	Surface roughness improved by 49%
Low alloy steel (DF2) [73]	Fr = 0.1 mm/rev; v = 50 m/min; d = 0.2 mm	F = 19 kHz; a = 15 μm	Reduction by 50%	Tool wear 20% less

2.2.4. Surface Defect Machining

It has been demonstrated that the surface defect machining (SDM) method harnesses the combined advantages of both porosity machining [74] and pulse laser pre-treatment machining [75], as shown in Figure 5a, by machining a workpiece initially by generating surface defects at depths less than the uncut chip thickness through either mechanical means and/or thermal means followed by a routine conventional machining operation. SDM enables ease of deformation by shearing the material at a reduced input energy [76,77]. Also, due to the large proportion of stress concentration in the cutting zone—rather than the sub-surface—a reduction in the associated residual stresses on the machined surface is enabled.

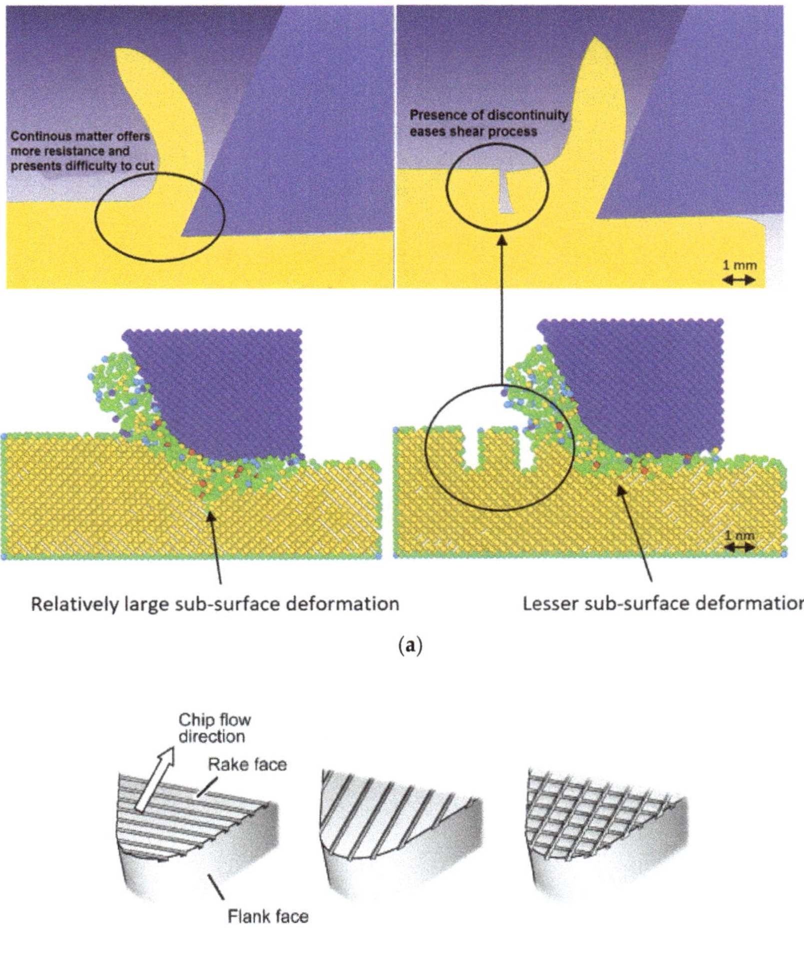

Figure 5. (**a**): Schematic diagram indicating the differences between the mode of deformation during conventional machining and surface defect machining (SDM) observed through an FEA (Finite Element Analysis) simulation of hard steel and MD (Molecular Dinamics) simulation of silicon carbide, respectively; (**b**) effect of providing nanogrooves on the tool [78]. (**a**) Surface defect machining. (**b**) Providing nanogrooves on the tool.

Figure 5b shows the provision of the structured surfaces on the rake face of the cutting tool. It has been implied particularly that when these structured surfaces are fabricated at a direction of 90° to the cut surface, they are helpful in reducing the extent of friction at the tool–chip interface. In contrast, probing these structures along the direction of cutting is rather disadvantageous as it destroys the integrity of the edge of the tool. Overall, a summary of these recent developments in this area of machining, as an add-on to the existing machines to improve the workpiece's machinability, is summarised in Table 2.

Table 2. Modified form of measures suggested for improved machinability [37].

S.No.	Theoretical Approach	Experimental Realization
	Modification of the process	
1	Reduction of chemical reaction rate between the tool and workpiece	Cryogenic turning [79]
2	Reduction of contact time between tool and workpiece	Vibration-assisted cutting [68,69,73,80–86]
3	Lowering of temperature rise and chemical contact	Usage of appropriate coolant [87,88]
	Modification of the cutting tool	
4	Building a diffusion barrier on the cutting tool	Use of protective coatings [89]
5	Modifying the cutting tool geometry	Providing nanogrooves on the cutting tool
6	Use of alternative cutting tool material	Use of cBN
	Workpiece modification	
7	Surface layer modification of the workpiece prior to cutting	Ion implantation
8	Surface defect machining	Pre-drilled laser holes in the workpiece reduce the shear strength, which is evident by observing a lower shear plane angle [90–92]

2.3. Protective Coatings

Several types of failure mechanisms, such as delamination, abrasion, oxidation, diffusion, etc., result from the surface of the cutting tool. Failure occurs due to the interactions at the interface between tool and workpiece or tool and ambient medium, respectively. This leads to the conclusion that it is possible to protect a tool by surface treatment, creating some additional interface or coating with individually designed features. Figure 6 shows the distribution of uncoated and coated cemented carbide cutting tools, where the growth of the latter is clearly visible in a timeline perspective.

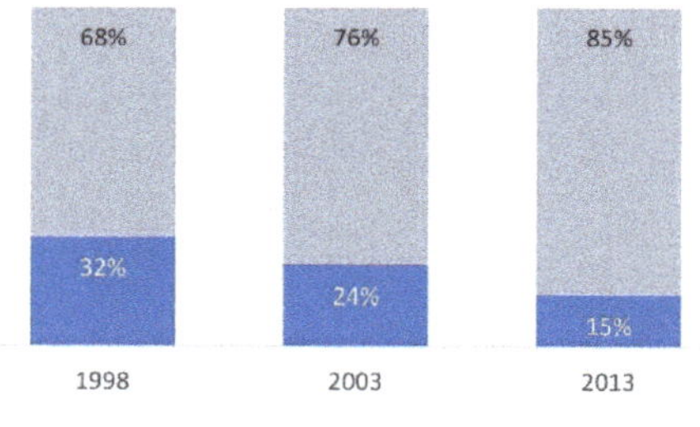

Figure 6. World market of cemented carbide cutting tools. Data from Dedalus Consulting, taken from [7].

Protective coatings may significantly increase the lifetime of tools and thus reduce the consumption of CRM content in bulk materials. Furthermore, most modern coatings are CRM-free. The evolution of the protective effect of coatings by the extension of tools' lifetimes and a comparison of coatings with available data from selected published works are shown in Figure 7. It is seen that coated cutting tools may have an extra lifetime of 200%–500% or more at the same cutting velocities. This can also lead to an increase in operational velocities (by 50% to 150%) with the same lifetime of cutting tools [93]. Coatings improve the endurance-related properties, and eventual wear failures may be dealt with through the wise application of the techniques outlined in Section 4.

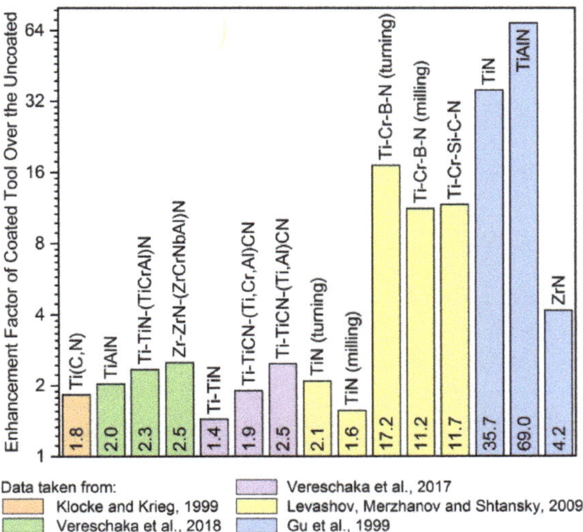

Figure 7. Comparison of enhancement factors of coated tools over uncoated tools (based on their lifetime). Analysed data are taken from selected published works [94–98]. The results strongly depend on operational conditions, such as the cutting method and speed, the workpiece material, and the thickness of the protective coating.

Each application area requires specific properties of protective coatings to reach maximum performance. The extreme conditions of service of cutting tools mean that working parts face high wear, high pressure, elevated temperatures caused by high machining speeds, oxidation, and corrosion from lubricants or cooling agents. The main criteria of the evaluation of protective coatings are summarised in Figure 8, even though a compromise among the different prerequisites must be achieved. Usually, the most successful in one of the criteria may perform poorly in many of the others. The enhancement of all necessary properties is still a challenge for modern materials science and a strong driving force for computational screening modelling efforts worldwide (see Section 4).

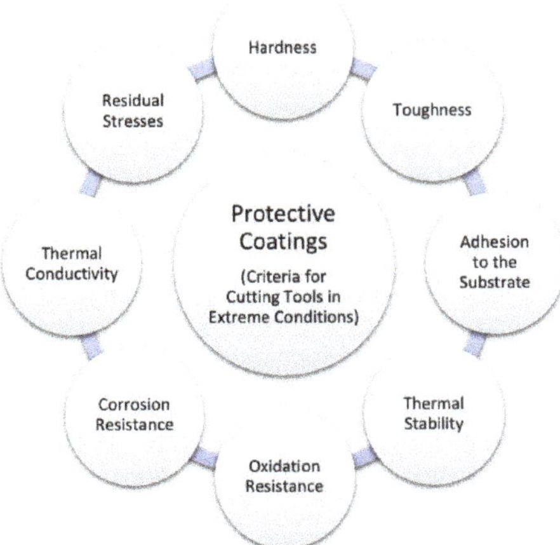

Figure 8. Criteria required for the success of the protective coatings of cutting tools.

Based on principal properties and functionality, protective coatings may be grouped mainly into hard coatings, coatings with enhanced thermal stability, coatings with high oxidation and corrosion resistance and thermal barrier coatings. High hardness is essential for cutting tool coatings. In addition to the usually extended tool lifetime, it helps to achieve a smooth high-quality surface and shape of machined parts. However, hard materials are often brittle and prone to cracking, which is why it is crucial for protective coatings to have both high hardness and toughness. Only a mixture of several parameters leads to high wear performance and the long lifetime of cutting tools under extreme conditions. In the next section, a brief description of the main properties of the hard coatings mainly used to coat WC–Co tools and other tool materials is reported.

2.3.1. Diamond and Diamond-Like Carbon (DLC) Coatings

Diamond is the intrinsically hardest material because of its strong nonpolar covalent C-C bonds (sp^3 bonding) and short bond length. It belongs to the class of ultrahard materials, with H = 70–100 GPa [99]. Diamond exhibits the highest room temperature thermal conductivity of ~20 W/cm·K and an extremely low coefficient of thermal expansion of $\sim 0.8 \times 10^{-6}$ K^{-1} at 300 K. At room temperature, it is inert to attack from acids and alkalis, and it is resistant to thermal shock. The topography of the diamond surface can be very different depending on the extent of polishing and orientation of the crystallographic planes, varying from an extremely smooth surface with a friction coefficient as low as 0.1 in air to a very rough surface with protruding edges [100]. Because of these properties, diamond has tremendous application in the field of tribology, especially as a protective coating for cutting tools. Polycrystalline diamond films are best synthesized by chemical vapour deposition (CVD) techniques (but not limited to this) [101,102]; the most common are hot filament-assisted CVD (HFCVD), direct current plasma-assisted CVD (DC PACVD), microwave plasma CVD (MPCVD) and combustion flame-assisted CVD (CFACVD) [100]. However, most of the metals and ceramics have a much higher thermal expansion coefficient than the very low one for diamond coatings. This often may cause residual stress and the further spallation of the coatings, thus limiting the number of workpiece materials for diamond-coated cutting tools. Tungsten carbide (WC) has the closest coefficient of thermal expansion to that of diamond, which allows diamond film deposition on WC substrates in almost stress-free conditions. Nevertheless, the diffusion of cobalt contained in WC tools promotes higher graphitisation at the diamond–carbide interface, which induces coating delamination during machining. To reduce such effects and improve adhesion, the deposition of an interlayer with a specific composition, dopants, the multilayer architecture of coatings, surface etching and other surface treatment techniques may be applied [103–108]. This should also help in the case of the deposition of a diamond coating on surfaces different from WC, with a higher coefficient of thermal expansion.

Diamond-like carbon (DLC), as is clear from the name, exhibits some of the properties intrinsic to diamond. Among them, the most important factors for application in cutting tools are its hardness at a super- and ultrahard level, high wear resistance and low coefficient of friction (~0.1). Most DLC films are structurally amorphous and can be synthesised by plasma-based PVD and CVD methods. Deposition method and the type of carbon source substantially influence the structural chemistry of the resulting films, which leads to large variations in their properties. Since the 1970s, DLC films have been well developed and discussed [109–111]. Structurally, they are made of sp^2- and sp^3-bonded carbon atoms and may be classified depending on their structure and hydrogen content as well as the presence of other dopants. DLC coatings with a lower hydrogen percentage (≤40%) and higher proportion of sp^3 bonding in the structure have a higher hardness. Thereby, hydrogen-free tetrahedral amorphous carbon (ta-C) with the highest sp^3 content (80%–88%) has an ultrahardness of 80 GPa. DLC films are not only hard but also smooth; furthermore, their tribological properties may be manipulated easily by introducing dopants such as nitrogen, silicon, etc. [112]. They also help to achieve high-quality shapes of the machined surface or cutting edge which are clean, smooth, and without chipping or rounding (see Figure 9). However, it is quite difficult to deposit thick coatings (of 2 μm and more) because of delamination due to the internal stress; in addition, there are limitations on the substrate choice with adherent film. Moreover, such coatings are prone to graphitise at temperatures above 300 °C with a further decrease of their hardness [100]. On the other hand, when the outer layer of DLC is turned to graphite when the tool is in service, both the friction and wear rates should decrease [113].

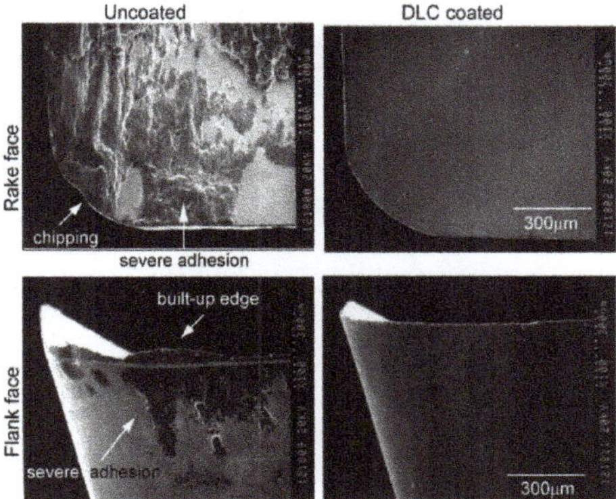

Figure 9. Scanning electron microscope (SEM) micrographs of the rake and flank face after the dry milling test for an $AlCu_{2.5}Si_{18}$ alloy [109].

2.3.2. Transition Metal Compounds

Transition metal nitrides (TMN), carbides (TMC) and borides (TMB) are largely employed as hard protective coatings in the cutting and forming tool industry. They attract interest due to their exceptional properties, such as their high hardness, chemical inertness, electronic properties, high melting point and thermal stability under harsh environments (oxidation, radiation, etc.). Such properties are mainly due to the variety of chemical bonding [114–116]. Based on the prevailing bonding type in TM nitrides, carbides and borides, Holleck [117] arranged their properties from low to high levels; some of these are shown in Figure 10. Depending on the priority properties and the suitable deposition technique, the coating which fits an application best can thus be chosen.

	low	medium	high
Hardness	Nitrides	Carbides	Borides
Brittleness	Borides	Carbides	Nitrides
Melting point	Nitrides	Borides	Carbides

Figure 10. Properties of transition metal nitrides, carbides and borides [117].

The first industrial CVD-coated tool coating on cemented carbide was TiC in 1969, and in 1980 TiN became the first PVD coating. Lower temperatures in the PVD process made the deposition of coatings on steel tools possible [7]. Since the 1970s–80s, and up to the present, TiN, TiC, TiB_2, CrN and ZrN have been the most frequently used binary coatings.

TiN has been the most widely studied TMN protective coating and has been in wide use since the late 1960s. However, it has some limitations and barely overcomes the modern challenges of thermal stability and oxidation resistance. At temperature above 500 °C, in fact, an oxide layer may be formed on the surface, which develops stress in the coating, which is high enough to damage or destroy the protective layer.

2.3.3. Multi-Elemental Compounds and High-Entropy Alloy Protective Coatings

To overcome the modern challenges related to performance in extreme conditions, new, more complex compositions of coatings, such as ternary and quaternary compounds, have been proposed, which exhibit some superior and specific properties (e.g., thermal stability and oxidation resistance). An important example of this are Al-containing TMN ternary compounds of the $TM_xAl_{1-x}N$ type, where TM is a transition metal, such as Ti, Cr, Zr, Nb, Hf, Ta, V, etc. The most popular are Ti-Al-N and Cr-Al-N systems, which have remained the "state-of-the art" coatings for a long time and are well-discussed in research and review papers [118–120]. Their exceptional hardness and oxidation resistance are mainly caused by the supersaturated solid-solution of hexagonal B4-structured AlN in the cubic B1 structure of TiN, with the resulting large volume mismatch, elastic strain energy and solid solution strengthening [119]. The mechanical and chemical properties of $TM_xAl_{1-x}N$ coatings usually improve with the increase of the Al fraction, but only up to a certain critical Al content (usually around 40%–50%). The formation of Al, Ti or Cr oxides increases the oxidation resistance at elevated temperatures. Moreover, metastable supersaturated films such as $TM_xAl_{1-x}N$ systems exhibit the phenomenon of age-hardening caused by decomposition with annealing temperature or time, leading to an increase in hardness (which is more significant in the Ti-Al-N system than in Cr-Al-N) [119]. This particular feature may have a high potential for the efficient enhancement of tool lifetimes.

Recently, the groups of T. Polcar and A. Cavaleiro performed studies on the addition of Cr to the Ti-Al-N system and its influence on thermal resistance, oxidation stability, tribological and cutting performance by the deposition of quaternary Ti-Al-Cr-N coatings [121,122]. A coating which demonstrates relatively low wear resistance at room temperature (Ti-Al-Cr-N system) in comparison to the other (in this case, the Ti-Al-N system) can exhibit much higher wear performance at elevated temperatures (650 °C) corresponding to realistic conditions for cutting tools' working temperatures.

Other ternary coatings which combine the superior properties of transition metal carbides and nitrides are transition metal carbonitrides (TMC_xN_{1-x}) such as TiC_xN_{1-x}, whose structure may be described as a TiN matrix with the substitution of N atoms by C atoms, which leads to distortion strengthening and increased resistance to dislocation motion [123]. Depending on the deposition method and deposition conditions, the film may also be composed of a mixture of TiN, TiC and C_3N_4 phases [124]. A transition metal nitride contributes to the strengthening and hardening of coatings, while carbon forms a graphite lubricating layer during work and thus substantially reduces the wear rate.

Apart from Al, other elements can be also added in multielement coatings to confer improved properties. Silicon is very often introduced in nitride coatings to form hard nanocomposite materials, where amorphous Si_3N_4 surrounds hard metal nitride grains [125], or a similar multi-layered configuration may be applied [126]. Refractory metals (Nb, Mo, Ta, W, Re) [127] significantly improve the thermal stability of the coatings and allow them to work under extreme conditions and high speed, but most of them are CRM elements (see Figure 2).

More complex compounds such as high-entropy alloy coatings can also be employed as protective coatings. High-entropy alloys (HEAs) are mostly identified as alloys which contain at least five principal elements with a concentration of each between 5 and 35 at.% and the possible inclusion of minor elements to modify the final properties [128,129], which finally depend on the material composition, microstructure, electronic structure and other features in complicated and sensitive ways. HEAs may be deposited as protective coatings as in [130], where Yeh and Lin experimented with the dry cutting of 304 steel with bare TiN, TiAlN, and high-entropy nitrides (HEN) $(Al_{0.34}Cr_{0.22}Nb_{0.11}Si_{0.11}Ti_{0.22})_{50}N_{50}$ coated WC-Co inserts and found that only the HEN-coated insert could produce long, curled chips, indicating that the cutting edge of the HEN-coated insert was still very sharp due to its superior oxidation resistance. Because of the great diversity of the possible involved elements in the composition, HEN coatings have high risk of containing CRMs; however, an appropriate screening method could allow the selection of HENs with a reduced or no critical raw material content (see Section 4).

2.3.4. Nanocomposite Super-Hard Coatings

Nanocomposite (nc) coatings are formed by nanometric-sized particles (usually MeN, MeC) embedded in amorphous or crystalline matrices and have attracted considerable interest because of their superior hardness which may allow the challenges of modern cutting tools to be overcome. The reason for their hardness resides in their grain size refinement. These so-called "third-generation ceramic coatings" represent a new class of materials that exhibit not only exceptional mechanical properties, but also excellent electronic, magnetic, and optical characteristics due to their nanoscale phase-separated domains of approximately 5–10 nm. The most recognised explanation for their improved performance is the increase of their grain boundary volume, because grain boundaries impede the movement and activation of dislocations. The so-called Hall–Petch strengthening [89,131] gives rise to high hardness H with a relatively low Young modulus E providing high toughness, enhanced wear resistance, high elastic recovery, resistance against crack formation and crack propagation, high thermal stability (up to 1100 °C) and reduced thermal conductivity [132]. Thus, grain size refinement allows the mechanical and tribological properties of the coatings to be controlled and optimised, and thus helps in increasing the tool's life. The opposite happens when grain sizes continue to decrease to values smaller than 10 nm, when the maximum hardness is reached. Overgrown volumes of boundaries and small grains cause a loss of material hardness. This phenomenon has been termed as the reverse or inverse Hall–Petch relation/effect.

There is now growing evidence that a critical H^3/E^2 ratio (>0.5 GPa) has to be satisfied for the coating to provide appropriate tribological protection [133] (Figure 11). In this respect, Plasma Enhanced Chemical Vapour Deposition (PECVD) and PVD metal-based nc coatings are very suitable candidates thanks to their optimal tribological properties which may be achieved by opportunely controlling the parameters of the deposition process (deposition time, target power or cathode current, bias voltage applied to the substrate, temperature, pressure, gasses flow, etc.) and the elemental composition of the coatings.

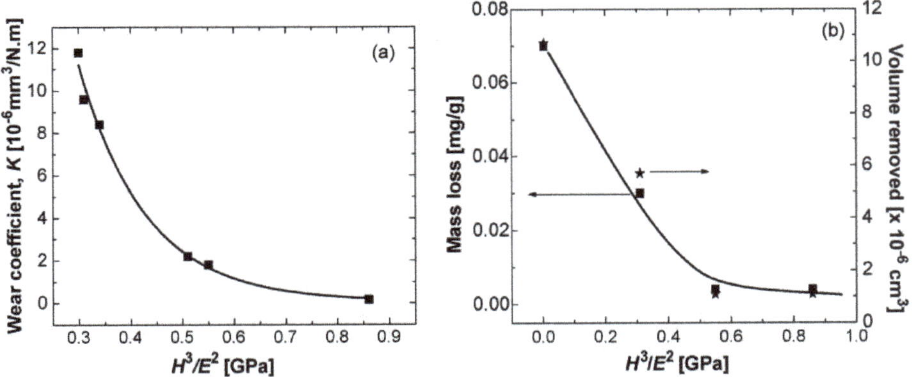

Figure 11. Tribological characteristics of TiN and nanocomposite nc-TiN/$SiN_{1.3}$ films as a function of the H^3/E^2 ratio: (**a**) wear coefficient; (**b**) erosion rate [132].

Nanocomposite coatings may provide superior performances than superlattices (multilayers in which a single layer has a thickness not greater than 10 nm), whose properties are strongly dependent on the precise thickness control of the single layers composing the multilayer stack, so that any error mismatch may affect the coating's performance.

Hard nanocomposite coatings for cutting tools may generally be classified into two groups: (1) nc-Me_1N/a-Me_2N (hard phase/hard phase composites), and (2) nc-Me_1N/Me_3 (hard phase/soft phase composites), where Me_1 = Ti, Zr, W, Ta, Cr, Mo, Al, etc. are the elements forming hard nitrides, Me_2 = Si, B, etc. and Me_3 = Cu, Ni, Ag, Au, Y, etc. [133]. The most used and interesting coatings for cutting tools are Ti-Al-Si-N with nc-(Ti_xAl_{1-x})N + a-Si_3N_4 phases [134–136], Cr-Al-Si-N of nc-(Cr_xAl_{x-1})N + a-Si_3N_4, TiZrSiN of c-(Ti,Zr)N solid solution + a-SiN_x [137], nc-TiC + a-C [138,139] Ti-Si-N [140, 141], Ti-Si-B-C [142–144], Ti-Si-B-C-N [145], AlTiN-Ni [146], ZrN/SiN_x [147], nc-W_2N/a-Si_3N_4 [148], (Zr-Ti-Cr-Nb)N [149,150], Mo_2BC [151], nc-AlN/a-SiO_2 [152]. In particular, several of these coatings, such as Ti-Al-Si-N, Ti-Si-N and AlTiN-Ni, were deposited and tested on cemented carbide WC–Co-based substrates like in [135,136,140,146]. Nanocomposite hard coatings are well discussed in the most recent and fundamental reviews of J. Musil [133], S. Veprek et al. [153,154], A.D. Pogrebnjak et al. [155,156], C.S. Kumar et al. [157].

2.3.5. Multi-Layered and Graded Coatings

Multilayer architecture is one of the most efficient and promising current approaches for the hardness and toughness enhancement of protective coatings. There are several paths to their design: depositing a set of films in a special order according to their functionality (e.g., substrate > adhesion film > superhard film > oxidation resistant film), alternating films with a similar crystal lattice for epitaxial growth, hard crystalline films with thin amorphous layers, alternating TMN, TMC or TMB films, nanocomposites, HEAs coatings, etc. Many research groups around the world have worked on this topic, and many papers have been published that review and evaluate the recent progress in this area [127,133,155,158–160]. Among the most recent and interesting multilayer solutions for protective coatings there are TiN/TiAlN [161,162], TiAlN/TaN [163], Ti(Al)N/Cr(Al)N [121], (TiAlSiY)N/MoN [164], CrN/AlSiN [165], AlCrN/TiAlTaN [166], TiSiC/NiC [167], TiN/MoN [168,169], TiN/WN [170], TiN/ZrN [171], Zr/ZrN [172], Ta/TaN [173], CrN/MoN [174–176], (TiZrNbHfTa)N/WN [177], and multilayer hard/soft DLC coatings [178]. In particular, the following multi-layered coatings were also deposited on cemented carbide substrates or machining tool inserts, including WC–Co-based inserts, for mechanical tests in laboratories or for industrial tests: TiN/TiAlN [162], TiAlN/TiSiN [179], CrAlSiN/TiVN [180], AlCrN/TiVN [181], TiVN/TiSiN [182], CrN/CrCN [183], AlTiCrSiYN/AlTiCrN [184] TiCrAlN/TiCrAlSiYN [185]. However, K.N. Andersen et al. [162] did not observe differences in the properties of the coatings when using the various substrates (e.g., high speed steel and cemented carbides). Moreover, it should be mentioned that cemented carbide-based substrates and tools usually may be exposed to higher operational temperatures.

At least three significant benefits of multi-layered coatings should be mentioned: the first is the possibility of building two-dimensional nanocomposite multi-layered films with a nanoscale thickness of each individual layer to enhance the mechanical and tribological properties [99]; the second is the adjustment of grain size by changing the bilayer thickness of the coatings, since the grain sizes decrease in thinner layers, as reported in many works; the third and the main intrinsic feature of the multilayer design of coatings is the ability to resist external forces and cracks and to interrupt their propagation toward the substrate on the interlayer interfaces, which prevents the direct impact of destructive factors on the workpiece material of cutting tools (Figure 12 [158]).

A combination of multilayer nc-coatings and cryo-machining was found to improve the cutting performances of the carbide cutting of the Inconel 718 superalloy [186].

Graded coatings are layers whose composition and microstructure are gradually tuned across the thickness from the interface with the substrate towards the outer surface. This kind of coatings, included in the so-called class of FGMs (functionally graded materials), represents an effective way to improve tool performance by enhancing coating/tool properties such as adhesion, toughness, thermal stability, resistance to corrosion, friction, fretting, wear, etc. [187], thus helping to increase the tool life and consequently reducing the consumption of CRMs contained within the tool or the coating itself. In FGMs, a proper gradient in the coating properties (Young's modulus, thermal expansion coefficient, etc.) can reduce residual stresses due to the lattice mismatch and different thermal characteristics of the substrate and coating, thus decreasing the possibility of coating debonding and delamination. Graded coatings based on nitride, carbide and carbonitride materials such as TiN, TiC, TiCN are commonly exploited for cutting tools based on cemented carbide or other materials containing CRMs such as W, Co and rare earth elements [188–190].

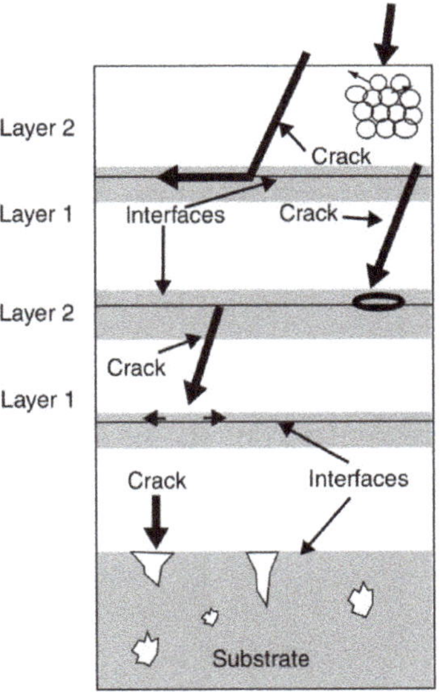

Figure 12. Toughening and strengthening mechanisms in multilayer coatings (taken from [158]).

Graded coatings are often used as intermediate layers to facilitate the bond between the substrate and a subsequent layer. For example, DLC coatings deposited directly on the surface of a component often result in the generation of high residual stress and consequent poor adhesion, thus hindering the excellent protection properties of DLC described in the previous paragraph. Thus, as shown in [191], an intermediate layer with a composition gradually changing from Ti to TiC (Ti/TiN/TiCN/TiC) can be deposited on a Ti-based alloy to promote the bonding of the final DLC coating. Similarly, a boron-doped graded layer diamond coating, acting as a transition layer between a non-graded boron-doped diamond coating and a nano-crystalline diamond coating, was studied in order to improve the machining performance of tungsten carbide cutting tools [103]. Other complex coating structures which take advantage of the gradient composition can be realized by different combinations of multiple layers, as in [192], where a layered coating was made of a lower fine periodic TiAlYN/CrN multilayer that graded into an upper amorphous TiAlY oxynitride layer in order to obtain enhanced oxidation resistance and reduced friction coefficient in tungsten carbide tools for high-speed cutting applications.

Graded layers can be also obtained by directly inducing modifications into the same surface of the tool; for example, by means of metal ion implantation or gas diffusion/reaction into the surface. In the first case, for instance, Cr ions can be implanted to create a graded layer and metal ion intermixing in the interface region to enhance the adhesion of protective CrN or TiAlN-based coatings on high-speed steels (HSS) [193,194], which are commonly used for cutting tools and contain a significant content of CMRs such as W, V or Co. In the second case, substrate surface modification is obtained by gas diffusion and/or reaction to create a graded layer. Plasma nitriding, carburising or carbonitriding are reported to be effective ways to improve the surface properties of working tool steels and cermets (containing, e.g., W, Co and V) and to enhance the adhesion of nitride or DLC coatings, thanks to the development of a graded diffusion layer and formation of intermediate interface compounds [195–198]. As an alternative, surface nitriding with a graded composition and microstructure can be achieved on the surface of the material directly during its production process. This technique is particularly useful to enhance the surface properties of WC–Co tools, such as in the case of the so-called functionally graded cemented carbides (FGCCs) or functionally graded hard metals (FGHM), which are a widely studied class of tool materials in which the graded layer is obtained by mixing and pressing powders with a suitable composition, followed by sintering steps under a controlled atmosphere. By properly tuning the different process parameters, the compositional and structural gradient of an FGCC can be tailored (Figure 13) according to the desired properties, improving the tribological properties of the material surface and strengthening the adhesion of additional protective coatings deposited afterwards [199–201].

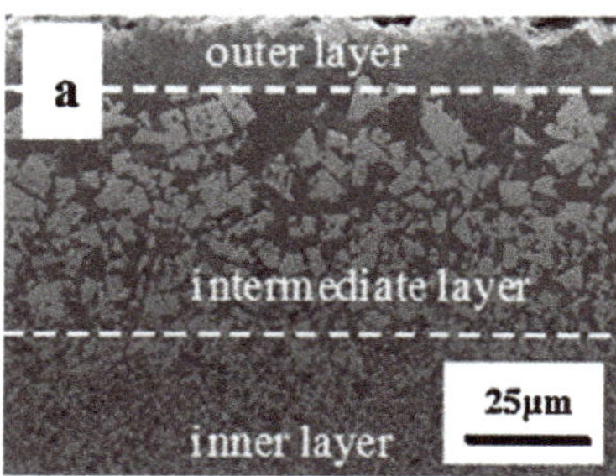

Figure 13. Cross-section SEM image of the surface of a graded cemented carbide material obtained in [201].

2.3.6. Thermal Barrier Coatings

A high amount of heat is generated in the cutting zone during machining. There are three different zones from where the heat flux comes into the cutting tool: the primary shear zone (plastic deformation and viscous dissipation), the secondary shear zone (frictional and plastic shearing energy), and the frictional rubbing of the cut surface on the tool insert flank. The diffusion of heat into the workpiece or tool body negatively influences the lifetime and work performance. There are many coatings with thermal barrier functionalities when applied to metallic surfaces which lead them to operate at elevated temperatures, but most of them exist for cases with no directly applied high mechanical load, such as gas turbines or aero-engine parts. In case of cutting tools, the influence of the coating on the heat distribution of the working interface is unknown or very poorly studied. It is not clear whether coatings influence the cutting process by an insulation effect (lower heat flux transmitted into the substrate) or rather by a tribological effect (lower level of heat generated by friction) [202]. It is quite difficult to perform in-situ or ex-situ experimental studies of thermal barrier coatings or to find thermal properties data in the literature due to the lack of a standard methodology to quantify these properties in the case of very thin layers. Thus, most of the existing approaches are based on simulation methods, and improvements may be expected by applying the techniques outlined in Section 4.

J. Rech et al. [202] proposed an analytical solution for heat transfer modelling to characterise the influence of a coating on the heat flow entering into the tool substrate. He showed that coatings have no capacity to insulate a substrate in continuous cutting applications, but in applications with a very short tool–chip contact duration, such as high-speed milling, coatings keep a large amount of heat in the interaction zone, which may improve the wear resistance of the tool. Experimental investigations made by the same authors were found in accordance with results from computational studies [203]. In addition, it was shown that the larger the coating thickness, the more it influences heat transfer. Moreover, it has been pointed out that the heat flux transmitted to a substrate is much more influenced by the tribological phenomenon at the cutting interface than by the thermal barrier properties of the coating [202].

M.A. Shalaby et al. [204] reported that the improvement of pure alumina ceramic tool performance in proportion to the cutting speed can be attributed to the thermal barrier properties of the ZrO_2 tribo-layer induced at a high cutting speed (temperature). In the case of SiAlON ceramic (Si_3N_4 + Al_2O_3), they pointed out that the high performance was due to the high amount of mullite (Al-Si-O) tribo-film formed on the tool face—a phase that reduces thermal conductivity and serves as a thermal barrier layer. W. Song et al. [205] reported the thermal barrier contribution to the tool wear resistance when coated by Ti-MoS_2/Zr. Gengler et al. [206] investigated thermal transport for Si-B-C-N ceramic films and concluded that their properties are ideal for thermal barrier applications for high-temperature protective systems in aircrafts, as well as for surfaces of cutting tools and optical devices.

Many protective coatings against wear and corrosion which may be applied for cutting tools are not considered here due to the huge variety of their types and infinite number of specific tasks; however, some general principles and tendencies were discussed, while more information is available in specialised papers and reviews [7,155,207–211]. Although chemical vapour deposition and physical vapour deposition are the two main methods of protective coating fabrication, it is suggested to look deeply into thematic papers and reviews to learn about different techniques [212,213] or find details on deposition methods in research papers concerned with specific materials and structures.

3. New Materials for Tools

High efficiency often translates into high cutting speeds. Ceramic tool materials are the basis of HSC (high-speed cutting) machining [212]. Ceramic tools show three to ten times more durability than sintered carbide tools, and they can work at least at several times higher cutting speeds. The development of "hard machining" and dry cutting technology is associated with ceramic materials. HSM (high-speed machining) was created as a result of the need to shorten the time taken in manufacturing elements, to eliminate inaccuracies resulting from the use of manual finishing treatments and to minimize manufacturing costs. HSM also enables the fast and efficient processing of hard materials (stainless steels, durable titanium alloys, tool steels), moulds and mould element processing with a high shape and surface precision with low roughness. All these advantages have an impact on simplifying construction work at the design stage. In addition, high-speed machining ensures high removal efficiency, shorter production times, lower cutting forces and reduced deformation of the workpiece as a result of significant heat dissipation through the chips. The above advantages require high rigidity and precision of the machine tool system: the tool, machine work with high spindle speeds and special cutting tools (appropriate tool materials and coatings). Thus, the materials from which the tools are made are one of the most important factors for machining at high speeds and must ensure the tools' long-term operation. The biggest problem is the negative impact of high cutting speed on the durability of the cutting blade. The most important features of tool materials for cutting blades are high hardness and abrasion resistance so that the tool does not require frequent regeneration. The necessary property of the cutting material is high resistance to dynamic loads and brittle fracture. The latter will protect the cutting blade against chipping (arising, for example, from the heterogeneity of the material's properties or the insufficient stability of the machine spindle). Due to the heating of the tool during operation, the material should retain its properties in a wide range of temperatures. For this reason, a favourable feature of the material is a high thermal conductivity coefficient, since in all cases of machining, it is possible to use cooling liquids. Cooling liquids are troublesome because they contain particles of often harmful elements that come from the workpiece. Sometimes, in the case of organic liquids (currently displaced by synthetics), these liquids can be bacterially contaminated. Nevertheless, an important feature of the tool material is the chemical resistance, excluding the possibility of chemical reaction with the material being processed, and thus its corrosive destruction. In addition to the features mentioned above, materials intended for cutting blades should be machined in an effective manner, ensuring proper shape and dimensions, especially for tool blades. All ceramic cutting tools have excellent wear resistance at high cutting speeds. There are a range of ceramic grades available for a variety of applications.

3.1. Basic Groups of Tool Materials Intended for Cutting

Ceramic tool materials have a stable position in terms of their range of application and range of species. It is estimated that they currently constitute 9% of all tool materials [9,213]. They are offered in the form of the mechanically fixed indexable inserts used by all major tool companies, regardless of several companies which specialise exclusively in tool ceramics (Figure 14).

Figure 14. Tool inserts made of ceramics.

The following groups of ceramic tool materials are distinguished here:

- Oxide ceramics are based on aluminium oxide (Al_2O_3). This material is chemically very stable, but lacks thermal shock resistance. Due to the low price and resistance to abrasion at high temperatures, it is used in the medium-fine machining of cast irons with a Brinell Hardness below HB235, carbon steels with a Rockwell Hardness the C scale lower than HRC38, as well as alloy steels. Most often, pure Al_2O_3 is used for machining parts made of grey cast iron for the automotive industry.
- Mixed ceramics Al_2O_3 with the addition of ZrO_2, TiC, TiN, or TaC, NbC, Mo_2C, Cr_3C_2: the most popular method of strengthening Al_2O_3 is the introduction of ZrO_2. The polymorphic ZrO_2 transformation occurs at 1150 °C and results in an increase in the volume of the zirconium-containing phase. The change in the volume of the ZrO_2 phase generates in the Al_2O_3 matrix the stresses which are able to absorb the energy of the cracks. Mixed ceramics are particle-reinforced through the addition of cubic carbides or carbonitrides. These additives improve the toughness and thermal conductivity of the material. The materials are used to anneal iron alloys with a hardness of 55–65 HRC, including cast irons, brittle materials such as composites on metal matrix-reinforced ceramics or intermetallic materials and high-density alloys based on tungsten. These materials can also be used in continuous and intermittent processing and milling and turning as well as roughing and fine machining conditions.
- Whisker-reinforced ceramics use silicon carbide whiskers (SiCw or Si_3N_4w) or single-leaf monocrystals, most commonly SiC. The critical stress intensity factor K_{Ic} for such materials is from 8 to 10 MPa $m^{1/2}$, and the bending strength is in the range of 600–900 MPa. These materials are used for machining with low cutting speeds of nickel alloys, hardened steels, non-metallic fragile materials and high hardness cast irons [214].
- Nitride ceramics: Si_3N_4 with additives to facilitate sintering, and SiAlON. Si_3N_4 elongated crystals form a self-reinforced material with high toughness. Silicon nitride grades are successful in grey cast iron machining, but their lack of chemical stability limits their use in other workpiece materials. Materials based on silicon nitride have a toughness similar to hard metals and the temperature resistance characteristic of oxides. This extends the scope of their applications and allows them to be used, for example, for the roughing and semi-finishing of cast iron castings with turning and milling as well as in the machining of special alloys with high nickel content. SiAlON grades combine the strength of a self-reinforced silicon nitride network with enhanced chemical stability. They are ideal for machining heat-resistant super alloys (HRSA).
- Superhard materials: diamond and regular boron nitride. These materials are designed for machining difficult-to-cut materials. The most commonly used are polycrystalline sintered

materials. Diamond is used for non-ferrous metals, and regular boron nitride is used for hardened steels.

Silicon metal and borate belong to the critical raw materials list, as natural graphite. However, various carbon sources can be used in the production of diamonds. Bearing in mind the limitation of the use of critical materials in materials intended for machining, the development of tools from Al_2O_3 and diamonds has been taken into account later in this chapter.

Ceramic materials have some disadvantages: they are not easily amenable to machining, and it is difficult to give them a complicated shape. Therefore, cutting inserts made of these materials usually have the shape of a circle, triangle, rhombus or square (Figure 14). Thanks to the advancement of materials engineering, there are many new materials for which machining technology needs to be developed. Most companies offering cutting inserts provide catalogue information in which the following materials are distinguished: unalloyed steels, low-alloy steels, high-alloy steels, stainless steels, tempered steels, cast irons, titanium alloys, nickel and cobalt, aluminium alloys, wood, and sometimes polymers and graphite.

3.2. *New Considerations for Cutting Tool Materials*

The development of civilization imposes the necessity of new solutions. New factors have to be taken into consideration which have not been considered properly to date. The basic factors are, of course, health and the human environment and the availability of resources; thus, the use of CRMs should be avoided as much as possible. Certain types of materials have been and are being used despite their harmful effects. The reason for this is the lack of substitutes. This is the driving force behind undertaking research work into new and better solutions. When searching for new substitutes, toxic, allergenic and carcinogenic materials or materials that can work only in the harmful environment of lubricating and cooling liquids should be avoided. Limiting the use of harmful cooling liquids is possible because of the higher thermal resistance of new cutting materials. Lower friction and wear materials reduce energy consumption and enhance the lifetime of used tools, thus saving materials and allowing a reduction of the used lubricants, or even enabling the use of dry machining. Coolant and lubricant costs account for 16% of the total machining cost, and tool costs only account for 4%. Therefore, the avoidance of cooling will have a high benefit and impact by the reduction of lubricant costs due to low-wear and low-friction ceramics. Human contact with harmful compounds may occur at various stages; e.g., material preparation or treatment with a blade made of this material (in the form of an inhaled aerosol), or as a result of the utilization of the tool material.

Ceramic tool materials usually have the ability to be machined at much higher cutting speeds compared to cemented carbide materials; e.g., the cutting speed should be properly selected to produce adequate heat in the cutting zone for chip plasticization, but it cannot be too high, or the chemical or phase composition or selected properties of the ceramics used will be changed. Higher feeds and cutting depths require a reduction in the cutting speed.

Ceramic materials can substitute cemented carbide tools. Currently, these cemented carbides account for about 50% of total tool production. Cemented carbides are popular as there is a large variety of commercial cutting tools made of them at a relatively low cost. In addition, they are easy to shape, and their operation parameters are consistent. However, without coating, they operate at lower machining speeds than materials with an Al_2O_3 matrix. When coated, the edge sharpness of the cutting tool may become an issue. Furthermore, they have low resistance to oxidation, they require intensive cooling, their cutting speed is lower than that of the ceramic materials, and their prices change frequently depending on the supply situation of their raw materials.

Today, there are several ways to develop new tool materials. One is the possibility of using new sintering techniques—for example, SPS/FAST (spark plasma sintering/field-assisted sintering technology) or ultra-high-pressure methods—guaranteeing the high density of sintered materials [156,157,215]. These methods allow us to obtain new materials which were previously not used as tool materials; this group includes carbides and high-melting borides or nanoceramics.

The conventional sintering of ceramic requires high temperatures and long times for densification; therefore, the lowest grain size achievable by this technique remains of about 0.5 mm. There is the need to investigate alternative sintering methods that enhance mass transport and make it possible to lower both the temperature and the time of consolidation and, thus, to control grain growth. For these techniques, the sintering of extremely refractory materials is possible, or, as an alternative, the lowering of the temperature of consolidation. SPS employs a pulsed DC current to activate and improve the sintering kinetics. Three mechanisms may contribute to field-assisted sintering: the activation of powder particles by a pulsed current, which leads to the cleaning and surface activation of powders, resistance sintering and pressure application. Thus, the densification of nanopowders occurs at temperatures significantly below those of larger-grained powders by up to several hundreds of degrees. Consequently, small final grain sizes may result, and sintering aids and undesirable phase transformation may be avoided [216]. Materials characterised by a nanometric particle size have better resistance to chipping and better mechanical properties. There are several publications documenting the great success in the field of synthesis of diamond and diamond composite materials sintering using SPS technology, which is cheaper than the high-pressure–high-temperature methods [217–220]. For example, the combination of an ultrafast heating rate of about 2000 °C/min and a short holding time for SPS was successful in fabricating diamond/cemented carbides. For most experiments, the size of the diamond was over 40 μm, and the diamond content was limited to up to 30 vol.%. Several experiments show that it is possible to prevent the diamond graphitization and improve its bonding strength with the matrix. The SPS/FAST method guarantees a reduction of the manufacturing cost, in comparison to the high-pressure–high-temperature sintering used for polycrystalline cBN sintering. The laser beam sintering of ceramic bodies is also a promising technique for new ceramic tools [221].

3.3. New Solutions for Tool Materials

The commercial ceramic tool materials presented above have many drawbacks and are imperfect. The fracture toughness of the materials is low because the dislocation movement is extremely limited by their ionic and/or covalent bonds. The brittleness and poor damage tolerance have so far limited their application as advanced engineering materials, especially in cutting applications. The problem is that the low thermal shock resistance of these materials implies using a huge amount of cooling liquid during the machining process. Good cutting properties depend on the thermal resistance of the material or the use of lubricant–coolant process media. Innovative solutions for cutting tool ceramics will result in the development of ceramics exhibiting higher hardness, fracture toughness, heat and wear resistance in comparison with previous solutions. New materials will have high thermal stability, high corrosion and wear resistance, high fracture toughness and will be obtained with a view to fully replacing the presently available commercial WC–Co hard materials. New Al_2O_3 matrix composites, with the inclusion of various novel reinforcing phases—for example, cBN (cubic boron nitride) [222] or Al_2O_3 fibres coated with graphene—will substitute hard metals (sintered carbides) [223,224]. It is evident from this discussion that the computational techniques outlined in Section 4 could play a major role in designing tool materials with improved mechanical properties.

A distinct change in properties is obtained by using metals, especially in the form of nanometric additives [201]. The simplest way of introducing metallic additives to Al_2O_3 is the metallization of powders by mechanical means through reduction or electrochemically. Cutting fluids are widely used in machining processes, especially for ceramic cutting tools. The main roles of cutting fluid are cooling, reducing friction, removing metal particles, and protecting the workpiece, the tool and the machine tool from corrosion. A new idea is to obtain self-lubricating ceramic cutting tool materials with the addition of solid lubricants. Solid sliding agents should be characterised by a low friction coefficient and resistance to oxidation at temperatures exceeding 800 °C [225–228].

One of the interesting methods to obtain self-lubricating ceramic cutting tool materials is the addition of metal-coated solid lubricant powders. Nickel coated CaF_2 composite powders with a core–shell structure were produced by the electroless plating technique [229].

For cutting tool edge applications, diamond polycrystals have been used since the early 1970s. Katzman and Libby have reported the liquid phase sintering of a diamond–cobalt system [227]. Hibbs and Wentorf have developed a method of cobalt infiltration into diamond layers under high-pressure conditions [230]. Diamond powders were sintered and bonded to a WC–Co substrate at the same time by the infiltration of Co from the substrate during the sintering process. The carbide substrate is very useful for tool producers because of the possibility of brazing to the tool body. The sufficiently high wettability of diamond materials by molten metal fillers is the principal requirement for successful brazing. The most popular commercial PCDs are two-layer materials with a cobalt phase. Cobalt provides a good wetting of diamond crystallites; this property allows the production of polycrystals characterized by a low amount (below 10 wt.%) of the bonding phase, resulting in high hardness. Cobalt-containing PCDs are chemically stable only up to 700 °C, while their working temperatures may rise even higher. The presence of a cobalt phase in the diamond layers has a strong influence on the decrease of thermal resistance. The thermal stability of a PCD material can be defined as the resistance to graphitization in an inert atmosphere at elevated temperatures. The fields of interest are diamond polycrystals with higher thermal resistance. The thermal resistance of diamond composites depends on the oxidation process more than on the graphitization process. The evolution of CO gas during diamond composite oxidation destroys the integrity of the composite microstructure. One of the possibilities to increase the thermal resistance of PCD materials is to reduce the cobalt bonding phase content by etching a metallic additive from the PCD layer after sintering [231]. A second method is the preparation of the materials with a non-metallic bonding phase, without the negative effects caused by the diamond–graphite solvent/catalyst on diamond graphitization [232,233]. It was confirmed that the introduction of ceramic phases, with high-temperature resistance, improved the resistance of polycrystalline diamond to oxidation and graphitization at high temperatures [234] (Figure 15).

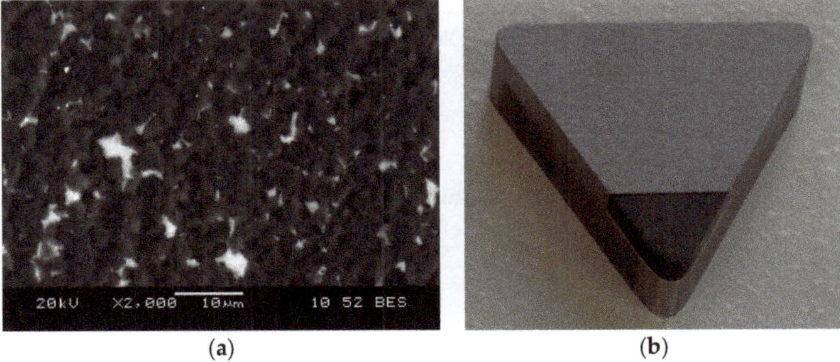

(a) (b)

Figure 15. (**a**) Microstructure of the samples for a diamond composite with 5 wt.% TiB_2 and (**b**) an insert with a diamond cutting edge.

The most popular method of obtaining bulk diamond compacts is the high-pressure–high-temperature sintering process. The development of multianvil apparatuses allows the use of pressures above 10 GPa even in commercial PCDs. Thus, it was possible to obtain a single-phase polycrystalline diamond by using direct conversion. This material is characterized by excellent temperature resistance and resistance to abrasion [235–237]. Other possible substitutes for cemented carbide tools could be materials based on intermetallics. Due to their interesting combination of properties, such as their relatively high fracture toughness and hardness, the anomalous temperature dependence of the yield strength [238] and oxidation resistance [239,240], aluminides of iron, nickel or titanium are the most promising and the most frequently mentioned candidates for the matrix of a composite tool material. Since aluminides form a natural oxide layer of aluminium oxide when exposed to corrosive or high-temperature environments, aluminium oxide seems to be the most suitable reinforcement for aluminides. The wear resistance of NiAl-Al_2O_3 composites prepared by self-propagating high-temperature synthesis (SHS) was studied in 2011 [241]. It was found that the composite has very good wear resistance in the case of the addition of 10 wt.% of Al_2O_3 short fibres. The wear behaviour was found to be comparable with AISI D2 tool steel, but it did not reach the performance of cemented carbides. In order to improve the properties, the more complex materials of NiTi-TiC-Al_2O_3 [242] and NiAl-AlN-Al_2O_3 were tested [243]. However, these materials were developed mostly as high-temperature alloys, and hence the wear behaviour is not known yet. On the other hand, the use of carbide reinforcement in nickel or iron aluminide provided highly interesting results. In the case of FeAl–WC composite tools, their successful application in the dry machining of copper bars has been reported [244]. If boron is added to a FeAl–WC composite, then it reaches a higher fracture toughness, higher hardness and wear resistance than WC–Co [245]. Composite materials of Ni_3Al–WC also achieve a comparable wear rate and fracture toughness to WC–Co [246]. This means that the cobalt in cemented carbides could be well substituted by iron or nickel aluminide. If we would also like to substitute the carbide phase, titanium carbide seems to be the best solution. FeAl-TiC has a fracture toughness approaching that of WC–Co (10.8 MPa·$m^{1/2}$) and a comparable hardness to the cemented carbide [247]. The wear resistance data for this material are not available yet, but it can be expected that they will be comparable to WC–Co due to the similarity of the other properties. Very good wear resistance, comparable to cemented carbides, was achieved in the case of Ni_3Al-TiC [248]. As presented above, aluminides could potentially substitute cobalt binders in cemented carbide and provide a similar wear behaviour as well as fracture toughness. However, there is one intermetallic phase which even has a much higher fracture toughness than aluminides: the NiTi phase. It is reported that its fracture toughness is around 40 MPa $m^{1/2}$ [249] in combination with a hardness of 303–362 HV [250]. The nitinol alloy, based on the NiTi phase, is used worldwide in medicine for stents or braces and in electronics for thermal switches due to its shape memory and superelasticity effects. It has been proved that it could be used also as a matrix for a composite, providing superior wear resistance and toughness, reaching the parameters of cemented carbide [251]. Since there are intermetallics with a high variety of properties, there are also ideas to develop a tool material based on the combination of softer intermetallics and harder ones, which act as the matrix and reinforcement, respectively. This approach was tested on the examples of TiAl-Ti_5Si_3 [252] and Ni_3Al-Ni_3V [253]. While the TiAl-Ti_5Si_3 exhibits very high hardness and wear resistance, but low fracture toughness [254], the Ni_3Al-Ni_3V has already been applied successfully as the tool for the friction stir welding of stainless steel and titanium [253]. High-entropy alloys, as novel candidate materials for cutting tools, are receiving increasing attention. Recent studies have investigated the substitution of Co binders in cemented carbide cutting tools by HEAs [255]. The search for Co substitutes is an active research field, which is demonstrated by the numerous papers recently reported [256–258]. The use of HEA binders with a reduced or no critical raw material content together with the opportune choice of the carbide phase eventually made by HEAs with reduced CRMs content may provide an alternative to WC–Co cemented carbides.

4. Modelling and Simulation

As stated in Section 1, the partial or total substitution of CRMs is one of the several approaches implemented within the RMI to try to deal with the scarcity and external dependence of these highly strategic materials. Even if the attainment of an evetual complete substitution is just an impractical dream in the most difficult cases and in the best of scenarios, it demands tremendous efforts from researchers and technologists [259,260]—the term "substitutability" has even been coined at Yale (Figure 16)—however, it is nevertheless worth trying to analyse in depth how to progress to a gradual and relative reduction of the most critical components of a given piece of machinery. In that regard, several computational modelling approaches have been considered in recent years to be the most appropriate by the experts in the field, all of which come under the umbrella of the multiscale, multiphysics and multidisciplinary modelling paradigm. Generally speaking, they may be divided into three broad categories: Data mining, computational screening and high-throughput machine learning-related tools.

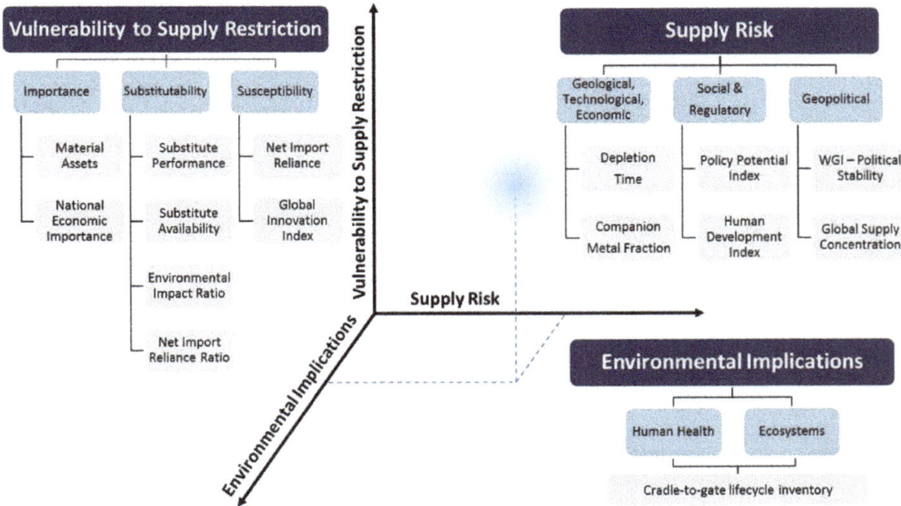

Figure 16. The methodology of criticality, showing the importance of substitution in the vertical axis (see [259]).

4.1. Data Mining

Given the immense amount of computational results that have been produced in the last 30 or more years, and especially beginning with the advent of the new century, several research groups have decided to join efforts by creating consortia responsible for developing and maintaining scientific research data repositories capable of storing the essential input/output results uploaded by materials scientists themselves, concurrently facilitating the cost-free retrieval of large amounts of information by interested researchers worldwide. The diversity of formats and quality of the available files and folders have led to an intense homogenisation strategy to include information which is as clear as possible on provenance and tracking. Due to the obvious links to original US (Materials Genome Initiative [261]), European (Psi-k [262], CoEs (Centres of Excellence in HPC applications [263], EXDCI (European Extreme Data & Computing Initiative, [264]) and Japanese (National Institute for Materials Science, NIMS [265]) undertakings, the databases and repositories are currently somewhat biased to electronic structure simulation data (NoMAD [266] and FAIR-DI [267], AiiDA [268], MaterialsCloud [269], CMR [270], MPDS [271], OMD [272], ESL [273], ESP [274], Materials Project [275] (Figure 17), AFLOW [276], OQMD [277], MatNavi [278]). However, several other initiatives are acquiring momentum as data mining techniques have become state-of-the-art frameworks in materials' design investigations; among them are several extensions of the above, as well as products—both open-source and proprietary—dealing with upper rungs of the multiscale ladder (CALPHAD and related [279–282], MMM@HPC [283], EUDAT [284], Kaye and Laby [285], MEDEA [286]), and more general (NanoHUB [287], pymatgen [288]) or more specific properties or systems (Imeall [289], MPInterfaces [290], pylada [291]), to give just a few examples of the impressively growing catalogue of the field. Obviously, extracting relevant information from these computations could be extremely hard without the highly computerized retrieval (data mining) software incorporated as another tool for the repositories; this benefits interested researchers, who may thus save huge amounts of CPU resources and time when discarding or focusing on the partial substitution of a certain element or material for another less critical one with similar or even improved specific properties tailored to a particular application.

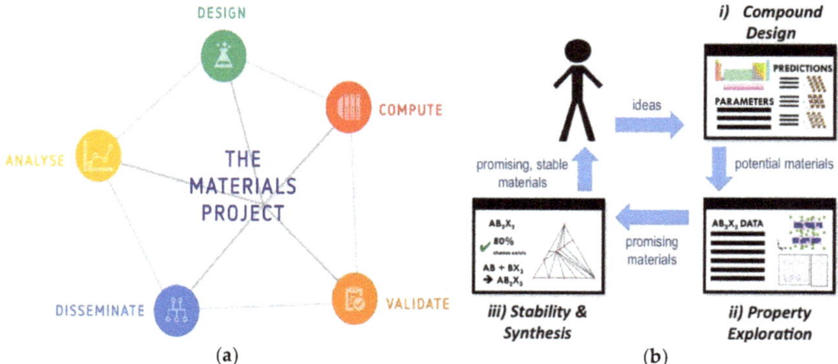

Figure 17. (**a**) Materials project thrusts. (**b**) In silico prototyping and iterative design steps of materials [292].

Figure 18 shows an example of data retrieval from two heavily used and pioneering repositories after searching for W–C–Co. Concerning computational screening (and its parallel approach, more experimentally biased, combinatorial screening, which has been used for a long time at least via a trial-and-error methodology, as this is essentially the procedure undertaken by humanity to create the very first alloys, namely bronze, CuSn, and brass, CuZn), the rapid advancement of computer capabilities and software has made possible the assessment of the performance of even slightly changed alloy compositions in the search for improved properties. For more than two decades, binary and ternary alloys using practically every element in the periodic table have been tracked and their phase diagrams described with unprecedented accuracy, both experimentally and computationally ([293–295]. The emergence of multi-component alloys (MCA) or (single-phase) high-entropy alloys (SP-HEA) in the present decade has boosted an immense and renewed interest in the stability, structural, electronic, elastic, magnetic, and thermodynamic properties of these combinations of several elements, none of which, in contrast to preceding alloying methodologies, can be said to be predominant [296–300].

Chemical formul	Space group	Basis set type	xc treatment	Code version	Upload date	Data access	References	Comment	Author(s)	Number of entri	Download
4*W3Co3C	Fd$\bar{3}$m	plane waves	GGA	VASP 5.3.2		Open	http://oqmd.org http://dx.doi.org/1		Hegde, Vinay Ward, Logan Wolverton, Chris	1	Show
WCo3C	Pm$\bar{3}$m	plane waves	GGA	VASP 5.3.5		Open			Botti, Silvana	2	Show
4*W4Co2C	R3m	plane waves	GGA	VASP 5.3.2		Open	http://oqmd.org http://dx.doi.org/1		Hegde, Vinay Ward, Logan Wolverton, Chris	1	Show
2*W6Co6C	R3m	plane waves	GGA	VASP 5.3.2		Open	http://oqmd.org http://dx.doi.org/1		Hegde, Vinay Ward, Logan Wolverton, Chris	1	Show
4*W3Co3C	R3m	plane waves	GGA	VASP 5.3.2		Open	http://oqmd.org http://dx.doi.org/1		Hegde, Vinay Ward, Logan Wolverton, Chris	1	Show
WCo slab	P1	plane waves	unknown	Quantum Espresso 5.1rc2 (svn rev.)		Open			Winther, Kirsten	1	Show
4*W4Co2C	Fd$\bar{3}$m	plane waves	GGA	VASP 5.2.11		Open			Huck, Patrick	2	Show
WCo2C	R3m	plane waves	GGA	VASP 5.3.3		Open			Gao, Qiang	6	Show
2*W6Co6C	Fd$\bar{3}$m	plane waves	GGA	VASP 5.3.2		Open	http://oqmd.org http://dx.doi.org/1		Hegde, Vinay Ward, Logan Wolverton, Chris	2	Show
2*W10Co3C4	Cmcm	plane waves	GGA	VASP 5.3.2		Open	http://oqmd.org http://dx.doi.org/1		Hegde, Vinay Ward, Logan Wolverton, Chris	2	Show

1 /2 › » [1 - 10 / 14]

(**a**)

Figure 18. *Cont.*

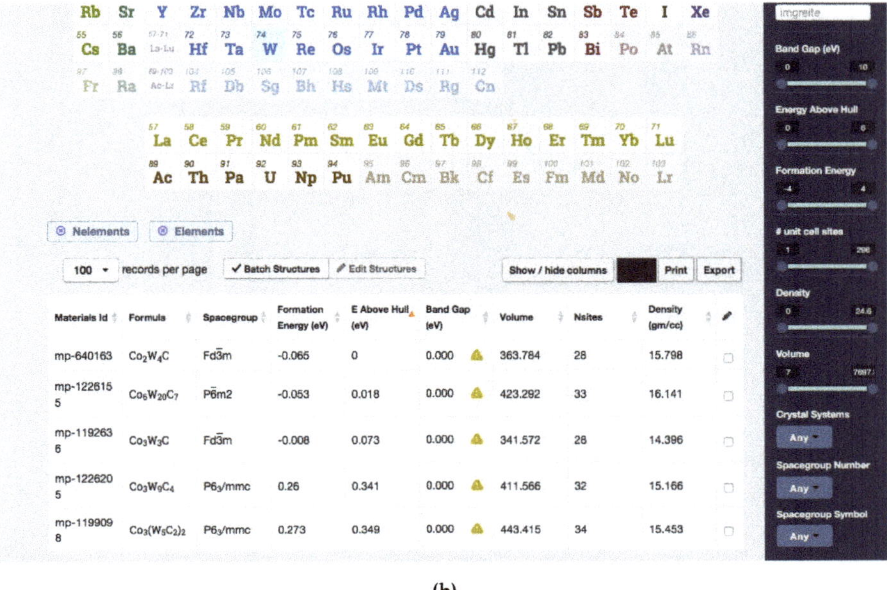

Materials Id	Formula	Spacegroup	Formation Energy (eV)	E Above Hull (eV)	Band Gap (eV)	Volume	Nsites	Density (gm/cc)
mp-640163	Co_2W_4C	$Fd\bar{3}m$	-0.065	0	0.000	363.784	28	15.798
mp-1226155	$Co_6W_{20}C_7$	$P\bar{6}m2$	-0.053	0.018	0.000	423.292	33	16.141
mp-1192636	Co_3W_3C	$Fd\bar{3}m$	-0.008	0.073	0.000	341.572	28	14.396
mp-1226205	$Co_3W_9C_4$	$P6_3/mmc$	0.26	0.341	0.000	411.566	32	15.166
mp-1199098	$Co_3(W_5C_2)_2$	$P6_3/mmc$	0.273	0.349	0.000	443.415	34	15.453

(b)

Figure 18. Search results for W–C–Co in two state-of-the-art computational data repositories. (**a**) In NOMAD [266]. (**b**) In the Materials Project [275].

4.2. Computational Screening

Genetic (USPEX [301], Figure 19a, GASP [302]), particle swarm optimization (CALYPSO [303]), cluster expansion (ATAT [304], UNCLE-MEDEA [305]), neural networks (RuNNer [306]), and random algorithms (AIRSS [307]), among others, have awakened the creativity of materials scientists, metallurgists, physicists and chemists alike in an unprecedented effort focused on the design of new materials containing smaller quantities of critical elements without compromising their ranges of applicability. As an example of the approach that may be pursued, the new superhard material WB_5, representing a possible alternative to tungsten carbide, although it still contains critical W, provides excellent performance even at high temperatures, as predicted by the global optimization code USPEX [308], as seen in Figure 19b.

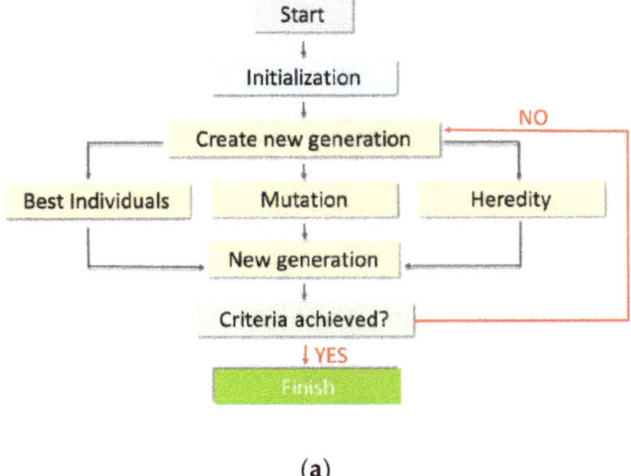

(a)

Figure 19. *Cont.*

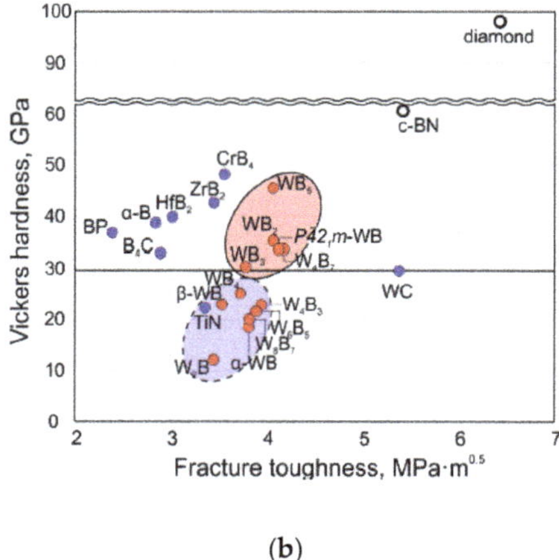

(b)

Figure 19. (**a**) Flowchart of a genetic algorithm as USPEX [309]. (**b**) Ashby plot of predicted new W-B phases (red points) compared to known superhard materials (blue points) [308].

4.3. Machine Learning

Machine learning is a cross-cutting methodology that has recently been developed in connection with the huge increase of computational power and advanced numerical and analytical techniques. In intimate connection with artificial intelligence concepts, appropriately configured computational programming and access to high-throughput resources allows a machine to extract patterns and learn from pre-existing data bases much faster and more accurately than ever before, iterating the processes until fully satisfying relationships and results have been obtained and, in doing so, reducing human intervention to a minimum. Interatomic potential fitting for ulterior MD modelling (Atomistica [310], Atomicrex [311], Potfit [312], OpenKIM [313]), hybrid DFT(Density Functional Theory)-MD simulations (Gaussian approximation potential (GAP) [314], SNAP [315]) or DFT-KMC [316,317] and finite element [318–320] multiscale modelling approaches are paradigmatic examples of advanced materials simulation methodologies that make use of machine learning-related techniques. The scaling up from ab-initio and atomistic time and lengths to mesoscopic and macroscopic scales benefits enormously from these kinds of procedures, as stated in several different recent reviews (Figure 20).

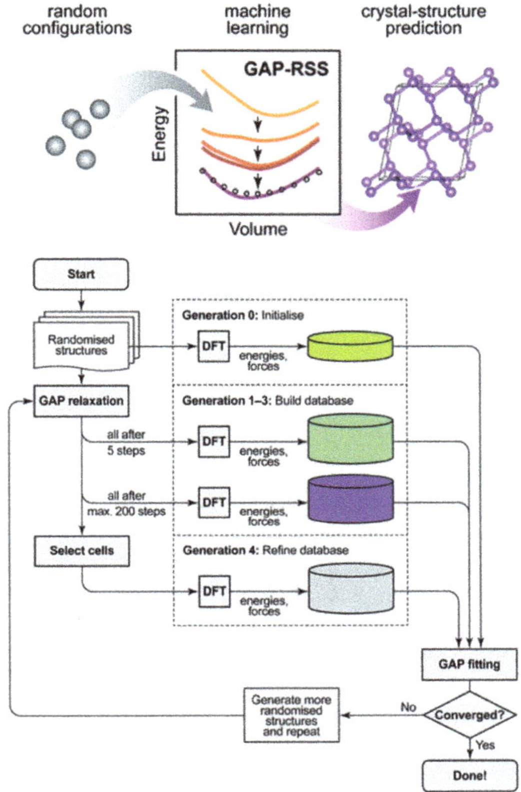

Figure 20. GAP-RSS (Gaussian approximation potential-based random structure searching) scheme and protocol [314].

4.4. Summary

In summary, the simultaneous application of the three aforementioned categories to a problem as intrinsically multiscale and multidisciplinary, as is the topic of the present review, provides innumerable advantages: one may envisage researchers willing to partially substitute a critical element (tungsten, cobalt, etc.) present in advanced cutting tool machinery. First, they explore the available similar compositions via computational materials repositories in a data mining process that builds on previous simulation results; secondly, once possible improvements in pre-existing alloy compositions have been identified, they wisely use computational screening techniques that will significantly reduce the number of possible variants, while crucially contributing to the identification of stable and metastable phases that deserve exploration while excellent working performance is not jeopardised; and thirdly, in the process, they build up a complete structural, electronic, elastic, magnetic database that in turn provides excellent quality input results to feed simulation techniques focused on higher lengths and times in an effort to advance to engineering scales and inform experimenters regarding the most industrially promising alloys with a lower critical materials content that could be amenable to sample synthesis in the laboratory and eventually reach the market.

5. Recycling

The recycling of hard metals and coating materials made of hard metals is an important element in the circular economy and environmental [321,322] and worker [323] safety. In terms of the life cycle of cutting tools, recyclability is also derived from a cost of ownership evaluation based on life cycle analysis tools [324].

Ishida et al. [325] have stated that if the volume of industrial production continues unaltered, global tungsten springs worldwide would be removed in about 40 years. Therefore, the recycling of tungsten from waste is becoming increasingly important. According to 2013 data from the International Tungsten Industry Association (ITIA), the recycling rate was 50% in Europe and the United States, in contrast to 30% in Japan. To improve the tungsten recycling rate in Japan, an efficient method could be to use scrap cemented carbide because the tungsten content in cemented carbide is more than 80%, which is higher than from other uses [326]. In fact, the Sumitomo Electric group focused on recycling tungsten from used carbide tools and succeeded in developing a technology and making a business. Various recycling processes for WC–Co cermets from cutting tools, such as chemical modification, thermal modification, the cold-stream method and the electrochemical method, have been investigated, and some of them are actually employed in industry. Presently, industrial recycling technology is committed to recycle tungsten carbide by two general methods. The first is the direct method, in which the binding metal is separated from the cemented carbide preserving the same composition; the second is an indirect method in which the dissolution of the binding metal takes place. In direct recycling techniques, such as the zinc process or the cold flow process, consumption is reduced and process costs are low, and the carbide scrap recycling rate is high. Instead, indirect recycling processes use acids and electrochemistry to dissolve the binder phase in cemented carbide waste. With this method, the energy consumption is high, process costs are high and the carbide scrap recycling rate is low. One of the main sources of the world supply of tungsten comes from the recycling of tungsten carbides, and it is believed that 30% of this comes from the recycling of tungsten carbide waste, mainly from tools such as cutters, turning, grinding and energy waste [325]. The used carbide inserts are also converted into powder by a process called the zinc process. In total, 60% of recycled tungsten comes from the recycling of high-speed steels. It is believed that tungsten recycling is much cheaper and more environmentally friendly than scrap disposal [326]. However, the industrial world continually requires the improvement of conventional methods and development of ever cheaper and more efficient recycling procedures. E. Altuncu et al. [327] investigated the applicability of the zinc-melt method (ZMM) for recycling WC–Co as a powder from cutting-tool scraps. It was proven that ZMM is an applicable technique for recovering the WC powder from the cutting tools. WC–Co powders are recovered and then spray-dried, sintered and obtained as a feedstock material for the thermal-spray coating processes. The zinc melt method can be used for the recycling of WC products, and the recycled powder with the addition of fresh powder can be used in the manufacturing of new samples, which shows that the best results are obtained with 70% of recycled WC powders. When identifying the need for customized tooling solutions using hard metal alloys, additive manufacturing can be chosen as a new production process thanks to its superior material and process flexibility.

Lee et al. developed a mechanochemical approach for WC recycling which involves a reaction with NaOH in a grinder [328]].

V.V. Popov et al. [329] studied the effect of powder recycling on Ti-6Al-4 V additive manufacturing. It was shown that as-printed samples produced from the recycled powder have dramatically decreased fatigue properties and a shortened lifetime. However, hot isostatic pressing (HIP) was suggested as a method to improve the quality of the sample and to achieve microstructural and mechanical properties very close to those of samples made of new powder.

Joost et al. suggested a recycling technique [330] which comprises a carbothermal process of simultaneous WC reactive sintering in a presence of graphite.

The electrochemical processing of WC–Co materials suggested by Malyshev et al. [331] offers good perspectives for the separation of the WC phase from hard metals and ore concentrates.

Acidic leaching is an up-to-date hydro-metallurgical approach for metal recycling including hard metals processing; however, more research is needed to increase the efficiency of reactions [332].

A microwave approach for recycling hard metal tools produced by chemical vapour deposition technique was proposed by Liu et al. [333].

A. C. Van Staden et al. [334] studied the SLM process using a cemented tungsten carbide powder for tools. The laser power, scan velocity, and hatch spacing of the SLM process were varied, and single powder layers were sintered accordingly. This was done to determine the influence of these parameter combinations on the melting behaviour of the material during sintering. It was found that a combination of high laser power, high hatch spacing, and low scan speed yielded the best results. It is hoped that tool manufacturing can soon be developed using additive techniques starting from recycled tungsten carbide powders.

6. Conclusions

The dependency of the EU on imports of CRMs—in particular, in the mechanical manufacturing industry—has raised a variety of concerns about the availability of raw materials and their rational use in tools manufacturing.

Due to the high import amount of CRMs needed for the manufacturing industry (such as tungsten, chromium, and niobium, to name just a few), it is vital to develop alternative solutions to face the high supply risk posed by the countries that are monopolistic suppliers of these CRMs.

Thus, several solutions must be pursued, such as improvements in strategies for longer-lasting tools (e.g., processing strategies and protective coatings), the development of alternative materials constituting the tools, the improved use of cutting-edge simulations and advanced use of recycling methods, as well as finding suitable candidates to partially or totally substitute the CRMs and increase their recycling.

European initiatives, such as the COST Action CA15102 "Solutions for Critical Raw Materials Under Extreme conditions", aim at developing strategies for the effective use of raw materials, including the reuse and recycling of end-of-life products containing CRMs.

The present review describes several strategies to face the CRMs issue in machining tools:

- The development of new cemented carbides based on environmentally harmless binders is a current research area. The substitution of cobalt in the cemented carbides is one of the research trends in the area of the environmental sustainability of industrial production and recycling processes.
- Different alternatives to the typical tungsten carbide material have been examined as constituents for the machining tools, such as ceramic materials, diamond-based systems, intermetallic systems and high entropy alloys, together with their related effective production techniques.
- We presented advanced machining techniques such as methods aided by laser, cryogenic temperatures, vibrations and surface defects with the aim of extending the tool life-span and thus reducing the amounts of CRMs used in the tools.
- We reviewed protecting coatings, which enable an increase in tool lifetimes under different machining situations.
- Additive manufacturing technologies along with the extensive use of advanced cost-effective fast-track computational methodologies facilitate the development of new materials by opening new ways of designing tools without or with only the partial use of CRMs with efficient strategies for the easy recycling of raw materials.
- Novel methodologies of tools manufacturing with geometric structures and challenging sub-structures and the potential to produce under-functionally different structures and optional materials with graded properties (e.g., the 4D-printing approach) provide a reasonable approach for decreasing the amounts of CRMs in tools.

In summary, every approach discussed in this review should result in the better quality of the machined parts, improve the machining performance and reduce the use of CRMs. However, in view of the conscious use of raw materials, it is clear that the different strategies should be combined together in a synergistic manner with a new way of thinking regarding the final products through the eco-friendly design of materials, tools and production methods to foster the integration of good practices and habits in the circular economy.

Author Contributions: A.R., M.L.G., and D.V. conceived the work. A.R. organized the work and focused on cryogenic machining and recycling; S.G. focused on advanced machining techniques; M.L.G., B.O.P., and D.V. focused on coatings and high entropy alloys; R.I. focused on modelling and simulation; L.J. focused on WC-Co and ceramic tools; P.N. focused on intermetallics; V.L. focused on recycling. B.O.P. made Figures 1, 2, 6–8 and 10. Writing—review and editing was performed by all the authors. All authors have read and agree to the published version of the manuscript.

Funding: This research received no external funding.

Acknowledgments: This article is based on work from COST Action CA15102 "Solutions for Critical Raw Materials under Extreme Conditions", supported by COST (European Cooperation in Science and Technology). AR wish to acknowledge the financial and scientific support received from the Italian National Projects SIADD ARS01_00806 MIUR nr. 1735, as well as the Department for Sustainability of ENEA for the equipment and expertise required to make this work possible. SG acknowledges that part of this work was summarised and carried out in the Centre for Doctoral Training in Ultra-Precision, which is supported by the UKRI via Grants No.: EP/K503241/1 and EP/L016567/1. Part of this work used Isambard Bristol, UK supercomputing service accessed by Resource Allocation Panel (RAP) grant. SG is particularly grateful to the fantastic financial support provided by the UKRI (Grant No. EP/S013652/1, EP/T001100/1 and EP/S036180/1), H2020 (Cost Actions (CA18125, CA18224 and CA16235) and EURAMET EMPIR A185 (2018)), Royal Academy of Engineering (Grant No. IAPP18-19\295), and Newton Fellowship award from the Royal Society (NIF\R1\191571) that allowed technical interactions across border. RI gratefully acknowledges financial and scientific support from the Spanish National Projects MAT2017-88258-R and ENE-2015-70300-C3-3-R. BP acknowledges that part of the contributions for this research was funded and supported by Portuguese Foundation for Science and Technology (FCT), grant number SFRH/BD/129614/2017, and by the Ukrainian state budget via research project number 0120U100475.

Conflicts of Interest: The authors declare that there is no conflict of interest.

References

1. Grilli, M.; Bellezze, T.; Gamsjäger, E.; Rinaldi, A.; Novak, P.; Balos, S.; Piticescu, R.; Ruello, M. Solutions for Critical Raw Materials under Extreme Conditions: A Review. *Materials* **2017**, *10*, 285. [CrossRef] [PubMed]
2. Commission of the European Communities. The Raw Materials Initiative: Meeting Our Critical Needs for Growth and Jobs in Europe, Brussels. 2008. Available online: https://eur-lex.europa.eu/legal-content/EN/TXT/?uri=CELEX:52008DC0699 (accessed on 17 March 2020).
3. Singla, A.K.; Singh, J.; Sharma, V.S. Processing of materials at cryogenic temperature and its implications in manufacturing: A review. *Mater. Manuf. Process.* **2018**, *33*, 1603–1640. [CrossRef]
4. European Commission. *DG Enterprise and Industry, Critical Raw Materials for the EU: Report of the Ad-hoc Working Group on Defining Critical Raw Materials*; European Commission: Brussels, Belgium, 2010.
5. European Commission. *Report on Critical Raw Materials for the EU: Report of the Ad hoc Working Group on Defining Critical Raw Materials*; European Commission: Brussels, Belgium, May 2014.
6. Communication from the Commission to the European Parliament, the Council, the European Economic and Social Committee and the Committee of the Regions, COM(2017) 490 Final. Available online: https://eur-lex.europa.eu/legal-content/en/ALL/?uri=CELEX:52017DC0490 (accessed on 17 March 2020).
7. Bobzin, K. High-performance coatings for cutting tools. *Cirp J. Manuf. Sci. Technol.* **2017**, *18*, 1–9. [CrossRef]
8. Byrne, G.; Ahearne, E.; Cotterell, M.; Mullany, B.; O'Donnell, G.E.; Sammler, F. High Performance Cutting (HPC) in the New Era of Digital Manufactoring–A Roadmap. *Procedia Cirp* **2016**, *46*, 1–6. [CrossRef]
9. Vanegas, P.; Durana, G.; Zubia, J.; De Ocariz, I.S. Advanced monitoring systems for smart tooling in aeronautical Industry 4.0. In Proceedings of the 14th Quantitative InfraRed Thermography Conference, Berlin, Germany, 25–29 June 2018. [CrossRef]
10. Cheng, K.; Chao, Z.; Wang, C.; Rakowski, R. Bateman Smart Cutting tools and smart machining: Development approaches and their implementation and application perspectives. *Chin. J. Mech. Eng.* **2017**, *30*. [CrossRef]
11. Paulsen, T.; Pecat, O.; Brinksmeier, E. Influence of different machining conditions on the subsurface properties of drilled TiAl6V4. *Procedia Cirp* **2016**, *46*, 472–475. [CrossRef]
12. Li, J.; Huang, Y.; Meng, X.; Xie, Y. A Review on High Entropy Alloys Coatings: Fabrication Processes and Property Assessment. *Adv. Eng. Mater.* **2019**, *21*, 1900343. [CrossRef]
13. Kurlov, S.A.; Gusev, A.I. *Tungsten Carbides: Structure, Properties and Application in Hardmetals*; Springer Series in Materials Science 184; Springer: Cham, Switzerland, 2013. [CrossRef]

14. Mills, B. Recent Developments in Cutting Tool Materials. *J. Mater. Process. Technol.* **1996**, *56*, 16–23. [CrossRef]
15. Agte, C.; Kohlermann, R. Hilfsmetallarme Hartmetallegierungen. *Die Tech.* **1957**, *10*, 686–689.
16. Richter, V. Hard sintered materials made of nano-sized powders. *Annu. Rep. Fraunhofer IKTS* **1995**, 44–45.
17. Li, S.K.; Li, J.Q.; Li, Y.; Liu, F.S.; Ao, W.Q. Dense pure binderless WC bulk material prepared by spark plasma sintering. *Mater. Sci. Technol.* **2015**, *31*, 1749–1756. [CrossRef]
18. Gubernat, A.; Rutkowski, P.; Grabowski, G.; Zientara, D. Hot pressing of tungsten carbide with and without sintering additives. *Int. J. Refract. Met. Hard Mater.* **2014**, *43*, 193–199. [CrossRef]
19. Sun, J.; Zhao, J.; Shen, X.; Huang, Z.; Yan, K.; Xing, J.; Gao, Y.; Jian, Y.; Yang, H.; Liat, B. A Review on Binderless Tungsten Carbide: Development and Application. *Nano-Micro Lett.* **2019**, *12*, 13. [CrossRef]
20. Gurland, J. A study of the effect of carbon content on the structure and properties of sintered WC–Co alloys. Trans. *AIME* **1954**, *200*, 285–290.
21. Gubedrechtrnat, L. Stobierski, Fractography of dense metal-like carbides sintered with carbon. *Key Eng. Mater.* **2009**, *409*, 287–290. [CrossRef]
22. Fox, R.T.; Nilsson, R. Binderless tungsten carbide carbon control with pressureless sintering. *Int. J. Refract. Met. Hard Mater.* **2018**, *76*, 82–89. [CrossRef]
23. Kim, H.C.; Shon, I.J.; Garay, J.E.; Munir, Z.A. Consolidation and properties of binderless sub-micron tungsten carbide by field-activated sintering. *Int. J. Refract. Met. Hard Mater.* **2004**, *22*, 257–264. [CrossRef]
24. Szutkowska, M.; Boniecki, M.; Cygan, S.; Kalinka, A.; Grilli, M.L.; Balos, S. Fracture behaviour of WC-Co partially substituted by titanium carbide. *Iop Conf. Ser. Mater. Sci. Eng.* **2018**, *329*. [CrossRef]
25. Tai, W.P.; Watanabe, T. Fabrication and Mechanical Properties of Al_2O_3-WC-Co Composites by Vacuum Hot Pressing. *J. Am. Ceram. Soc.* **1998**, *81*, 1673–1676. [CrossRef]
26. Basu, B.; Lee, J.H.; Kim, D.Y. Development of WC–ZrO_2 Nanocomposites by Spark Plasma Sintering. *Am. Ceram. Soc.* **2004**, *87*, 317–319. [CrossRef]
27. Uvarova, I.; Babutina, T.; Konchakovskay, I.; Timofeeva, I.; Petuchov, A. Nanostructure Composition of Diamond-WC-Co. In *Advanced Science and Technology of Sintering*; Stojanović, B.D., Skorokhod, V.V., Nikolić, M.V., Eds.; Springer: Berlin/Heidelberg, Germany, 1999.
28. Martínez, V.; Echeberria, J. Hot isostatic pressing of cubic boron nitride-tungsten carbide/cobalt (cBN-WC/Co) composites: Effect of cBN particle size and some processing parameters on their microstructure and properties. *J. Am. Ceram Soc.* **2007**, *90*, 415–424. [CrossRef]
29. Rong, H.Y.; Pen, Z.J.; Ren, X.Y.; Wang, C.B.; Fu, Z.Q.; Qi, L.H.; Miao, H.Z. Microstructure and mechanical properties of ultrafine WC-Ni-VC-TaC-cBN cemented carbides fabricated by spark plasma sintering. *Int. J. Refract. Met. Hard. Mater.* **2011**, *29*, 733–738. [CrossRef]
30. Rosinski, M.; Michalski, A. WCCo/cBN composites produced by pulse plasma sintering method. *J. Mater. Sci.* **2012**, *47*, 7064–7071. [CrossRef]
31. Mao, C.; Ren, Y.; Gan, H.; Zhang, M.; Zhang, J.; Tang, K. Microstructure and mechanical properties of cBN-WC-Co composites used for cutting tools. *Int. J. Adv. Manuf. Technol.* **2015**, *76*, 2043–2049. [CrossRef]
32. Bengisu, M.; Inal, O.T. Whisker toughening of ceramics, toughening mechanisms, fabrication, and composite properties. *Annu. Rev. Mater. Sci.* **1994**, *24*, 83–124. [CrossRef]
33. Chao, Y.J.; Liu, J. Study of WC ceramic tool material by SiC whisker toughening. *Rare Met. Cem. Carbides* **2005**, *33*, 13–16.
34. Chen, K.; Xiao, W.; Li, Z.; Wu, J.; Hong, K.; Ruan, X. Effect of Graphene and Carbon Nanotubes on the Thermal Conductivity of WC–Co Cemented Carbide. *Met. Open Access Metall. J.* **2019**, *9*, 377. [CrossRef]
35. Díaz-Álvarez, J.; Criado, V.; Miguélez, H.; Cantero, J. PCBN Performance in High Speed Finishing Turning of Inconel 718. *Metals* **2018**, *8*, 582. [CrossRef]
36. Fortunato, A.; Valli, G.; Liverani, E.; Ascari, A. Additive Manufacturing of WC-Co Cutting Tools for Gear Production. *Lasers Manuf. Mater. Process.* **2019**, *6*, 247–262. [CrossRef]
37. Goel, S.; Luo, X.; Agrawal, A.; Reuben, R.L. Diamond machining of silicon: A review of advances in molecular dynamics simulation. *Int. J. Mach. Tools Manuf.* **2015**, *88*, 131–164. [CrossRef]
38. Komanduri, R.; Lee, M.; Flom, D.G.; Thompson, R.A.; Jones, M.G.; Douglas, R.J. General Electric Co, 1982. Pulse laser pretreated machining. U.S. Patent 4,356,376, 26 October 2020.
39. Wu, J.F.; Guu, Y.B. Laser assisted machining method and device. Google Patents US20040104207A1, 21 Feburary 2006.
40. Patten, J. Micro Laser Assisted Machining. Google Patents US8933366B2, 13 January 2015.

41. Shin, Y.C. Laser Assisted Machining Process with Distributed Lasers. Google Patents US8698041B2, 15 April 2014.
42. Shin, Y.C. Machining apparatus and process. Google Patents US20110048183A1, 3 March 2011.
43. Dahotre, N.B.; Santhanakrishnan, S. Laser-assisted machining (lam) of hard tissues and bones. Google Patents US9387041B2, 12 July 2016.
44. Santner, J.S.; Sciammarella, F.M.; Kyselica, S. Laser assisted machining system for ceramics and hard materials. Google Patents US20130134141A1, 30 May 2013.
45. Abdulghani, O.; Sobih, M.; Youssef, A.; El-Batahgy, A.M. Modeling and Simulation of Laser Assisted Turning of Hard Steels. *Modeling Numer. Simul. Mater. Sci.* **2013**, *3*, 106–113. [CrossRef]
46. Razavykia, A.; Delprete, C.; Baldissera, P. Correlation between Microstructural Alteration, Mechanical Properties and Manufacturability after Cryogenic Treatment: A Review. *Materials* **2019**, *12*, 3302. [CrossRef] [PubMed]
47. Varghese, V.; Akhil, K.; Ramesh, M.R.; Chakradhar, D. Investigation on the performance of AlCrN and AlTiN coated cemented carbide inserts during end milling of maraging steel under dry, wet and cryogenic environments. *J. Manuf. Process.* **2019**, *43*, 136–144. [CrossRef]
48. Goel, S.; Martinez, F.D.; Chavoshi, S.Z.; Khatri, N.; Giusca, C. Molecular dynamics simulation of the elliptical vibration-assisted machining of pure iron. *J. Micromanuf.* **2018**, *1*, 6–19. [CrossRef]
49. Shokrani, A.; Dhokia, V.; Munos-Escalona, P.; Newmann, S.T. State-of-art cryogenic machining and processing. *Int. J. Comput. Integr. Manuf.* **2013**, *26*, 616–648. [CrossRef]
50. Rakesh, S.; Nirmal, K. Cryogenic Treatment of Tool Materials: A Review. *Mater. Manuf. Process.* **2010**, *25*, 1077–1100. [CrossRef]
51. Yildiz, Y.; Nalbant, M. A review of cryogenic cooling in machining processes. *Int. J. Mach. Tools Manuf.* **2008**, *48*, 947–964. [CrossRef]
52. Gill, S.S.; Singh, R.; Singh, H.; Singh, J. Investigation on wear behavior of cryogenically treated TiAlN coated tungsten carbide inserts in turning. *Int. J. Mach. Tools Manuf.* **2011**, *51*, 25–33. [CrossRef]
53. Seah, K.H.W.; Rahaman, M.; Yong, K.H. Performance evaluation of cryogenically treated tungsten carbide cutting tool inserts Proceeding of the Institution of Mechanical Engineers-Part B. *J. Eng. Manuf.* **2003**, *217*, 29–43. [CrossRef]
54. Yong, A.Y.L.; Seah, K.H.W.M. Rahman Performance of cryogenically treated tungsten carbide tools in milling operations. *Int. J. Adv. Manuf. Technol.* **2006**, *32*, 638–643. [CrossRef]
55. Sreeramareddy, T.V.; Sornakumar, T.; VenkataramaReddy, M.; Venkatram, R. Machining of C45 steel with deep cryogenic treated tungsten carbide cutting tool inserts. *Int. J. Refract. Met. Hard Mater.* **2009**, *27*, 181–185. [CrossRef]
56. Stewart, H.A. Cryogenic treatment of tungsten carbide reduces tool wear when machining medium density fiberboard. *For. Prod. J.* **2004**, *54*, 53–56.
57. Bryson, W.E. *Cryogenics*; Carl Hanser Verlag GmbH & Co.: Cincinnati, OH, USA, 1999; pp. 81–107. ISBN 978-1569902745.
58. Thakur, D.G.; Ramamoorthy, B.; Vijayaraghavan, L. Influence of different post treatments on tungsten carbide-cobalt inserts. *Mater. Lett.* **2008**, *62*, 4403–4406. [CrossRef]
59. Sert, A.; Celik, O.N. Characterization of the mechanism of cryogenic treatment on the microstructural changes in tungsten carbide cutting tools. *Mater. Charact.* **2019**, *150*, 1–7. [CrossRef]
60. Reddy, T.V.S.; Ajaykumar, B.S.; Reddy, M.V.; Venkataram, R. Machining performance of low temperature treated P-30 tungsten carbide cutting tool inserts. *Cryogenic* **2008**, *48*, 458–461.
61. Vadivel, K.; Rudramoorthy, R. Performance analysis of cryogenically treated coated carbide inserts. *Int. J. Adv. Manuf. Technol.* **2009**, *42*, 222–232. [CrossRef]
62. Swamini, A.; Chopra, V.; Sargade, G. Metallurgy behind the Cryogenic Treatment of Cutting Tools: An Overview. *Mater. Today Proceed.* **2015**, *2*, 1814–1824. [CrossRef]
63. Ahmed, M.I.; Ismail, A.F.; Abakr, Y.A.; Amin, A.K.M.N. Effectiveness of cryogenic machining with modified tool holder. *J. Mater. Process. Technol.* **2007**, *185*, 91–96. [CrossRef]
64. Dhananchezian, M. Study the machinability characteristics of Nicked based Hastelloy C-276 under cryogenic cooling. *Measurement* **2019**, *136*, 694–702. [CrossRef]
65. VakkasYıldırım, Ç. Experimental comparison of the performance of nanofluids, cryogenic and hybrid cooling in turning of Inconel 625. *Tribol. Int.* **2019**, *137*, 366–378.

66. Biswal, B.; Sarkar, B.; Mahanta, P. (Eds.) Characteristics During Hard Turning of Alloy Steel with Untreated and Cryotreated Cermet Inserts. In *Advances in Mechanical Engineering; Lecture Notes in Mechanical Engineering*; Springer: Singapore, 2018.
67. Sui, H.; Zhang, X.; Zhang, D.; Jiang, X.; Wu, R. Feasibility study of high-speed ultrasonic vibration cutting titanium alloy. *J. Mater. Process. Technol.* **2017**, *247*, 111–120. [CrossRef]
68. Bulla, B.; Kloche, F.; Dambon, O.; Hunter, M. Ultrasonic Assisted Diamond turning of hardned steel for mould manufacturing. *Key Eng. Mater.* **2012**, *516*, 437–442. [CrossRef]
69. Muhammad, R.; Hussain, M.S.; Maurotto, A.; Siemers, C.; Roy, A.; Silberschmidt, V.V. Analysis of a free machining $\alpha+\ \beta$ titanium alloy using conventional and ultrasonically assisted turning. *J. Mater. Process. Technol.* **2014**, *214*, 906–915. [CrossRef]
70. Ultrasonically Assisted Machining of Titanium Alloys. Available online: https://hdl.handle.net/2134/15909 (accessed on 17 March 2020).
71. Joshi, S.; Tewari, A.; Joshi, S.S. Microstructural characterization of chip segmentation under different machining environments in orthogonal machining of Ti6Al4V. *J. Eng. Mater. Technol.* **2015**, *137*, 011005. [CrossRef]
72. Maurotto, A.; Muhammad, R.; Roy, A. Comparing machinability of Ti-15-3-3-3 and Ni-625 alloys in UAT. *Procedia Cirp* **2012**, *1*, 330–335. [CrossRef]
73. Nath, C.; Rahman, M.; Andrew, S. A study on ultrasonic vibration cutting of low alloy steel. *J. Mater. Process. Technol.* **2007**, *192*, 159–165. [CrossRef]
74. Tutunea-Fatan, O.R.; Fakhri, M.A.; Bordatchev, E.V. Porosity and cutting forces: From macroscale to microscale machining correlations. Proceedings of the Institution of Mechanical Engineers. *Part B J. Eng. Manuf.* **2011**, *225*, 619–630. [CrossRef]
75. Pulse Laser Pretreated Machining. US Patent No. 4356376 against Application No. 263235. Available online: https://patents.google.com/patent/US4356376A/en (accessed on 18 March 2020).
76. Goel, S.; Rashid, W.B.; Luo, X.; Agrawal, A.; Jain, V. A theoretical assessment of surface defect machining and hot machining of nanocrystalline silicon carbide. *J. Manuf. Sci.Eng.* **2014**, *136*, 021015. [CrossRef]
77. Brinksmeier, E.; Glabe, R. Advances in Precision Machining of Steel. *Cirp Ann. Manuf. Tecnol.* **2001**, *50*, 385–388. [CrossRef]
78. Kawasegi, N.; Sugimori, H.; Morimoto, H.; Morita, N.; Hori, I. Development of cutting tools with microscale and nanoscale textures to improve frictional behavior. *Precis. Eng.* **2009**, *33*, 248–254. [CrossRef]
79. Evans, C.; Bryan, J.B. Cryogenic Diamond Turning of Stainless Steel. *Cirp Ann. Manuf. Technol.* **1991**, *40*, 571–575. [CrossRef]
80. Chang, W.; Sun, J.; Luo, X.; Ritchie, J.M.; Mack, C. Investigation of microstructured milling tool for deferring tool wear. *Wear* **2011**, *271*, 2433–2437. [CrossRef]
81. Kazuhiro, F.; Hideo, Y.; Naomichi, F.; Yutaka, Y.; Takashi, T.; Ryutaro, H.; Akitake, M.; Toshiro, H. Development of ultra-fine-grain binderless cBN tool for precision cutting of ferrous materials. *J. Mater. Process. Technol.* **2009**, *209*, 5646–5652. [CrossRef]
82. Fang, F.Z.; Chen, Y.H.; Zhang, X.D.; Hu, X.T.; Zhang, G.X. Nanometric cutting of single crystal silicon surfaces modified by ion implantation. *Cirp Ann. Manuf. Technol.* **2011**, *60*, 527–530. [CrossRef]
83. To, S.; Wang, H.; Jelenković, E.V. Enhancement of the machinability of silicon by hydrogen ion implantation for ultra-precision micro-cutting. *Int. J. Mach. Tools Manuf.* **2013**, *74*, 50–55. [CrossRef]
84. Muhammad, R.; Maurotto, A.; Demiral, M.; Roy, A.; Silberschmidt, V.V. Thermally enhanced ultrasonically assisted machining of Ti alloy. *Cirp J. Manuf. Sci. Technol.* **2014**, *7*, 159–167. [CrossRef]
85. Patil, S.; Joshi, S.; Tewari, A.; Joshi, S.S. Modelling and simulation of effect of ultrasonic vibrations on machining of Ti6Al4V. *Ultrasonics* **2014**, *54*, 694–705. [CrossRef]
86. Muhammad, R.; Mistry, A.; Khan, W.S.; Ahmed, N.; Roy, A.; Silberschmidt, V.V. Analysis of tool wear in ultrasonically assisted turning of Iranica. *B Transac. Engg.* **2016**, *23*, 1800–1810. [CrossRef]
87. Yan, J.; Zhang, Z.; Kuriyagawa, T. Effect of Nanoparticle Lubrication in Diamond Turning of Reaction-Bonded SiC. *Int. J. Autom. Technol.* **2011**, *5*, 307–312. [CrossRef]
88. Inada, A.; Min, S.; Ohmori, H. Micro cutting of ferrous materials using diamond tool under ionized coolant with carbon particles. *Cirp Ann. Manuf. Technol.* **2011**, *60*, 97–100. [CrossRef]
89. Zareena, A.R.; Veldhuis, S.C. Tool wear mechanisms and tool life enhancement in ultra-precision machining of titanium. *J. Mater. Process. Technol.* **2012**, *212*, 560–570. [CrossRef]

90. Rashid, W.B.; Goel, S. Advances in the surface defect machining (SDM) of hard steels. *J. Manuf. Process.* **2016**, *23*, 37–46. [CrossRef]
91. Rashid, W.B.; Goel, S.; Luo, X.; Ritchie, J.M. An experimental investigation for the improvement of attainable surface roughness during hard turning process. Proceedings of the Institution of Mechanical Engineers. *Part B J. Eng. Manuf.* **2013**, *227*, 338–342. [CrossRef]
92. Rashid, W.B.; Goel, S.; Luo, X.; Ritchie, J.M. The development of a surface defect machining method for hard turning processes. *Wear* **2013**, *302*, 1124–1135. [CrossRef]
93. Tamerabeta, Y.; Briouaa, M.; Tamerabeta, M.; Khoualdia, S. Experimental Investigation on Tool Wear Behavior and Cutting Temperature during Dry Machining of Carbon Steel SAE 1030 Using KC810 and KC910 Coated Inserts. *Tribol. Ind.* **2018**, *40*, 52–65. [CrossRef]
94. Klocke, F.; Krieg, T. Coated Tools for Metal Cutting–Features and Applications. *Cirp Ann.* **1999**, *48*, 515–525. [CrossRef]
95. Vereschaka, A.; Kataeva, E.; Sitnikov, N.; Aksenenko, A.; Oganyan, G.; Sotova, C. Influence of Thickness of Multilayered Nano-Structured Coatings Ti-TiN-(TiCrAl)N and Zr-ZrN-(ZrCrNbAl)N on Tool Life of Metal Cutting Tools at Various Cutting Speeds. *Coatings* **2018**, *8*, 44. [CrossRef]
96. Levashov, E.A.; Merzhanov, A.G.; Shtanskv, D.V. Advanced technologies, materials and coatings developed in scientific-educational center of SHS. *Galvanotechnik* **2009**, *100*, 2102–2114.
97. Gu, J.; Barber, G.; Tung, S.; Gu, R.-J. Tool life and wear mechanism of uncoated and coated milling inserts. *Wear* **1999**, *225–229*, 273–284. [CrossRef]
98. Vereschaka, A.A.; Grigoriev, S.N.; Sitnikov, N.N.; Oganyan, G.V.; Batako, A. Working efficiency of cutting tools with multilayer nano-structured Ti-TiCN-(Ti,Al)CN and Ti-TiCN-(Ti,Al,Cr)CN coatings: Analysis of cutting properties, wear mechanism and diffusion processes. *Surf. Coat. Technol.* **2017**, *332*, 198–213. [CrossRef]
99. Veprek, S. Recent search for new superhard materials: Go nano! *J. Vac. Sci. Technol. A Vac. Surf. Film* **2013**, *31*, 050822. [CrossRef]
100. Roy, M. Protective Hard Coatings for Tribological Applications. In *Materials Under Extreme Condition*; Elsevier: Amsterdam, The Netherlands, 2017; pp. 259–292.
101. Polini, R.; Barletta, M.; Rubino, G.; Vesco, S. Recent Advances in the Deposition of Diamond Coatings on Co-Cemented Tungsten Carbides. *Adv. Mater. Sci. Eng.* **2012**, *2012*, 1–14. [CrossRef]
102. Kuo, C.; Wang, C.; Ko, S. Wear behaviour of CVD diamond-coated tools in the drilling of woven CFRP composites. *Wear* **2018**, *398–399*, 1–12. [CrossRef]
103. Ramasubramanian, K.; Arunachalam, N.; Rao, M.S.R. Wear performance of nano-engineered boron doped graded layer CVD diamond coated cutting tool for machining of Al-SiC MMC. *Wear* **2019**, *426–427*, 1536–1547. [CrossRef]
104. Poulon-Quintin, A.; Faure, C.; Teulé-Gay, L.; Manaud, J.P. A multilayer innovative solution to improve the adhesion of nanocrystalline diamond coatings. *Appl. Surf. Sci.* **2015**, *331*, 27–34. [CrossRef]
105. Linnik, S.A.; Gaydaychuk, A.V.; Okhotnikov, V.V. Improvement to the adhesion of polycrystalline diamond films on WC-Co cemented carbides through ion etching of loosely bound growth centers. *Surf. Coat. Technol.* **2018**, *334*, 227–232. [CrossRef]
106. Ye, F.; Li, Y.; Sun, X.; Yang, Q.; Kim, C.-Y.; Odeshi, A.G. CVD diamond coating on WC-Co substrate with Al-based interlayer. *Surf. Coat. Technol.* **2016**, *308*, 121–127. [CrossRef]
107. Wang, T.; Zhang, S.; Jiang, C.; Handschuh-Wang, S.; Chen, G.; Zhou, X.; Tang, Y. TiB_2 barrier interlayer approach for HFCVD diamond deposition onto cemented carbide tools. *Diam. Relat. Mater.* **2018**, *83*, 126–133. [CrossRef]
108. Chandran, M.; Sammler, F.; Uhlmann, E.; Akhvlediani, R.; Hoffman, A. Wear performance of diamond coated WC-Co tools with a CrN interlayer. *Diam. Relat. Mater.* **2017**, *73*, 47–55. [CrossRef]
109. An Mahmud, K.A.H.; Kalam, M.A.; Masjuki, H.H.; Mobarak, H.M.; Zulkifli, N.W.M. An updated overview of diamond-like carbon coating in tribology. *Crit. Rev. Solid State Mater. Sci.* **2015**, *40*, 90–118. [CrossRef]
110. Fukui, H.; Okida, J.; Omori, N.; Moriguchi, H.; Tsuda, K. Cutting performance of DLC coated tools in dry machining aluminum alloys. *Surf. Coat. Technol.* **2004**, *187*, 70–76. [CrossRef]
111. Erdemir, A.; Donnet, C. Tribology of diamond-like carbon films: Recent progress and future prospects. *J. Phys. D. Appl. Phys.* **2006**, *39*, R311–R327. [CrossRef]

112. Huang, L.; Yuan, J.; Li, C.; Hong, D. Microstructure, tribological and cutting performance of Ti-DLC/α-C:H multilayer film on cemented carbide. *Surf. Coat. Technol.* **2018**, *353*, 163–170. [CrossRef]
113. Liu, Y.; Meletis, E.I. Evidence of graphitization of diamond-like carbon films during sliding wear. *J. Mater. Sci.* **1997**, *32*, 3491–3495. [CrossRef]
114. Chen, J.G. Carbide and Nitride Overlayers on Early Transition Metal Surfaces: Preparation, Characterization, and Reactivities. *Chem. Rev.* **1996**, *96*, 1477–1498. [CrossRef]
115. Han, Y.; Dai, Y.; Shu, D.; Wang, J.; Sun, B. Electronic and bonding properties of TiB2. *J. Alloy. Compd.* **2007**, *438*, 327–331. [CrossRef]
116. Rasaki, S.A.; Zhang, B.; Anbalgam, K.; Thomas, T.; Yang, M. Synthesis and application of nano-structured metal nitrides and carbides: A review. *Prog. Solid State Chem.* **2018**, *50*, 1–15. [CrossRef]
117. Holleck, H. Material selection for hard coatings. *J. Vac. Sci. Technol. A Vacuum Surfaces Film* **1986**, *4*, 2661–2669. [CrossRef]
118. Grilli, M.L.; Valerini, D.; Piticescu, R.R.; Bellezze, T.; Yilmaz, M.; Rinaldi, A.; Cuesta-López, S.; Rizzo, A. Possible alternatives to critical elements in coatings for extreme applications. *IOP Conf. Ser. Mater. Sci. Eng.* **2017**, *329*, 012005. [CrossRef]
119. Mayrhofer, P.H.; Rachbauer, R.; Holec, D.; Rovere, F.; Schneider, J.M. Protective Transition Metal Nitride Coatings. In *Comprehensive Materials Processing*; Elsevier: Amsterdam, The Netherlands, 2014; pp. 355–388. [CrossRef]
120. Kalss, W.; Reiter, A.; Derflinger, V.; Gey, C.; Endrino, J.L. Modern coatings in high performance cutting applications. *Int. J. Refract. Met. Hard Mater.* **2006**, *24*, 399–404. [CrossRef]
121. Fernandes, F.; Danek, M.; Polcar, T.; Cavaleiro, A. Tribological and cutting performance of TiAlCrN films with different Cr contents deposited with multilayered structure. *Tribol. Int.* **2018**, *119*, 345–353. [CrossRef]
122. Danek, M.; Fernandes, F.; Cavaleiro, A.; Polcar, T. Influence of Cr additions on the structure and oxidation resistance of multilayered TiAlCrN films. *Surf. Coat. Technol.* **2017**, *313*, 158–167. [CrossRef]
123. Mi, P.; He, J.; Qin, Y.; Chen, K. Nanostructure reactive plasma sprayed TiCN coating. *Surf. Coat. Technol.* **2017**, *309*, 1–5. [CrossRef]
124. Kumar, T.S.; Jebaraj, A.V.; Sivakumar, K.; Shankar, E.; Tamiloli, N. Characterization of ticn coating synthesized by the plasma enhanced physical vapour deposition process on a cemented carbide tool. *Surf. Rev. Lett.* **2018**, *25*, 1950028. [CrossRef]
125. Patscheider, J. Nanocomposite Hard Coatings for Wear Protection. *MRS Bull.* **2003**, *8*, 180–183. [CrossRef]
126. Beake, B.D.; Vishnyakov, V.M.; Valizadeh, R.; Colligon, J.S. Influence of mechanical properties on the nanoscratch behaviour of hard nanocomposite TiN/Si3N4coatings on Si. *J. Phys. D: Appl. Phys.* **2006**, *39*, 1392–1397. [CrossRef]
127. Pogrebnjak, A.; Smyrnova, K.; Bondar, O. Nanocomposite Multilayer Binary Nitride Coatings Based on Transition and Refractory Metals. *Struct. Prop. Coat.* **2019**, *9*, 155. [CrossRef]
128. Yeh, J.-W. Recent progress in high-entropy alloys. *Ann. Chim. Sci. Des Matériaux.* **2006**, *31*, 633–648. [CrossRef]
129. Yeh, J.-W.; Chen, S.-K.; Lin, S.-J.; Gan, J.-Y.; Chin, T.-S.; Shun, T.-T.; Tsau, C.-H.; Chang, S.-Y. Nanostructured High-Entropy Alloys with Multiple Principal Elements: Novel Alloy Design Concepts and Outcomes. *Adv. Eng. Mater.* **2004**, *6*, 299–303. [CrossRef]
130. Yeh, J.-W.; Lin, S.-J. Breakthrough applications of high-entropy materials. *J. Mater. Res.* **2018**, *33*, 3129–3137. [CrossRef]
131. Yip, S. The strongest size. *Nature* **1998**, *391*, 532–533. [CrossRef]
132. Martinu, L.; Zabeida, O.; Klemberg-Sapieha, J.E. Plasma Enhanced Chemical Vapor Deposition of Functional Coatings. *Handb. Depos. Technol. Film. Coat.* **2010**, 392–465. [CrossRef]
133. Musil, J. Hard nanocomposite coatings: Thermal stability, oxidation resistance and toughness. *Surf. Coat. Technol.* **2012**, *207*, 50–65. [CrossRef]
134. Uhlmann, E.; Fuentes, J.A.O.; Gerstenberger, R.; Frank, H. nc-AlTiN/a-Si3N4 and nc-AlCrN/a-Si3N4 nanocomposite coatings as protection layer for PCBN tools in hard machining. *Surf. Coat. Technol.* **2013**, *237*, 142–148. [CrossRef]
135. Ma, Q.; Li, L.; Xu, Y.; Gu, J.; Wang, L.; Xu, Y. Effect of bias voltage on TiAlSiN nanocomposite coatings deposited by HiPIMS. *Appl. Surf. Sci.* **2017**, *392*, 826–833. [CrossRef]

136. Settineri, L.; Faga, M.G. Laboratory tests for performance evaluation of nanocomposite coatings for cutting tools. *Wear* **2006**, *260*, 326–332. [CrossRef]
137. Abadias, G.; Daniliuk, A.Y.; Solodukhin, I.A.; Uglov, V.V.; Zlotsky, S.V. Thermal Stability of TiZrAlN and TiZrSiN Films Formed by Reactive Magnetron Sputtering, Inorg. *Mater. Appl. Res.* **2018**, *9*, 418–426. [CrossRef]
138. Musil, J.; Novák, P.; Čerstvý, R.; Soukup, Z. Tribological and mechanical properties of nanocrystalline-TiC/a-C nanocomposite thin films. *J. Vac. Sci. Technol. A Vacuum Surfaces Film* **2010**, *28*, 244–249. [CrossRef]
139. El Mel, A.A.; Gautron, E.; Christien, F.; Angleraud, B.; Granier, A.; Souček, P.; Vašina, P.; Buršíková, V.; Takashima, M.; Ohtake, N.; et al. Titanium carbide/carbon nanocomposite hard coatings: A comparative study between various chemical analysis tools. *Surf. Coat. Technol.* **2014**, *246*, 41–46. [CrossRef]
140. Qiu, L.; Du, Y.; Wang, S.; Li, K.; Yin, L.; Wu, L.; Zhong, Z.; Albir, L. Mechanical properties and oxidation resistance of chemically vapor deposited TiSiN nanocomposite coating with thermodynamically designed compositions. *Int. J. Refract. Met. Hard Mater.* **2019**, *80*, 30–39. [CrossRef]
141. Schwaller, P.; Haug, F.-J.; Michler, J.; Patscheider, J. Nanocomposite Hard Coatings: Deposition Issues and Validation of their Mechanical Properties. *Adv. Eng. Mater.* **2005**, *7*, 318–322. [CrossRef]
142. Mahato, P.; Nyati, G.; Singh, R.J.; Mishra, S.K. Nanocomposite TiSiBC Hard Coatings with High Resistance to Wear, Fracture and Scratching. *J. Mater. Eng. Perform.* **2016**, *25*, 3774–3782. [CrossRef]
143. Verma, D.; Banerjee, D.; Mishra, S.K. Effect of Silicon Content on the Microstructure and Mechanical Properties of Ti-Si-B-C Nanocomposite Hard Coatings. *Met. Mater. Trans. A* **2019**, *50*, 894–904. [CrossRef]
144. Mahato, P.; Singh, R.J.; Mishra, S.K. Nanocomposite Ti–Si–B–C hard coatings deposited by magnetron sputtering: Oxidation and mechanical behaviour with temperature and duration of oxidation. *Surf. Coat. Technol.* **2016**, *288*, 230–240. [CrossRef]
145. Mahato, P.; Singh, R.J.; Pathak, L.C.; Mishra, S.K. Effect of nitrogen on mechanical, oxidation and structural behaviour of Ti-Si-B-C-N nanocomposite hard coatings deposited by DC sputtering. *Surf. Interface Anal.* **2016**, *48*, 1080–1089. [CrossRef]
146. Yi, J.; Chen, S.; Chen, K.; Xu, Y.; Chen, Q.; Zhu, C.; Liu, L. Effects of Ni content on microstructure, mechanical properties and Inconel 718 cutting performance of AlTiN-Ni nanocomposite coatings. *Ceram. Int.* **2019**, *45*, 474–480. [CrossRef]
147. Saladukhin, I.A.; Abadias, G.; Uglov, V.V.; Zlotski, S.V.; Michel, A.; Van Vuuren, A.J. Thermal stability and oxidation resistance of ZrSiN nanocomposite and ZrN/SiNx multilayered coatings: A comparative study. *Surf. Coat. Technol.* **2017**, *332*, 428–439. [CrossRef]
148. Anwar, S.; Islam, A.; Bajpai, S.; Anwar, S. Structural and mechanical studies of W2N embedded Si3N4 nanocomposite hard coating prepared by reactive magnetron sputtering. *Surf. Coat. Technol.* **2017**, *311*, 268–273. [CrossRef]
149. Pogrebnjak, A.D.; Postol'nyi, B.A.; Kravchenko, Y.A.; Shipilenko, A.P.; Sobol', O.V.; Beresnev, V.M.; Kuz'menko, A.P. Structure and properties of (Zr-Ti-Cr-Nb)N multielement superhard coatings. *J. Superhard Mater.* **2015**, *37*, 101–111. [CrossRef]
150. Bondar, O.V.; Postolnyi, B.O.; Kravchenko, Y.A.; Shypylenko, A.P.; Sobol, O.V.; Beresnev, V.M.; Kuzmenko, A.P.; Zukowski, P. Fabrication and Research of Superhard (Zr-Ti-Cr-Nb)N Coatings. *Acta Phys. Pol. A* **2015**, *128*, 867–871. [CrossRef]
151. Gleich, S.; Breitbach, B.; Peter, N.J.; Soler, R.; Bolvardi, H.; Schneider, J.M.; Dehm, G.; Scheu, C. Thermal stability of nanocomposite Mo2BC hard coatings deposited by magnetron sputtering. *Surf. Coat. Technol.* **2018**, *349*, 378–383. [CrossRef]
152. Kawasaki, M.; Nose, M.; Onishi, I.; Shiojiri, M. Structural Investigation of AlN/SiOx Nanocomposite Hard Coatings Fabricated by Differential Pumping Cosputtering. *Microsc. Microanal.* **2016**, *22*, 673–678. [CrossRef] [PubMed]
153. Veprek, S.; Zhang, R.F.; Veprek-Heijman, M.G.J.; Sheng, S.H.; Argon, A.S. Superhard nanocomposites: Origin of hardness enhancement, properties and applications. *Surf. Coat. Technol.* **2010**, *204*, 1898–1906. [CrossRef]
154. Veprek, S.; Veprek-Heijman, M.G.J. Limits to the preparation of superhard nanocomposites: Impurities, deposition and annealing temperature. *Thin Solid Film.* **2012**, *522*, 274–282. [CrossRef]
155. Pogrebnjak, A.D.; Bagdasaryan, A.A.; Pshyk, A.; Dyadyura, K. Adaptive multicomponent nanocomposite coatings in surface engineering. *Phys. Uspekhi.* **2017**, *60*, 586–607. [CrossRef]
156. Pogrebnyak, A.D.; Shpak, A.P.; Azarenkov, N.A.; Beresnev, V.M. Structures and properties of hard and superhard nanocomposite coatings. *Phys. Uspekhi* **2009**, *52*, 29–54. [CrossRef]

157. Kumar, C.S.; Patel, S.K. Application of surface modification techniques during hard turning: Present work and future prospects. *Int. J. Refract. Met. Hard Mater.* **2018**, *76*, 112–127. [CrossRef]
158. Ziebert, C.; Stüber, M.; Leiste, H.; Ulrich, S.; Holleck, H. Nanoscale PVD Multilayer Coatings. In *Encyclopedia Material Science Technology*; Elsevier: Amsterdam, The Netherlands, 2011; pp. 1–8. [CrossRef]
159. Inspektor, A.; Salvador, P.A. Architecture of PVD coatings for metalcutting applications: A review. *Surf. Coat. Technol.* **2014**, *257*, 138–153. [CrossRef]
160. Khadem, M.; Penkov, O.V.; Yang, H.-K.; Kim, D.-E. Tribology of multilayer coatings for wear reduction: A review. *Friction* **2017**, *5*, 248–262. [CrossRef]
161. Wang, J.; Yazdi, M.A.P.; Lomello, F.; Billard, A.; Kovács, A.; Schuster, F.; Guet, C.; White, T.J.; Sanchette, F.; Dong, Z. Influence of microstructures on mechanical properties and tribology behaviors of TiN/Ti X Al 1−X N multilayer coatings. *Surf. Coat. Technol.* **2017**, *320*, 441–446. [CrossRef]
162. Andersen, K.N.; Bienk, E.J.; Schweitz, K.O.; Reitz, H.; Chevallier, J.; Kringhøj, P.; Bøttiger, J. Deposition, microstructure and mechanical and tribological properties of magnetron sputtered TiN/TiAlN multilayers. *Surf. Coat. Technol.* **2000**, *123*, 219–226. [CrossRef]
163. Contreras, E.; Bejarano, G.; Gómez, M. Synthesis and microstructural characterization of nanoscale multilayer TiAlN/TaN coatings deposited by DC magnetron sputtering. *Int. J. Adv. Manuf. Technol.* **2019**, *101*, 663–673. [CrossRef]
164. Pshyk, A.V.; Kravchenko, Y.; Coy, E.; Kempiński, M.; Iatsunskyi, I.; Załęski, K.; Pogrebnjak, A.D.; Jurga, S. Microstructure, phase composition and mechanical properties of novel nanocomposite (TiAlSiY)N and nano-scale (TiAlSiY)N/MoN multifunctional heterostructures. *Surf. Coat. Technol.* **2018**, *350*, 376–390. [CrossRef]
165. Illana, A.; Almandoz, E.; Fuentes, G.G.; Pérez, F.J.; Mato, S. Comparative study of CrAlSiN monolayer and CrN/AlSiN superlattice multilayer coatings: Behavior at high temperature in steam atmosphere. *J. Alloy. Compd.* **2019**, *778*, 652–661. [CrossRef]
166. Seidl, W.M.; Bartosik, M.; Kolozsvári, S.; Bolvardi, H.; Mayrhofer, P.H. Mechanical properties and oxidation resistance of Al-Cr-N/Ti-Al-Ta-N multilayer coatings. *Surf. Coat. Technol.* **2018**, *347*, 427–433. [CrossRef]
167. Braic, M.; Balaceanu, M.; Parau, A.C.; Dinu, M.; Vladescu, A. Investigation of multilayered TiSiC/NiC protective coatings. *Vacuum* **2015**, *120*, 60–66. [CrossRef]
168. Pogrebnjak, A.D.; Bondar, O.V.; Abadias, G.; Eyidi, D.; Beresnev, V.M.; Sobol, O.V.; Postolnyi, B.O.; Zukowski, P. Investigation of Nanoscale TiN/MoN Multilayered Systems, Fabricated Using Arc Evaporation. *Acta Phys. Pol. A* **2015**, *128*, 836–841. [CrossRef]
169. Postolnyi, B.O.; Konarski, P.; Komarov, F.F.; Sobol, O.V.; Kyrychenko, O.V.; Shevchuk, D.S. Study of elemental and structural phase composition of multilayer nanostructured TiN/MoN coatings, their physical and mechanical properties. *J. Nano- Electron. Phys.* **2014**, *6*, 04016.
170. Buchinger, J.; Koutná, N.; Chen, Z.; Zhang, Z.; Mayrhofer, P.H.; Holec, D.; Bartosik, M. Toughness enhancement in TiN/WN superlattice thin films. *Acta Mater.* **2019**, *172*, 18–29. [CrossRef]
171. Pogrebnjak, A.; Ivashchenko, V.; Bondar, O.; Beresnev, V.; Sobol, O.; Załęski, K.; Jurga, S.; Coy, E.; Konarski, P.; Postolnyi, B. Multilayered vacuum-arc nanocomposite TiN/ZrN coatings before and after annealing: Structure, properties, first-principles calculations. *Mater. Charact.* **2017**, *134*, 55–63. [CrossRef]
172. Major, L.; Major, R.; Kot, M.; Lackner, J.M.; Major, B. Ex situ and in situ nanoscale wear mechanisms characterization of Zr/Zr x N tribological coatings. *Wear* **2018**, *404–405*, 82–91. [CrossRef]
173. Pogrebnjak, A.D.; Kravchenko, Y.O.; Bondar, O.V.; Zhollybekov, B.; Kupchishin, A.I. Kupchishin, Structural Features and Tribological Properties of Multilayer Coatings Based on Refractory Metals. *Prot. Met. Phys. Chem. Surf.* **2018**, *54*, 240–258. [CrossRef]
174. Postolnyi, B.; Bondar, O.; Opielak, M.; Rogalski, P.; Araújo, J.P. Structural analysis of multilayer metal nitride films CrN/MoN using electron backscatter diffraction (EBSD). In Proceedings of the SPIE 10010, Advanced Topics in Optoelectronics, Microelectronics, and Nanotechnologies VIII, 100100E, Constanta, Romania, 14 December 2016. [CrossRef]
175. Pogrebnjak, A.D.; Beresnev, V.M.; Bondar, O.V.; Postolnyi, B.O.; Zaleski, K.; Coy, E.; Jurga, S.; Lisovenko, M.O.; Konarski, P.; Rebouta, L.; et al. Superhard CrN/MoN coatings with multilayer architecture. *Mater. Des.* **2018**, *153*, 47–59. [CrossRef]

176. Postolnyi, B.O.; Bondar, O.V.; Zaleski, K.; Coy, E.; Jurga, S.; Rebouta, L.; Araujo, J.P. Multilayer Design of CrN/MoN Superhard Protective Coatings and Their Characterisation. In *Advances in Thin Films, Nanostructured Materials, and Coatings*; Pogrebnjak, A.D., Novosad, V., Eds.; Springer: Singapore, 2019; pp. 17–29. [CrossRef]
177. Bagdasaryan, A.A.; Pshyk, A.V.; Coy, L.E.; Kempiński, M.; Pogrebnjak, A.D.; Beresnev, V.M.; Jurga, S. Structural and mechanical characterization of (TiZrNbHfTa)N/WN multilayered nitride coatings. *Mater. Lett.* **2018**, *229*, 364–367. [CrossRef]
178. Lin, Y.; Zia, A.W.; Zhou, Z.; Shum, P.W.; Li, K.Y. Development of diamond-like carbon (DLC) coatings with alternate soft and hard multilayer architecture for enhancing wear performance at high contact stress. *Surf. Coat. Technol.* **2017**, *320*, 7–12. [CrossRef]
179. Zha, X.; Jiang, F.; Xu, X. Investigating the high frequency fatigue failure mechanisms of mono and multilayer PVD coatings by the cyclic impact tests. *Surf. Coat. Technol.* **2018**, *344*, 689–701. [CrossRef]
180. Chang, Y.-Y.; Chiu, W.-T.; Hung, J.-P. Mechanical properties and high temperature oxidation of CrAlSiN/TiVN hard coatings synthesized by cathodic arc evaporation. *Surf. Coat. Technol.* **2016**, *303*, 18–24. [CrossRef]
181. Chang, Y.-Y.; Weng, S.-Y.; Chen, C.-H.; Fu, F.-X. High temperature oxidation and cutting performance of AlCrN, TiVN and multilayered AlCrN/TiVN hard coatings. *Surf. Coat. Technol.* **2017**, *332*, 494–503. [CrossRef]
182. Chang, Y.-Y.; Chang, H.; Jhao, L.-J.; Chuang, C.-C. Tribological and mechanical properties of multilayered TiVN/TiSiN coatings synthesized by cathodic arc evaporation. *Surf. Coat. Technol.* **2018**, *350*, 1071–1079. [CrossRef]
183. Kong, Y.; Tian, X.; Gong, C.; Chu, P.K. Enhancement of toughness and wear resistance by CrN/CrCN multilayered coatings for wood processing. *Surf. Coat. Technol.* **2018**, *344*, 204–213. [CrossRef]
184. Beake, B.D.; Fox-Rabinovich, G.S. Progress in high temperature nanomechanical testing of coatings for optimising their performance in high speed machining. *Surf. Coat. Technol.* **2014**, *255*, 102–111. [CrossRef]
185. Chowdhury, S.; Beake, B.; Yamamoto, K.; Bose, B.; Aguirre, M.; Fox-Rabinovich, G.; Veldhuis, S. Improvement of Wear Performance of Nano-Multilayer PVD Coatings under Dry Hard End Milling Conditions Based on Their Architectural Development. *Coatings* **2018**, *8*, 59. [CrossRef]
186. Kursuncu, B.; Caliskan, H.; Guven, S.Y.; Panjan, P. Improvement of cutting performance of carbide cutting tools in milling of the Inconel 718 superalloy using multilayer nanocomposite hard coating and cryogenic heat treatment. *Int. J. Adv. Manuf. Technol.* **2018**, *97*, 467–479. [CrossRef]
187. Rezapoor, M.; Razavi, M.; Zakeri, M.; Rahimipour, M.R.; Nikzad, L. Fabrication of functionally graded Fe-TiC wear resistant coating on CK45 steel substrate by plasma spray and evaluation of mechanical properties. *Ceram. Int.* **2018**, *44*, 22378–22386. [CrossRef]
188. Narasimhan, K.; Boppana, S.P.; Bhat, D.G. Development of a graded TiCN coating for cemented carbide cutting tools—a design approach. *Wear* **1995**, *188*, 123–129. [CrossRef]
189. Miao, H.; Shi, F.; Peng, Z.; Yang, S.; Liu, C.; Qi, L. Nanometer grain titanium carbonitride coatings with continuously graded interface onto silicon nitride cutting tools by pulsed high energy density plasma. *Mater. Sci. Eng. A* **2004**, *384*, 202–208. [CrossRef]
190. Damerchi, E.; Abdollah-zadeh, A.; Poursalehi, R.; Mehr, M.S. Effects of functionally graded TiN layer and deposition temperature on the structure and surface properties of TiCN coating deposited on plasma nitrided H13 steel by PACVD method. *J. Alloy. Compd.* **2019**, *772*, 612–624. [CrossRef]
191. Bell, T.; Dong, H.; Sun, Y. Realising the potential of duplex surface engineering. *Tribol. Int.* **1998**, *31*, 127–137. [CrossRef]
192. Lembke, M.I.; Lewis, D.B.; Titchmarsh, J.M. Joint Second Prize Significance of Y and Cr in TiAlN Hard Coatings for Dry High Speed Cutting. *Surf. Eng.* **2001**, *17*, 153–158. [CrossRef]
193. Schönjahn, C.; Ehiasarian, A.P.; Lewis, D.B.; New, R.; Münz, W.-D.; Twesten, R.D.; Petrov, I. Optimization of in situ substrate surface treatment in a cathodic arc plasma: A study by TEM and plasma diagnostics. *J. Vac. Sci. Technol. A Vac. Surf. Film.* **2001**, *19*, 1415–1420. [CrossRef]
194. Lattemann, M.; Ehiasarian, A.P.; Bohlmark, J.; Persson, P.Å.O.; Helmersson, U. Investigation of high power impulse magnetron sputtering pretreated interfaces for adhesion enhancement of hard coatings on steel. *Surf. Coat. Technol.* **2006**, *200*, 6495–6499. [CrossRef]
195. Stoiber, M.; Wagner, J.; Mitterer, C.; Gammer, K.; Hutter, H.; Lugmair, C.; Kullmer, R. Plasma-assisted pre-treatment for PACVD TiN coatings on tool steel. *Surf. Coat. Technol.* **2003**, *174–175*, 687–693. [CrossRef]
196. El-Hossary, F.M.; Negm, N.Z.; El-Rahman, A.M.A.; Hammad, M. Duplex treatment of 304 AISI stainless steel using rf plasma nitriding and carbonitriding. *Mater. Sci. Eng. C* **2009**, *29*, 1167–1173. [CrossRef]

197. Zheng, Y.; Zhong, J.; Lv, X.; Zhao, Y.; Zhou, W.; Zhang, Y. Microstructure and performance of functionally graded Ti(C,N)-based cermets prepared by double-glow plasma carburization. *Int. J. Refract. Met. Hard Mater.* **2014**, *44*, 109–112. [CrossRef]
198. Yang, Y.; Yan, M.F.; Zhang, Y.X. Tribological behavior of diamond-like carbon in-situ formed on Fe3C-containing carburized layer by plasma carburizing. *Appl. Surf. Sci.* **2019**, *479*, 482–488. [CrossRef]
199. Glühmann, J.; Schneeweiß, M.; Van den Berg, H.; Kassel, D.; Rödiger, K.; Dreyer, K.; Lengauer, W. Functionally graded WC–Ti(C,N)–(Ta,Nb)C–Co hardmetals: Metallurgy and performance. *Int. J. Refract. Met. Hard Mater.* **2013**, *36*, 38–45. [CrossRef]
200. Garcia, J.; Pitonak, R. The role of cemented carbide functionally graded outer-layers on the wear performance of coated cutting tools. *Int. J. Refract. Met. Hard Mater.* **2013**, *36*, 52–59. [CrossRef]
201. Chen, J.; Deng, X.; Gong, M.; Liu, W.; Wu, S. Research into preparation and properties of graded cemented carbides with face center cubic-rich surface layer. *Appl. Surf. Sci.* **2016**, *380*, 108–113. [CrossRef]
202. Rech, J.; Battaglia, J.L.; Moisan, A. Thermal influence of cutting tool coatings. *J. Mater. Process. Technol.* **2005**, *159*, 119–124. [CrossRef]
203. Rech, J.; Kusiak, A.; Battaglia, J. Tribological and thermal functions of cutting tool coatings. *Surf. Coat. Technol.* **2004**, *186*, 364–371. [CrossRef]
204. Shalaby, M.A.; Veldhuis, S.C. Wear and Tribological Performance of Different Ceramic Tools in Dry High Speed Machining of Ni-Co-Cr Precipitation Hardenable Aerospace Superalloy. *Tribol. Trans.* **2019**, *62*, 62–77. [CrossRef]
205. Song, W.; Wang, Z.; Deng, J.; Zhou, K.; Wang, S.; Guo, Z. Cutting temperature analysis and experiment of Ti–MoS2/Zr-coated cemented carbide tool. *Int. J. Adv. Manuf. Technol.* **2017**, *93*, 799–809. [CrossRef]
206. Gengler, J.J.; Hu, J.; Jones, J.G.; Voevodin, A.A.; Steidl, P.; Vlček, J. Thermal conductivity of high-temperature Si–B–C–N thin films. *Surf. Coat. Technol.* **2011**, *206*, 2030–2033. [CrossRef]
207. Bouzakis, K.-D.; Michailidis, N.; Skordaris, G.; Bouzakis, E.; Biermann, D.; M'Saoubi, R. Cutting with coated tools: Coating technologies, characterization methods and performance optimization. *Cirp Ann.* **2012**, *61*, 703–723. [CrossRef]
208. Chen, Y.; Wang, J.; Chen, M. Enhancing the machining performance by cutting tool surface modifications: A focused review. *Mach. Sci. Technol.* **2019**, 1–33. [CrossRef]
209. Oliaei, S.N.B.; Karpat, Y.; Davim, J.P.; Perveen, A. Micro tool design and fabrication: A review. *J. Manuf. Process.* **2018**, *36*, 496–519. [CrossRef]
210. Swan, S.; Abdullah, M.S.B.; Kim, D.; Nguyen, D.; Kwon, P. Tool Wear of Advanced Coated Tools in Drilling of CFRP. *J. Manuf. Sci. Eng.* **2018**, *140*, 111018. [CrossRef]
211. Volosova, M.; Grigoriev, S.; Metel, A.; Shein, A. The Role of Thin-Film Vacuum-Plasma Coatings and Their Influence on the Efficiency of Ceramic Cutting Inserts. *Coatings* **2018**, *8*, 287. [CrossRef]
212. Mitterer, C. PVD and CVD Hard Coatings. In *Compr Hard Mater*; Elsevier: Amsterdam, The Netherlands, 2014; pp. 449–467. [CrossRef]
213. Rokni, M.R.; Nutt, S.R.; Widener, C.A.; Champagne, V.K.; Hrabe, R.H. Review of Relationship Between Particle Deformation, Coating Microstructure, and Properties in High-Pressure Cold Spray. *J. Spray Technol.* **2017**, *26*, 1308–1355. [CrossRef]
214. Abukhshim, N.A.; Mativenga, P.T.; Sheikh, M.A. Heat generation and temperature prediction in metal cutting: A review and implications for high speed machining. *Int. J. Mach. Tools Manuf.* **2006**, *46*, 782–800. [CrossRef]
215. Jaworska, L.; Cyboron, J.; Cygan, S.; Laszkiewicz-Lukasik, J.; Podsiadlo, M.; Novak, P.; Holovenko, Y. New materials through a variety of sintering methods. *IOP Conf. Series Mater. Sci. Eng.* **2018**, *329*, 012004. [CrossRef]
216. Billman, E.R.; Mehrotra, P.K.; Shuster, A.F.; Beechly, C.W. Machining with Al_2O_3-Sic Whisker Cutting Tools. In Proceedings of the 12th Annual Conference on Composites and Advanced Ceramic Materials: Ceramic Engineering and Science Proceedings, Cocoa Beach, FL, USA, 1 January 1988; Wachtman, J.B.J., Ed.; John Wiley & Sons, Inc.: Hoboken, NJ, USA; pp. 543–552.
217. McMillan, P.F. New materials from high-pressure experiments. *Nat. Mater.* **2002**, *1*, 19–25. [CrossRef]
218. Wang, L.; Zhang, J.; Jiang, W. Recent development in reactive synthesis of nanostructured bulk materials by spark plasma sintering. *Int. J. Refract. Met. Hard Mater.* **2013**, *39*, 103–112. [CrossRef]

219. Grasso, S.; Hu, C.; Maizza, G.; Sakka, Y. Spark Plasma Sintering of Diamond Binderless WC Composites. *J. Am. Ceram. Soc.* **2012**, *95*, 2423–2428. [CrossRef]
220. Zhou, X.; Wang, Y.; Li, T.; Li, X.; Cheng, X.; Dong, L.; Yuan, Y.; Zang, J.; Lu, J.; Yu, Y.; et al. Fabrication of diamond–SiC–TiC composite by a spark plasma sintering-reactive synthesis method. *J. Eur. Ceram. Soc.* **2015**, *35*, 69–76. [CrossRef]
221. Sing, S.L.; Yeong, W.Y.; Wiria, F.E.; Tay, B.Y.; Zhao, Z.; Zhao, L.; Tian, Z.; Yang, S. Direct selective laser sintering and melting of ceramics: A review. *Rapid Prototyp. J.* **2017**, *23*, 611–623. [CrossRef]
222. Klimczyk, P.; Cura, M.E.; Vlaicu, A.M.; Mercioniu, I.; Wyżga, P.; Jaworska, L.; Hannula, S.-P. Al_2O_3 –cBN composites sintered by SPS and HPHT methods. *J. Eur. Ceram. Soc.* **2016**, *36*, 1783–1789. [CrossRef]
223. Wozniak, J.; Cygan, T.; Petrus, M.; Cygan, S.; Kostecki, M.; Jaworska, L.; Olszyna, A. Tribological performance of alumina matrix composites reinforced with nickel-coated grapheme. *Ceram. Int.* **2018**, *44*, 9728–9732. [CrossRef]
224. Cygan, T.; Wozniak, J.; Kostecki, M.; Petrus, M.; Jastrzębska, A.; Ziemkowska, W.; Olszyna, A. Mechanical properties of graphene oxide reinforced alumina matrix composites. *Ceram. Int.* **2017**, *43*, 6180–6186. [CrossRef]
225. Broniszewski, K.; Wozniak, J.; Kostecki, M.; Czechowski, K.; Jaworska, L.; Olszyna, A. Al_2O_3–V cutting tools for machining hardened stainless steel. *Ceram. Int.* **2015**, *41*, 14190–14196. [CrossRef]
226. Benedicto, E.; Carou, D.; Rubio, E.M. Technical, Economic and Environmental Review of the Lubrication/Cooling Systems Used in Machining Processes. *Procedia Eng.* **2017**, *184*, 99–116. [CrossRef]
227. Putyra, P.; Figiel, P.; Podsiadło, M.; Klimczyk, P. Alumina composites with solid lubricant participations, sintered by SPS-method. *Kompozyty* **2011**, *11*, 107–110.
228. Deng, J.; Cao, T. Self-lubricating mechanisms via the in situ formed tribofilm of sintered ceramics with CaF2 additions when sliding against hardened steel. *Int. J. Refract. Met. Hard Mater.* **2007**, *25*, 189–197. [CrossRef]
229. Deng, J.; Can, T.; Sun, J. Microstructure and mechanical properties of hot-pressed Al_2O_3/TiC ceramic composites with the additions of solid lubricants. *Ceram. Int.* **2005**, *31*, 249–256. [CrossRef]
230. Wu, G.; Xu, C.; Xiao, G.; Yi, M.; Chen, Z.; Xu, L. Self-lubricating ceramic cutting tool material with the addition of nickel coated CaF2 solid lubricant powders. *Int. J. Refract. Met. Hard Mater.* **2016**, *56*, 51–58. [CrossRef]
231. Katzman, H.; Libby, W.F. Sintered Diamond Compacts with a Cobalt Binder. *Science* **1971**, *172*, 1132–1134. [CrossRef]
232. Hibbs, L.E., Jr.; Wentorf, R.E., Jr. High pressure sintering of diamond by cobalt infiltration. *High Temp.-High Press.* **1974**, *6*, 409–413.
233. Griffin, N.D.; Hughes, P.R. Polycrystalline Diamond Partially Depleted of Catalizing Material, US6592985B2, 2003. Available online: https://patents.google.com/patent/US6592985B2/en (accessed on 17 March 2020).
234. Bertagnolli, K.E.; Vail, M.A. Polycrystalline Diamond Compact (PDC) Cutting Element Having Multiple Catalytic Elements, US8342269B1, 2013. Available online: https://patents.google.com/patent/US8342269 (accessed on 17 March 2020).
235. Osipov, A.S.; Klimczyk, P.; Cygan, S.; Melniychuk, Y.A.; Petrusha, I.A.; Jaworska, L.; Bykov, A.I. Diamond-CaCO 3 and diamond-Li 2 CO 3 materials sintered using the HPHT method. *J. Eur. Ceram. Soc.* **2017**, *37*, 2553–2558. [CrossRef]
236. Jaworska, L.; Szutkowska, M.; Klimczyk, P.; Sitarz, M.; Bucko, M.; Rutkowski, P.; Figiel, P.; Lojewska, J. Oxidation, graphitization and thermal resistance of PCD materials with the various bonding phases of up to 800 °C. *Int. J. Refract. Met. Hard Mater.* **2014**, *45*, 109–116. [CrossRef]
237. Sumiya, H. Novel Development of High-Pressure Synthetic Diamonds Ultra-Hard Nano-Polycrystalline Diamonds. Sei Tech. Rev. 2012, 15–23. Available online: https://global-sei.com/technology/tr/bn74/pdf/74-03.pdf (accessed on 17 March 2020).
238. Morris, D.G.; Muñoz-Morris, M.A. The stress anomaly in FeAl–Fe3Al alloys. *Intermetallics* **2005**, *13*, 1269–1274. [CrossRef]
239. Novák, P.; Nová, K. Oxidation Behavior of Fe–Al, Fe–Si and Fe–Al–Si Intermetallics. *Materials* **2019**, *12*, 1748. [CrossRef] [PubMed]
240. Grabke, H.J. Oxidation of Aluminides. *Mater. Sci. Forum* **1997**, *251–254*, 149–162. [CrossRef]
241. Novák, P.; Šotka, D.; Novák, M.; Michalcová, A.; Šerák, J.; Vojtěch, D. Production of NiAl–matrix composites by reactive sintering. *Powder Metall.* **2011**, *54*, 308–313. [CrossRef]

242. Sheng, L.Y.; Yang, F.; Xi, T.F.; Guo, J.T. Investigation on microstructure and wear behavior of the NiAl–TiC–Al_2O_3 composite fabricated by self-propagation high-temperature synthesis with extrusion. *J. Alloy. Compd.* **2013**, *554*, 182–188. [CrossRef]
243. Choo, H.; Nash, P.; Dollar, M. Mechanical properties of NiAl–AlN–Al_2O_3 composites. *Mater. Sci. Eng. A* **1997**, *239–240*, 464–471. [CrossRef]
244. Furushima, R.; Shimojima, K.; Hosokawa, H.; Matsumoto, A. Oxidation-enhanced wear behavior of WC-FeAl cutting tools used in dry machining oxygen-free copper bars. *Wear* **2017**, *374–375*, 104–112. [CrossRef]
245. Mottaghi, M.; Ahmadian, M. Comparison of the wear behavior of WC/(FeAl-B) and WC-Co composites at high temperatures. *Int. J. Refract. Met. Hard Mater.* **2017**, *67*, 105–114. [CrossRef]
246. Ahmadian, M.; Wexler, D.; Chandra, T.; Calka, A. Abrasive wear of WC–FeAl–B and WC–Ni3Al–B composites. *Int. J. Refract. Met. Hard Mater.* **2005**, *23*, 155–159. [CrossRef]
247. Subramanian, R.; Schneibel, J.H. FeAl–TiC and FeAl–WC composites—melt infiltration processing, microstructure and mechanical properties. *Mater. Sci. Eng. A* **1998**, *244*, 103–112. [CrossRef]
248. Buchholz, S.; Farhat, Z.N.; Kipouros, G.J.; Plucknett, K.P. The reciprocating wear behaviour of TiC–Ni3Al cermets. *Int. J. Refract. Met. Hard Mater.* **2012**, *33*, 44–52. [CrossRef]
249. Robertson, S.W.; Pelton, A.R.; Ritchie, R.O. Mechanical fatigue and fracture of Nitinol. *Int. Mater. Rev.* **2013**, *57*, 1–37. [CrossRef]
250. Baumann, M.A. Nickel-titanium: Options and challenges. *Dent. Clin. N. Am.* **2004**, *48*, 55–67. [CrossRef] [PubMed]
251. Novák, P.; Kristianová, E.; Valalik, M.; Darme, C.; Salvetr, P. New Composite Materials Based on NiTi. *Manuf. Technol.* **2015**, *15*, 644–647.
252. Novák, P.; Kříž, J.; Průša, F.; Kubásek, J.; Marek, I.; Michalcová, A.; Voděrová, M.; Vojtěch, D. Structure and properties of Ti–Al–Si-X alloys produced by SHS method. *Intermetallics* **2014**, *39*, 11–19. [CrossRef]
253. Mochizuki, N.; Takasugi, T.; Kaneno, Y.; Oki, S.; Hirata, T. Friction stir welding of 430 stainless steel and pure titanium using Ni3Al-Ni3V dual two-phase intermetallic alloy tool. In *Proceedings of the 1st International Joint Symposium on Joining and Welding*; Woodhead Publishing: Sawston, Cambridge, UK, 2013; pp. 465–471.
254. Knaislová, A.; Novák, P.; Cabibbo, M.; Průša, F.; Paoletti, C.; Jaworska, L.; Vojtěch, D. Combination of reaction synthesis and Spark Plasma Sintering in production of Ti-Al-Si alloys. *J. Alloy. Compd.* **2018**, *752*, 317–326. [CrossRef]
255. Holmström, E.; Linder, D.; Salmasi, A.; Wang, W.; Kaplan, B.; Mao, H.; Larsson, H.; Vitos, L. High entropy alloys: Substituting for cobalt in cutting edge technology. *Appl. Mater. Today* **2008**, *12*, 322–329. [CrossRef]
256. Arragó, J.M.; Ferrari, C.; Reig, B.; Coureaux, D.; Schneider, L.; Llanes, L. Mechanics and mechanisms of fatigue in a WC–Ni hardmetal and a comparative study with respect to WC–Co hard metals. *Int. J. Fatigue* **2015**, *70*, 252–257. [CrossRef]
257. Viswanadham, R.K.; Lindquist, R.G. Transformation-Toughening in Cemented Carbides: Part I. Binder Composition Control. *Met. Trans. A* **1987**, *18*, 2163–2173. [CrossRef]
258. Schubert, W.D.; Fugger, B.M.; Wittmann, R. Useldinger, Aspects of sintering of cemented carbides withFe-based binders. *Int. J. Refract. Met. Hard Mater.* **2015**, *49*, 110–123. [CrossRef]
259. Graedel, T.E.; Harper, E.M.; Nassar, N.T.; Reck, B.K. On the materials basis of modern society. *Proc. Natl. Acad. Sci.* **2015**, *112*, 6295–6300. [CrossRef] [PubMed]
260. Provis, J.L. Grand Challenges in Structural Materials. *Front. Mater.* **2015**, *2*. [CrossRef]
261. Materials Genome Initiative Homepage. Available online: https://www.mgi.gov (accessed on 4 March 2020).
262. Psi-k Homepage. Available online: http://psi-k.net/ (accessed on 4 March 2020).
263. Overview of the EU funded Centres of Excellence for Computing Applications. Available online: https://ec.europa.eu/programmes/horizon2020/en/news/overview-eu-funded-centres-excellence-computing-applications (accessed on 4 March 2020).
264. EXDCI Homepage. Available online: https://exdci.eu (accessed on 4 March 2020).
265. NIMS. 2019. Available online: https://www.nims.go.jp/eng/index.html (accessed on 4 March 2020).
266. The Novel Materials Discovery (NOMAD) Laboratory Homepage. Available online: https://nomad-coe.eu/ (accessed on 4 March 2020).
267. FAIR-DI (Data Infrastructure). Pillar, A, Computational Materials Science–NOMAD. Available online: https://fairdi.eu/index.php?page=pillar-a (accessed on 4 March 2020).

268. AiiDA (Automated Interactive Infrastructure and Database for Computational Science) Homepage. Available online: http://www.aiida.net/ (accessed on 4 March 2020).
269. MaterialsCloud Homepage. Available online: https://www.materialscloud.org/ (accessed on 4 March 2020).
270. CMR (Computational Materials Repository) Homepage. Available online: https://cmr.fysik.dtu.dk/ (accessed on 4 March 2020).
271. MPDS (Materials Platform for Data Science) Homepage. Available online: https://mpds.io/ (accessed on 4 March 2020).
272. OMD (Open Materials Database) Homepage. Available online: http://openmaterialsdb.se/ (accessed on 4 March 2020).
273. ESL (Electronic Structure Library) Homepage. Available online: https://esl.cecam.org/ (accessed on 4 March 2020).
274. ESP (Electronic Structure Project) Homepage. Available online: http://gurka.fysik.uu.se/ESP/ (accessed on 4 March 2020).
275. Materials Project Homepage. Available online: https://www.materialsproject.org/ (accessed on 4 March 2020).
276. AFLOW (Automatic-FLOW for Materials Discovery) Homepage. Available online: http://www.aflow.org/ (accessed on 4 March 2020).
277. OQMD (Open Quantum Materials Database) Homepage. Available online: http://oqmd.org/ (accessed on 4 March 2020).
278. MatNavi (NIMS Materials Database) Homepage. Available online: https://mits.nims.go.jp/index_en.html (accessed on 4 March 2020).
279. factsage Homepage. Available online: http://www.factsage.com/ (accessed on 4 March 2020).
280. Computherm Homepage. Available online: https://computherm.com/ (accessed on 4 March 2020).
281. Thermocalc Homepage. Available online: https://www.thermocalc.com/ (accessed on 4 March 2020).
282. Opencalphad Homepage. Available online: http://www.opencalphad.com/ (accessed on 4 March 2020).
283. MMM@HPC (Multiscale Materials Modelling on High Performance Computing Architectures) Homepage. Available online: http://www.multiscale-modelling.eu/ (accessed on 4 March 2020).
284. EUDAT Homepage. Available online: https://eudat.eu/ (accessed on 4 March 2020).
285. Kaye and Laby. Available online: http://www.npl.co.uk/resources (accessed on 4 March 2020).
286. MEDEA Homepage. Available online: https://www.materialsdesign.com/ (accessed on 4 March 2020).
287. NanoHUB Homepage. Available online: http://nanohub.org/ (accessed on 4 March 2020).
288. Pymatgen Homepage. Available online: http://pymatgen.org/ (accessed on 4 March 2020).
289. Imeall Homepage. Available online: https://github.com/kcl-tscm/imeall (accessed on 4 March 2020).
290. MPInterfaces Homepage. Available online: https://github.com/henniggroup/MPInterfaces (accessed on 4 March 2020).
291. pylada Homepage. Available online: https://github.com/pylada/pylada-defects (accessed on 4 March 2020).
292. Jain, A.; Ong, S.P.; Hautier, G.; Chen, W.; Richards, W.D.; Dacek, S.; Cholia, S.; Gunter, D.; Skinner, D.; Ceder, G.; et al. Commentary: The Materials Project: A materials genome approach to accelerating materials innovation. *APL Mater.* **2013**, 011002. [CrossRef]
293. Massalski, T.B.; Okamoto, H.; Subramanian, P.R.; Kacprzak, L. (Eds.) *Binary Alloy Phase Diagrams*, 2nd ed.; ASM International: Cleveland, OH, USA, 1990.
294. Villars, P.; Prince, A.; Okamoto, H. (Eds.) *Handbook of Ternary Alloy Phase Diagrams*; ASM International: Cleveland, OH, USA, 1995.
295. NIST-ASM (National Institute of Standards and Technology – Materials Data Repository) Homepage. Available online: https://materialsdata.nist.gov (accessed on 4 March 2020).
296. Ye, Y.F.; Wang, Q.; Lu, J.; Liu, C.T.; Yang, Y. High-entropy alloy: Challenges and prospects. *Mater. Today* **2016**, *19*, 349–362. [CrossRef]
297. Fu, X.; Schuh, C.A.; Olivetti, E.A. Materials selection considerations for high entropy alloys, Scr. *Mater.* **2017**, *138*, 145–150. [CrossRef]
298. Abu-Odeh, A.; Galvan, E.; Kirk, T.; Mao, H.; Chen, Q.; Mason, P.; Malak, R.; Arróyave, R. Efficient exploration of the High Entropy Alloy composition-phase space. *Acta Mater.* **2018**, *152*, 41–57. [CrossRef]
299. Ikeda, Y.; Grabowski, B.; Körmann, F. Ab initio phase stabilities and mechanical properties of multicomponent alloys: A comprehensive review for high entropy alloys and compositionally complex alloys. *Mater. Charact.* **2019**, *147*, 464–511. [CrossRef]

300. Gorbachev, I.; Popov, V.; Katz-Demyanetz, A.; Popov, V.V., Jr.; Eshed, E. Prediction of the Phase Composition of High-Entropy Alloys Based on Cr–Nb–Ti–V–Zr Using the Calphad Method. *Phys. Met. Metallogr.* **2019**, *120*, 378–386. [CrossRef]
301. USPEX (Universal Structure Predictor: Evolutionary Xtallography) Homepage. Available online: http://uspex-team.org/en/ (accessed on 4 March 2020).
302. GASP (Genetic Algorithm for Structure and Phase Prediction) Homepage. Available online: http://gasp.mse.ufl.edu/ (accessed on 4 March 2020).
303. CALYPSO (Crystal structure AnaLYsis by Particle Swarm Optimization) Homepage. Available online: http://www.calypso.cn/ (accessed on 4 March 2020).
304. ATAT (Alloy Theoretic Automated Toolkit) Homepage. Available online: https://www.brown.edu/Departments/Engineering/Labs/avdw/atat/ (accessed on 4 March 2020).
305. UNCLE-MEDEA. Available online: https://www.materialsdesign.com/products (accessed on 4 March 2020).
306. RuNNer Homepage. Available online: http://www.uni-goettingen.de/de/560580.html (accessed on 4 March 2020).
307. AIRSS (Ab Initio Random Structure Searching) Homepage. Available online: https://www.mtg.msm.cam.ac.uk/Codes/AIRSS (accessed on 4 March 2020).
308. Kvashnin, A.G.; Zakaryan, H.A.; Zhao, C.; Duan, Y.; Kvashnina, Y.A.; Xie, C.; Dong, H.; Oganov, A.R. New tungsten borides, their stability and outstanding mechanical properties. *J. Phys. Chem. Lett.* **2018**, *9*, 3470. [CrossRef]
309. Mikhail Kuklin. Available online: https://wiki.aalto.fi/display/IMM/USPEX+guide (accessed on 4 March 2020).
310. Atomistica Homepage. Available online: http://www.atomistica.org/ (accessed on 4 March 2020).
311. Atomicrex Homepage. Available online: https://atomicrex.org/ (accessed on 4 March 2020).
312. Potfit Homepage. Available online: https://www.potfit.net (accessed on 4 March 2020).
313. OpenKIM (Knowledgebase of Interatomic Models) Homepage. Available online: https://openkim.org/ (accessed on 4 March 2020).
314. GAP (Gaussian Approximation Potentials) Homepage. Available online: http://www.libatoms.org/Home/Software (accessed on 4 March 2020).
315. SNAP (Spectral Neighbor Analysis Potential) Homepage. Available online: https://github.com/materialsvirtuallab/snap (accessed on 4 March 2020).
316. González, C.; Panizo-Laiz, M.; Gordillo, N.; Guerrero, C.L.; Tejado, E.; Munnik, F.; Piaggi, P.; Bringa, E.; Iglesias, R.; Perlado, J.M.; et al. H trapping and mobility in nanostructured tungsten grain boundaries: A combined experimental and theoretical approach. *Nucl. Fusion.* **2015**, *55*, 113009. [CrossRef]
317. Valles, G.; Panizo-Laiz, M.; González, C.; Martin-Bragado, I.; González-Arrabal, R.; Gordillo, N.; Iglesias, R.; Guerrero, C.L.; Perlado, J.M.; Rivera, A. Influence of grain boundaries on the radiation-induced defects and hydrogen in nanostructured and coarse-grained tungsten. *Acta Mater.* **2017**, *122*, 277–286. [CrossRef]
318. The Minerals, Metals & Materials Society (TMS). *Advanced Computation and Data in Materials and Manufacturing: Core Knowledge Gaps and Opportunities*; TMS: Pittsburgh, PA, USA, 2018.
319. Bessa, M.A.; Glowacki, P.; Houlder, M. Bayesian Machine Learning in Metamaterial Design: Fragile Becomes Supercompressible. *Adv. Mater.* **2019**, *31*, 1904845. [CrossRef] [PubMed]
320. White Paper on Gaps and Obstacles in Materials Modelling. Available online: https://emmc.info/wp-content/uploads/2019/12/EMMC-CSA-whitepaper-DRAFTV20191220v2.pdf (accessed on 4 March 2020).
321. Shin, S.-H.; Kim, H.-Y.; Rim, K.-Y. Worker Safety in the Rare Earth Elements Recycling Process from the Review of Toxicity and Issues. *Saf. Health Work* **2019**, *10*, 409–419. [CrossRef] [PubMed]
322. Dominik, B.; Kleinert, F.; Imiela, J.; Westkämper, E. Life Cycle Management of Cutting Tools: Comprehensive Acquisition and Aggregation of Tool Life Data. *Procedia Cirp* **2017**, *61*, 311–316.
323. Ishida, T.; Itakura, T.; Moriguchi, H.; Ikegaya, A. Development of technologies for recycling cemented carbidescrap and reducing tungstenuse in cemented carbide tools. *SEI Tech. Rev.* **2012**, *75*, 38–46.
324. Hayashi, T.; Sato, F.; Sasaya, K.; Ikegaya, A. Industrialization of Tungsten Recovering from Used Cemented Carbide Tools. *SEI Tech. Rev.* **2016**, *82*, 33–38.
325. Freemantle, C.; Sacks, N. Recycling of cemented tungsten carbide mining tool scrap. *J. South. Afr. Inst. Min. Metall.* **2015**, *115*, 1207–1213. [CrossRef]
326. Angelo, P.C.; Subramanian, R. *Powder Metallurgy: Science, Technology and Applications*; PHI Learning Pvt. Ltd.: New Delhi, India, 2008.

327. Altuncu, E.; Ustel, F.; Turk, A.; Ozturk, S.; Erdogan, G. Cutting-tool recycling process with the zinc-melt method for obtaining thermal-spray feedstock powder (wc-co) mtaec9. *UDK* **2013**, *47*, 115.
328. Lee, J.; Kim, S.; Kim, B. A New Recycling Process for Tungsten Carbide Soft Scrap That Employs a Mechanochemical Reaction with Sodium Hydroxide. *Metals* **2017**, *7*, 230.
329. Popov, V., Jr.; Katz-Demyanetz, A.; Garkun, A.; Bamberger, M. The effect of powder recycling on the mechanical properties and microstructure of electron beam melted Ti-6Al-4 V specimens. *Addit. Manuf.* **2018**, *22*, 834–843. [CrossRef]
330. Joost, R.J.; Pirso, J.; Letunovitš, S.; Juhani, K. Recycling of WC-Co Hardmetals by Oxidation and Carbothermal Reduction in Combination with Reactive Sintering. *Est. J. Eng.* **2012**, *18*, 127. [CrossRef]
331. Malyshev, V.; Shakhnin, D.; Gab, A.; Uskova, N.; Gudymenko, O.; Glushakov, V.; Kushchevska, N. Tungsten Resource-Saving: Cobalt Cermets Wastes Recycling and Concentrates Extraction. *J. Env. Sci. Eng. B* **2015**, *4*. [CrossRef]
332. Gregor, K.; Luidold, S.; Czettl, C.; Storf, C. Successful Control of the Reaction Mechanism for Semi-Direct Recycling of Hard Metals. *Int. J. Refract. Met. Hard Mater.* **2020**, *86*, 105131.
333. Liu, H.; Hanyu, H.; Murakami, Y.; Kamiya, S.; Saka, M. Recycling Technique for CVD Diamond Coated Cutting Tools. *Surf. Coat. Technol.* **2001**, *137*, 246–248. [CrossRef]
334. Van Staden, A.C.; Hagedorn-Hansen, D.; Oosthuizen, G.A.; Sacks, N. Characteristics of single layer selective laser melted tool grade cemented tungsten carbide. In Proceedings of the Competitive Manufacturing, International Conference on Competitive Manufacturing (COMA '16), Stellenbosch University, Stellenbosch, South Africa, 27–29 January 2016; Stellenbosch University: Stellenbosch, South Africa.

Article

High Strength X3NiCoMoTi 18-9-5 Maraging Steel Prepared by Selective Laser Melting from Atomized Powder

Angelina Strakosova *, Jiří Kubásek, Alena Michalcová, Filip Průša, Dalibor Vojtěch and Drahomír Dvorský

Department of Metals and Corrosion Engineering, Faculty of Chemical Technology, University of Chemistry and Technology, Technická 5, Praha 6—Dejvice, 166 28 Prague, Czech Republic; Jiri.Kubasek@vscht.cz (J.K.); Alena.Michalcova@vscht.cz (A.M.); Filip.Prusa@vscht.cz (F.P.); Drahomir.Dvorsky@vscht.cz (D.V.); dalibor.vojtech@vscht.cz (D.D.)

* Correspondence: strakosa@vscht.cz

Received: 21 November 2019; Accepted: 10 December 2019; Published: 12 December 2019

Abstract: Maraging steels are generally characterized by excellent mechanical properties, which make them ideal for various industrial applications. The application field can be further extended by using selective laser melting (SLM) for additive manufacturing of shape complicated products. However, the final mechanical properties are strongly related to the microstructure conditions. The present work studies the effect of heat treatment on the microstructure and mechanical properties of 3D printed samples prepared from powder of high-strength X3NiCoMoTi 18-9-5 maraging steel. It was found that the as-printed material had quite low mechanical properties. After sufficient heat treatment, the hardness of the material increased from 350 to 620 HV0.1 and the tensile yield strength increased from 1000 MPa up to 2000 MPa. In addition, 3% ductility was maintained. This behavior was primarily affected by strong precipitation during processing.

Keywords: maraging steel; atomized powder; selective laser melting; heat treatment; precipitation hardening

1. Introduction

Selective laser melting (SLM) is one of the best-known methods of additive production of materials. In everyday life, the most commonly encountered name for this method is 3D printing [1]. The aim of this method is to produce a three-dimensional metal part. The principle of manufacturing these parts lies in the use of a laser for melting metallic powders and subsequent layer-by-layer application of these powders. Due to the high cooling rate, rapid solidification and material transformations occur in the material. The advantage of this method is that the product is already made into the desired shape and size without the necessity for secondary machining [1]. The most useful applications of SLM technology are found in medicine, aviation, the automotive industry, and also in the work of architects [2].

The main advantages of 3D printing over conventional methods of manufacturing metallic materials are as follows. First, it provides the ability to produce components with various complex surface and volume shapes. This means that details can be produced during a single operation. Secondly, it requires minimal surface machining of the finished product and thus minimizes waste [1]. Third, SLM technology allows us to work with a wide range of materials. The most extensively studied are steels, Al alloys, Ti alloys, and Ni superalloys [1,3]. The microstructure of 3D printed material is almost non-porous, reaching up to 99.9% of theoretical density [1]. It is also very different from the microstructure of the materials produced by other methods. Thanks to the high cooling rate and solidification of the alloy during the SLM process, we get a material with a very fine-grained cell microstructure. Grain size reduction has a beneficial effect on the mechanical properties of 3D printed products [4–8].

Maraging steels are widely used in the aerospace industry. Also, they are widely used in tooling applications and in the production of weapons. Maraging steels are also well-known thanks to their excellent mechanical properties. The most important properties are very high hardness, good weldability, high ductility, and easy machinability after solution annealing. Spatial stability during aging is one more positive feature of the maraging steels [8–13]. Their properties and range of applications are one of the reasons for the interest in their production by using modern technologies, including 3D printing.

Maraging steels exert these very good mechanical properties after the application of heat treatment involving solution annealing and aging. The first state after quenching is the martensitic structure of the material. This structure may be caused by different means. In maraging steels, the creation of martensite is supported by the high Ni content in an alloy. In contrast to this, in other steels, the martensitic structure is due to a relatively high carbon content. Aging at a temperature between 480 °C to 510 °C causes intermetallic precipitation [13,14]. Several phases were identified in maraging steels like $Ni_3(Ti, Mo)$ [13], Ni_3X (X = Ti, Al, Mo) [10], $(Fe, Ni, Co)_3(Ti, Mo)$, $(Fe, Ni, Co)_3(Mo, Ti)$, and $(Fe, Ni, Co)_7Mo_6$ [15].

Several major works have been carried out to get a better idea of the properties of X3NiCoMoTi 18-9-5 maraging steel made by 3D printing. Kempen et al. [1] investigated how impact and tensile properties depend on aging treatment and process parameters. Tensile, fracture, and fatigue crack growth of maraging steel was studied by Suryawanshi et al. [4]. Some researchers [7,16] have investigated the effect of different heat treatments on microstructure and mechanical properties. In all the work mentioned above, the aging treatment significantly improved the properties due to the formation of precipitates. However, only limited information is known about the precipitates which form during the treatment of 3D printed steel. For example, the formation of a Ni3X (X = Ti, Al, Mo) precipitates after heat treatment was described by references [10,15].

The aim of this work is to investigate the impact of the 3D printing process and two heat treatment modes on microstructure and mechanical properties X3NiCoMoTi 18-9-5 maraging steel.

2. Materials and Methods

Samples in the shape of a dog bone (Figure 1) were produced by 3D printing on SLM solution 280HL (NETME Center, Brno, Czech Republic) from the powder X3NiCoMoTi 18-9-5 alloy. The chemical composition (Table 1) was found using XRF analysis on spectrometer ARL 9400.

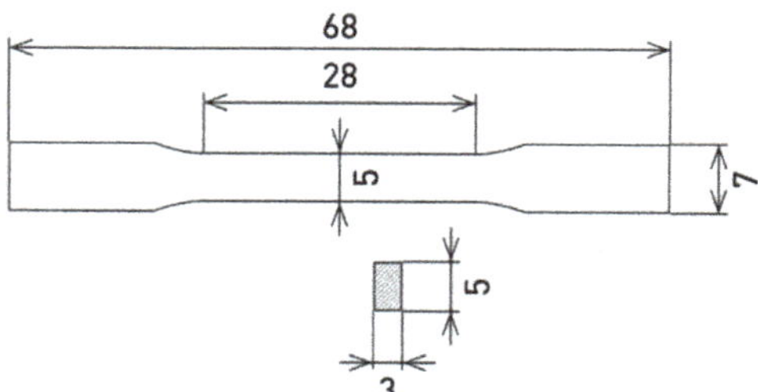

Figure 1. Dimensions of 3D-printed sample for a tensile test.

Table 1. Chemical composition of X3NiCoMoTi 18-9-5 maraging steel.

Element	Ni	Co	Mo	Ti	C	Al	Cr	Mn	Si	Fe
Wt/%	19	9.3	5	0.64	≤0.03	0.06	0.08	0.04	0.07	Bal.

The powder morphology structure and size distribution are shown in Figure 2. Particle size analysis was performed using a scanning electron microscope (TESCAN VEGA 3 LMU, Brno, Czech Republic) and the ImageJ program.

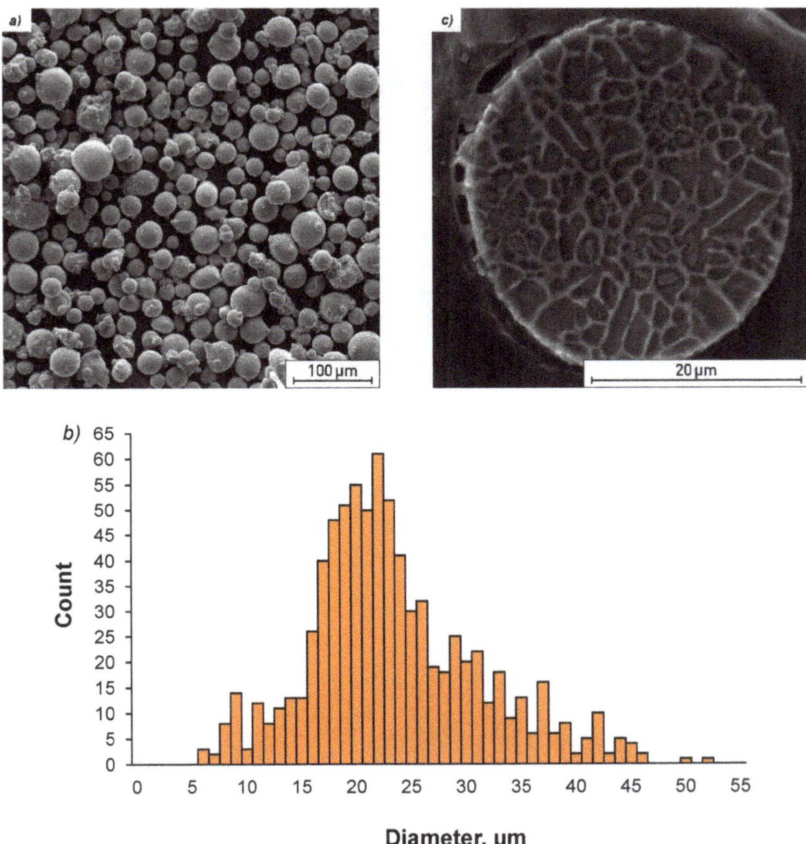

Figure 2. Characterization of X3NiCoMoTi 18-9-5 alloy powder; (**a**) powder morphology, (**b**) particle size distribution, (**c**) the microstructure of the powder.

Two heat treatment regimes were used for 3D printed specimens. The first regime was solution annealing at 820 °C for 1 h, air cooling, and subsequent aging at 490 °C for 6 h. The second one included only aging at 490 °C for 6 h. Electric resistance furnaces were used for heat treatment.

Metallographic samples were made by grinding on SiC abrasive papers with grit size P280-P4000. The samples were polished using a diamond paste (D 2 µm). Final polishing was done on an Eposil F suspension. Metallographic light microscope Olympus PME3 (LM) was used to observe the microstructure of the material. The samples were etched in Nital 2 solution (2 mL HNO_3 + 98 mL ethanol) to visualize the microstructure. Scanning (TESCAN VEGA 3 LMU, Brno, Czech Republic, equipped by EDS-OXFORD Instruments, High Wycombe, UK) and transmission electron microscopy (JEOL 2200 FS equipped by EDS - OXFORD Instruments, High Wycombe, UK) technologies were used for a more detailed study of the structural components of the examined material. Samples for TEM analysis were prepared through grinding and polishing. A thin plate, that was cut from the studied sample, was thinned with SiC abrasive paper (P4000). Discs with a diameter of 3 mm were punched from this thinned plate, which were then polished. Polishing was carried out on PIPs GATAN in an Ar atmosphere with an ion gun setting of 5 keV and rotation of 3 rpm. Phase composition study was performed by X-ray diffraction on PANanalytical X'Pert PRO with Kα radiation on Co lamp ($\lambda = 0.17929$ nm) and generator setting of 40 mA and 35 kV in a 2θ range of 6–110° using a step size of 0.033° and scan step duration of 81.28 s.

The Vickers microhardness with a load of 100 g and holding time of 10 seconds was measured on a FUTURE TECH FM-700 machine. The microhardness was measured in the horizontal section and in the vertical section, of each sample. The final value was calculated from 10 measurements. A tensile strength test was performed at room temperature using the universal test machine LabTest 5.250SP1-VM (LaborTech, Opava, Czech Republic). The tensile test was performed only in the building direction of specimens and tensile yield strength corresponding to the 0.2 proof stress ($TYS_{0.2}$), ultimate tensile strength (UTS), and elongation (A) were evaluated.

3. Results and Discussion

3.1. Microstructure Characterization

As can be seen from Figure 2a, the X3NiCoMoTi 18-9-5 alloy powder particles are characterized by different sizes. Most particles have almost a spherical shape that is preferred for the SLM process. Several photos of the studied powder were made using SEM and the size of eight hundred particles was calculated based on image analysis using the ImageJ software. Each particle was measured in three different directions. The particle size distribution is shown in Figure 2b. We can see that the majority of the metal particles (73%) have a size between 15 and 30 μm. The microstructure of the powder particles in the cross-section shown in Figure 2c demonstrated that the powder has a characteristic cellular structure. This cellular structure was preferred due to the high cooling rate of the powder particles during their production caused by atomization. Therefore, the alloying elements segregate at the cell boundaries (Figure 3). The phase composition of the powder of maraging steel, based on the results of XRD analysis which will be shown below, consisted of both α—martensite, and γ—austenite as a consequence of a rapid cooling rate.

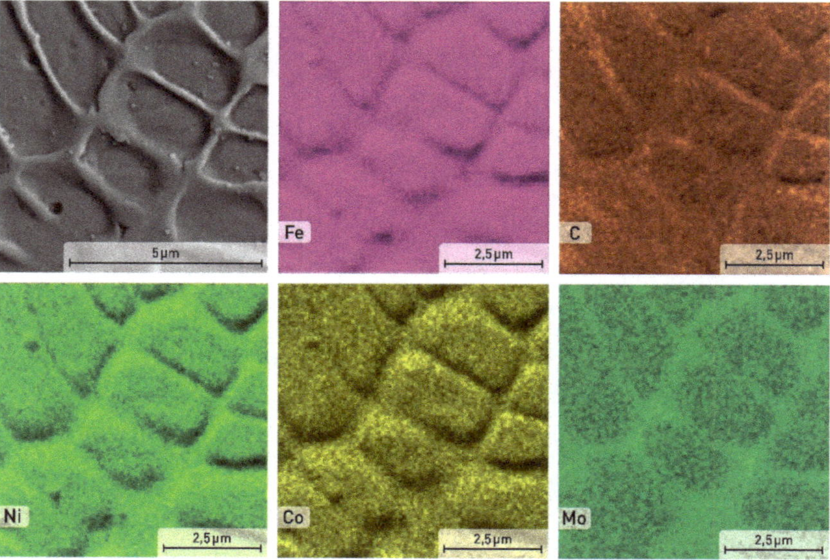

Figure 3. SEM image and EDS element distribution maps in the powder particle of the X3NiCoMoTi 18-9-5 maraging steel.

The microstructure of the printed samples was examined before and after the application of the heat treatment regimes. As shown in Figure 4a, the sample microstructure shows the presence of very fine cellular martensite which fully corresponds to the structure of 3D printed materials as has been reported also in [1,4,10,15]. There are also visible melting pools, which show how the molten material has deposited and solidified layer-by-layer. These fine cells were formed thanks to a high cooling rate up to 10^4–10^6 K/s [14,16] and a difference in temperature gradients during the 3D printing.

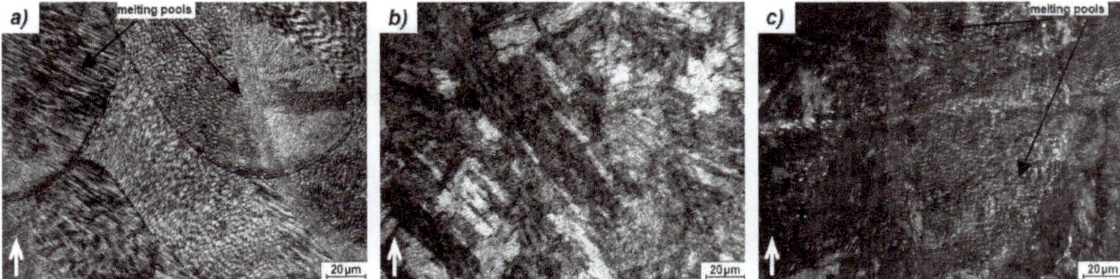

Figure 4. LM micrographs of the material structure; (**a**) as-printed, (**b**) solution annealed and aged, (**c**) aged; ↑—building direction.

A detailed distribution of alloying elements of the as-printed sample is shown in Figure 5. The cell boundaries are enriched by C, other elements like Fe, Ni, Co, and Mo are homogeneously distributed within the cells.

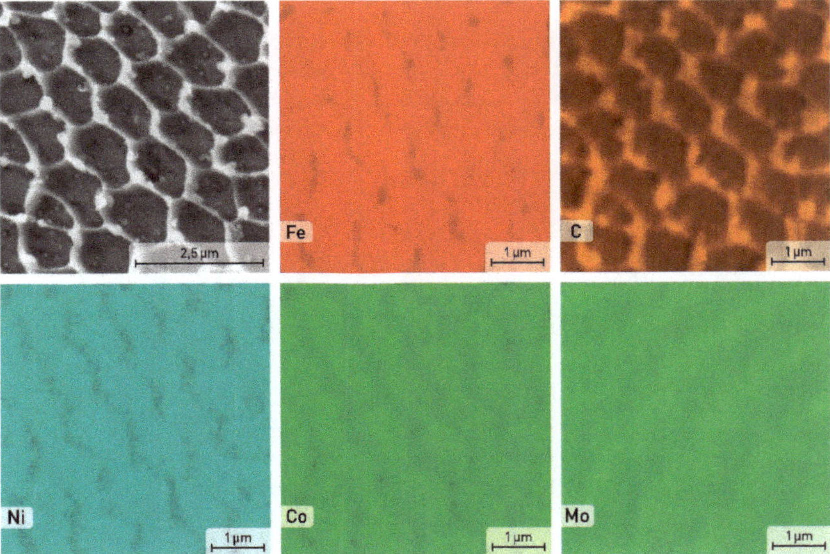

Figure 5. SEM image and EDS element distribution maps in the as-printed sample of the X3NiCoMoTi 18-9-5 maraging steel.

After the application of solution annealing (820 °C/1 h) and aging (490 °C/6 h), a complete change in the material microstructure can be observed (Figure 4). General characteristics of the 3D print structural components disappeared, as shown in Figure 4b. The fine-grained cell structure was transformed into a coarser needle-like martensite structure. This conversion occurred as a result of heating the material to 820 °C. At this temperature, the α-Fe solid solution became supersaturated and homogenized [10]. A more detailed view of this structure is shown in Figure 6. It can be seen that the material lost its cellular structure and was characterized by a uniform distribution of alloying elements. In addition, martensite needles are indicated by arrows in Figure 6.

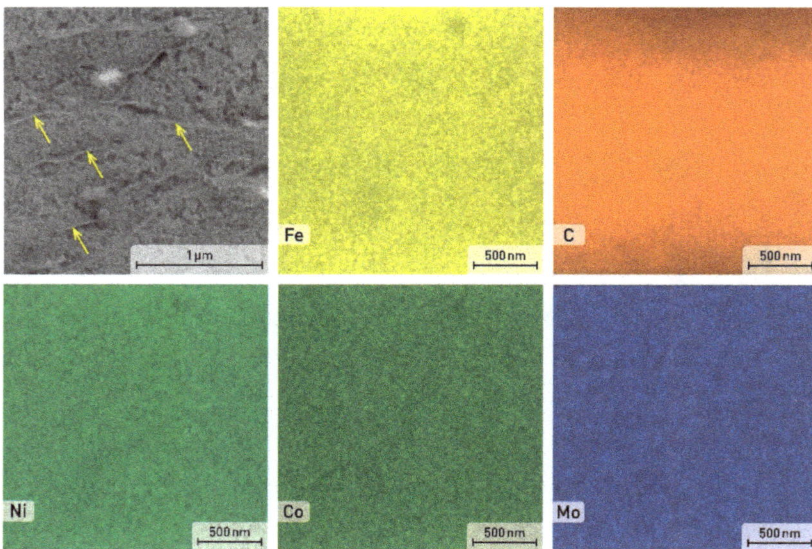

Figure 6. SEM image and EDS element distribution maps in the solution annealed and aged sample.

After aging at 490 °C for 6 h, the microstructure (Figure 4c) becomes similar to the microstructure of the as-printed sample with a specific structure. The alloying elements segregation is also located at the cell boundary (Figure 7). This phenomenon can be justified by the low aging temperature, which is not sufficient for diffusion processes to homogenize the material composition.

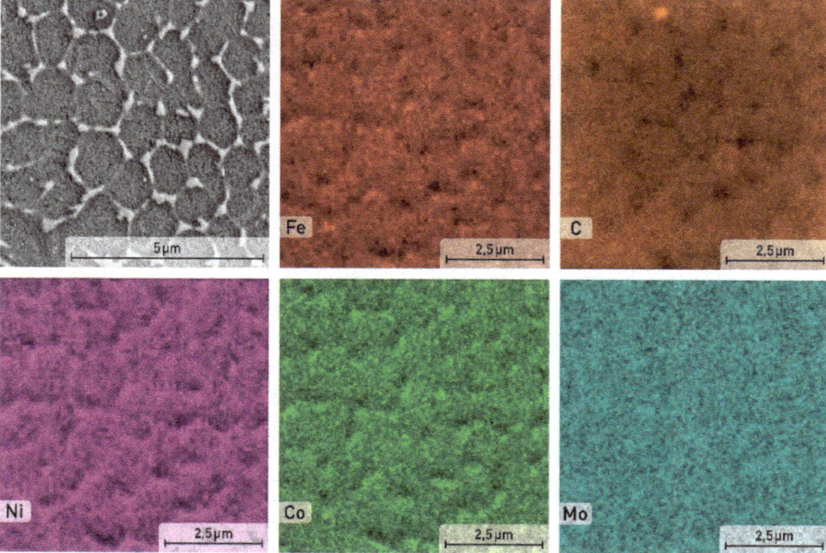

Figure 7. SEM image and EDS element distribution maps in the aged sample of the X3NiCoMoTi 18-9-5 maraging steel.

As shown in Figure 8a, grains, grain boundaries, and also defects (mainly dislocations) are clearly visible in the microstructure of the as-printed sample by TEM. Both gas atomization process and 3D printing are characterized by high cooling rates which favor the formation of various defects like dislocations or vacancies. They also provide very high internal stresses. Based on the TEM of the as-printed X3NiCoMoTi 18-9-5 alloy, no precipitates were observed in the microstructure. The chemical composition within the cells is similar to in Table 1.

The microstructure of the heat-treated samples is different compared to the as-printed sample (Figure 8). As shown in Figure 8b,c, precipitates were formed in both solution annealed and aged samples, and only aged samples. The chemical composition of the heat-treated materials was examined using EDS analysis. Light and dark zones were analyzed with the assumption that the dark ones correspond to the precipitates (Figure 8b,c). It was found that light zones consist of 69% Fe, 15% Ni, 9.5% Co, 3% Mo and 0.7% Ti. Dark areas consist of 62% Fe, 20% Ni, 8.5% Co, 6.5% Mo, and 1% Ti. Precipitates contain a slightly higher amount of Ni, Mo, Ti, and therefore we assume the formation of Ni_3(Ti, Mo) [10,13].

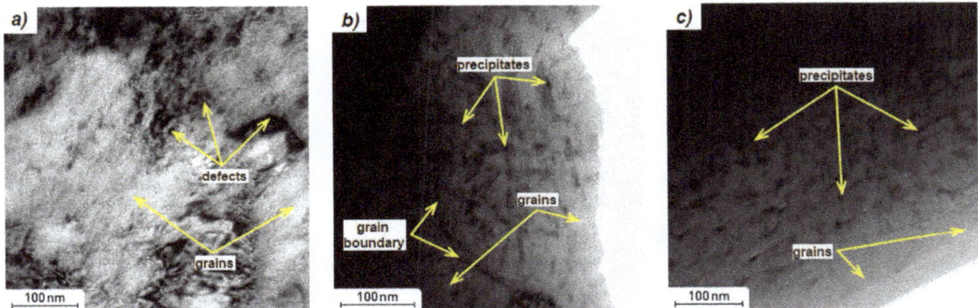

Figure 8. TEM images of X3NiCoMoTi 18-9-5 maraging steel; (**a**) as-printed, (**b**) solution annealed and aged, (**c**) aged.

The X3NiCoMoTi 18-9-5 maraging steel in four different states (powder, as-printed, aged, and solution annealed and aged was also subjected to X-ray diffraction analysis. The results are shown in Figure 9. As can be seen, the sample that has been annealed and aged contained 100% of α-phase. This confirms the martensitic structure of the material (see Figure 4b). In contrast, the retained austenite (ɣ-phase) was found in the powder, as-printed and in the aged samples. This phenomenon can be explained by methods involving studied powder or built samples. As for the aged sample, the aging temperature was not high enough to convert the retained ɣ-phase into the α-phase. According to the literature [10], the temperature necessary for this conversion is, considering the actual chemical composition, around 600 °C.

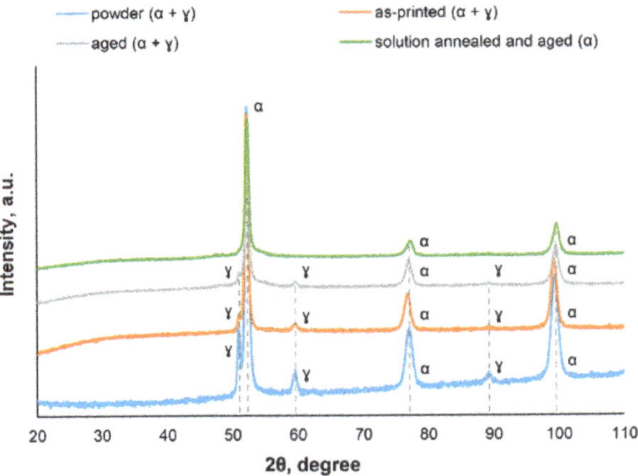

Figure 9. XRD patterns of X3NiCoMoTi 18-9-5 maraging steel.

3.2. Mechanical Properties

Figure 10 and Table 2 give us an overview of the dependence of mechanical properties on the heat treatment regimes. As can be seen in Figure 10a, the direction of hardness examination has no effect on the microhardness of the material. When either material is examined in the horizontal section or in the vertical section, the microhardness is almost the same. Thus, since the tests did not prove any anisotropic behavior with respect to the print direction, several tests were done on samples in the building direction. The hardness of the as-printed samples was relatively low. This is due to the fact that since as-printed samples have a low carbon content in martensite, martensite cannot have a really hard phase in this case.

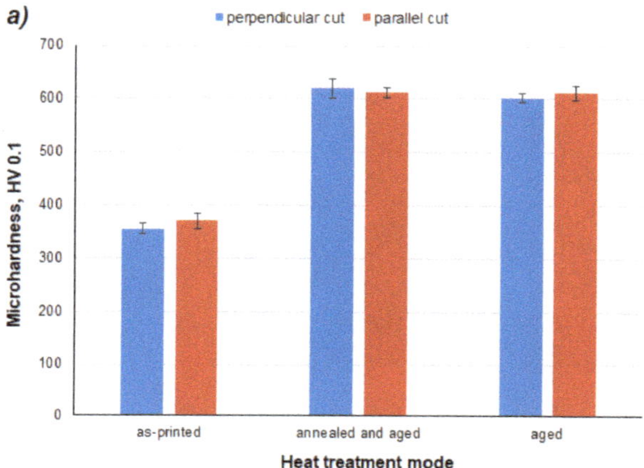

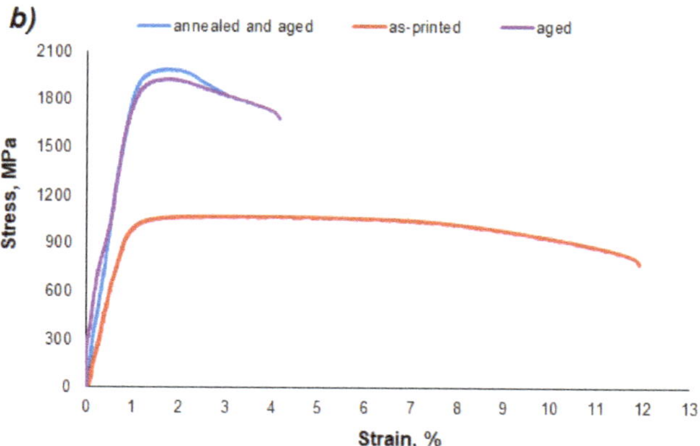

Figure 10. Mechanical tests of the X3NiCoMoTi 18-9-5 maraging steel depending on the heat treatment regime; (**a**) microhardness HV 0.1, (**b**) tensile test.

Table 2. Summary of X3NiCoMoTi 18-9-5 alloy properties depending on heat treatment.

	A	UTS, MPa	TYS0,2, MPa
As-printed	11.3	1080	999
Annealed and aged	2.5	1992	1943
Aged	3.5	1944	1867

Two regimes of heat treatment were carried out in the presented work. An interesting fact is that the resulting microhardness values were almost the same in both cases (Figure 10a). The increase in the hardness of the material after solution annealing and aging can be explained by the formation of a completely martensitic structure after solution annealing and precipitation hardening after the aging process (Figure 8b). Due to the low carbon content, martensite is not a reinforcing element. Therefore, precipitates become responsible for hardening.

After the application of aging alone, the original microstructure of the material was retained, but the temperature was sufficient to cause precipitation hardening (Figure 8c). As in the case of solution annealed and aged samples, the aged samples have a high hardness thanks to the precipitates.

Tensile tests (Figure 10b, Table 2) of 3D printed and heat-treated X3NiCoMoTi 18-9-5 maraging steel show the same trend as hardness. As can be seen in Figure 10b, the as-printed sample is characterized by low yield strength, but high ductility. In contrast, the strength characteristics of the material after heat treatment are nearly twice as high as that of the printed sample, but low ductility is observed. One can see that the annealed and aged sample has slightly higher strength characteristics, but a lower ductility compared to the sample which was only aged (Table 2). This phenomenon can be explained by phase composition (Figure 9). The aged material contains the α-phase in addition to the γ-phase. It is well known that the γ-phase is soft, tough, and ductile. In contrast, the annealed and aged alloy has a 100% α-phase and, as shown in Figure 4b, it has a martensitic structure. During the solution treatment of the as-printed sample, th cellular structure is removed and the distribution of alloying elements is homogenized. This particular leads to stronger precipitation during aging, which causes the increase of TYS and UTS, as well as the decrease of elongation.

The mechanical properties of the 3D printed maraging steel were compared with the properties of the same steel produced by conventional methods. It has been found that the hardness, ultimate tensile strength, and elongation of wrought alloys are slightly higher than of 3D printed ones [17–19]. This difference is justified by the presence of residual porosity in additively produced materials.

4. Conclusions

The effect of solution annealing (820 °C/1 h) plus aging (490 °C/6 h), and also only aging (490°C/ 6 h) on the microstructure and mechanical properties of 3D printed X3NiCoMoTi 18-9-5 maraging steel was investigated in this work. The as-printed X3NiCoMoTi 18-9-5 maraging steels were characterized by a martensitic structure with residual austenite. The material was composed of specific cells containing supersaturated solid solution and a high quantity of defects. After the application of solution annealing and aging, the cell structure disappeared and a coarse martensitic structure was formed. In addition, precipitates enriched by Ti and Mo were also formed. Aging of as-printed material caused precipitation of Ti and Mo enriched phases, but the cellular structure of the as-printed sample remained. Application of two regimes of heat treatment occurred, such as solution annealing and aging, and also only aging. The mechanical properties of the maraging steel were significantly improved after thermal processing, including annealing and aging or only aging. Hardness and tensile yield strength increased almost twice at the expense of slightly impaired ductility. Base on the observed properties, the solution annealing stage seems not to be necessary to improve the mechanical properties of 3D-printed samples. For these purposes, it is sufficient to apply only the aging regime.

Author Contributions: Investigation, A.S., J.K., A.M. and F.P.; Methodology, D.V.; Supervision, D.V.; writing—original draft, A.S. and D.D.; Writing—review & editing, A.S., J.K. and D.V.

Funding: Financial support from specific university research (MSMT No 21-SVV/2019). The authors acknowledge the assistance in TEM sample preparation provided by the Research Infrastructure NanoEnviCz, supported by the Ministry of Education, Youth, and Sports of the Czech Republic under Project No. LM2015073.

Conflicts of Interest: The authors declare no conflict of interest.

References

1. Herzog, D.; Seyda, V.; Wycisk, E.; Emmelmann, C. Additive manufacturing of metals. *Acta Mater.* **2016**, *117*, 371–392. [CrossRef]
2. Wong, K.V.; Hernandez, A. A Review of additive manufacturing. *ISRN Mech. Eng.* **2012**, *2012*, 208760. [CrossRef]
3. Murr, L.E.; Martinez, E.; Amato, K.N.; Gaytan, S.M.; Hernandez, J.; Ramirez, D.A.; Shindo, P.W.; Medina, F.; Wicker, R.B. Fabrication of metal and alloy components by additive manufacturing: Examples of 3D materials science. *J. Mater. Res. Technol.* **2012**, *1*, 42–54. [CrossRef]
4. Suryawanshi, J.; Prashanth, K.G.; Ramamurty, U. Tensile, fracture, and fatigue crack growth properties of a 3D printed maraging steel through selective laser melting. *J. Alloys Compd.* **2017**, *725*, 355–364. [CrossRef]
5. Afkhami, S.; Dabiri, M.; Habib Alavi, S.; Björk, T.; Salminen, A. Fatigue characteristics of steels manufactured by selective laser melting. *Int. J. Fatigue* **2019**, *122*, 72–83. [CrossRef]
6. Casati, R.; Lemke, J.; Vedani, M. Microstructure and fracture behavior of 316L austenitic stainless steel produced by selective laser melting. *J. Mater. Process. Technol.* **2016**, *32*, 738–744. [CrossRef]
7. Casati, R.; Lemke, J.N.; Tuissi, A.; Vedani, M. Aging behaviour and mechanical performance of 18-Ni 300 steel processed by selective laser melting. *Metals* **2016**, *6*, 218. [CrossRef]
8. Kempen, K.; Yasa, E.; Thijs, L.; Kruth, J.P.; Van Humbeeck, J. Microstructure and mechanical properties of selective laser melted 18Ni-300 steel. *Phys. Procedia* **2011**, *12*, 255–263. [CrossRef]
9. Tan, C.; Zhu, K.; Tong, X.; Huang, Y.; Li, J.; Ma, W.; Li, F.; Kuang, T. Microstructure and mechanical properties of 18Ni-300 maraging steel fabricated by selective laser melting. In Proceedings of the 2016 6th International Conference on Advanced Design and Manufacturing Engineering (ICADME 2016), Zhuhai, China, 23–24 July 2017.
10. Tan, C.; Zhou, K.; Ma, W.; Zhang, P.; Liu, M.; Kuang, T. Microstructural evolution, nanoprecipitation behavior and mechanical properties of selective laser melted high-performance grade 300 maraging steel. *Mater. Des.* **2017**, *134*, 23–34. [CrossRef]
11. Xu, X.; Ganguly, S.; Ding, J.; Guo, S.; Williams, S.; Martina, F. Microstructural evolution and mechanical properties of maraging steel produced by wire + arc additive manufacture process. *Mater. Char.* **2018**, *143*, 152–162. [CrossRef]
12. Shamantha, C.R.; Narayanan, R.; Iyer, K.J.L.; Radhakrishnan, V.M.; Seshadri, S.K.; Sundararajan, S.; Sundaresan, S. Microstructural changes during welding and subsequent heat treatment of 18Ni (250-grade) maraging steel. *Mater. Sci. Eng.* **2000**, *287*, 43–51. [CrossRef]
13. Tewari, R.; Mazumder, S.; Batra, I.S.; Dey, G.K.; Banerjee, S. Precipitation in 18 wt% Ni maraging steel of grade 350. *Acta Mater.* **2000**, *48*, 1187–1200. [CrossRef]
14. Jägle, E.A.; Sheng, Z.; Kürnsteiner, P.; Ocylok, S.; Weisheit, A.; Raabe, D. Comparison of maraging steel micro- and nanostructure produced conventionally and by laser additive manufacturing. *Materials* **2017**, *10*, 8. [CrossRef] [PubMed]
15. Jägle, E.A.; Choi, P.P.; van Humbeeck, J.; Raabe, D. Precipitation and austenite reversion behavior of a maraging steel produced by selective laser melting. *J. Mater. Res.* **2014**, *29*, 2072–2079. [CrossRef]
16. Bai, Y.; Wang, D.; Yang, Y.; Wang, H. Effect of heat treatment on the microstructure and mechanical properties of maraging steel by selective laser melting. *Mater. Sci. Eng.* **2019**, *760*, 105–117. [CrossRef]
17. Tariq, F.; Naz, N.; Baloch, R.A. Effect of cyclic aging on mechanical properties and microstructure of maraging steel 250. *J. Mater. Eng. Perform.* **2010**, *19*, 1005–1014. [CrossRef]
18. SAE Standard. *AMS 6514H, Steel, Maraging, Bars, Forgings, Tubing, and Rings 18.5Ni-9.0Co-4.9Mo-0.65Ti-0.10Al Consumable Electrode Vacuum Melted, Annealed;* SAE International: Warrendale, PA, USA, 2012.
19. ASM International Handbook Committee. *Properties and Selection: Iron Steels and High Performance Alloy, ASM Handbook;* Materials Information Company: Materials Park, OH, USA, 1991; pp. 1872–1873.

MDPI
St. Alban-Anlage 66
4052 Basel
Switzerland
Tel. +41 61 683 77 34
Fax +41 61 302 89 18
www.mdpi.com

Materials Editorial Office
E-mail: materials@mdpi.com
www.mdpi.com/journal/materials

www.ingramcontent.com/pod-product-compliance
Lightning Source LLC
LaVergne TN
LVHW070110170726
843515LV00007B/1652

* 9 7 8 3 0 3 9 3 6 5 2 3 4 *